T0305799

Nonlinear Field Theories
and Unexplained
Phenomena in Nature

NONLINEAR FIELD THEORIES AND UNEXPLAINED PHENOMENA IN NATURE

Alexander S Rabinowitch

HSE University, Russia

World Scientific

NEW JERSEY · LONDON · SINGAPORE · BEIJING · SHANGHAI · HONG KONG · TAIPEI · CHENNAI · TOKYO

Published by

World Scientific Publishing Co. Pte. Ltd.

5 Toh Tuck Link, Singapore 596224

USA office: 27 Warren Street, Suite 401-402, Hackensack, NJ 07601

UK office: 57 Shelton Street, Covent Garden, London WC2H 9HE

Library of Congress Control Number: 2023006253

British Library Cataloguing-in-Publication Data
A catalogue record for this book is available from the British Library.

NONLINEAR FIELD THEORIES AND UNEXPLAINED PHENOMENA
IN NATURE

ISBN 978-981-126-411-5 (hardcover)
ISBN 978-981-126-412-2 (ebook for institutions)
ISBN 978-981-126-413-9 (ebook for individuals)

For any available supplementary material, please visit
https://www.worldscientific.com/worldscibooks/10.1142/13078#t=suppl

Desk Editors: Nimal Koliyat/Lai Fun Kwong

Typeset by Stallion Press
Email: enquiries@stallionpress.com

Printed in Singapore

Dedicated to my wife Vera Chaykovskaya

Preface

The monograph is devoted to a number of topical problems of modern mathematical and theoretical physics. In it, new nonlinear generalizations of the classical theories of physical fields are studied and applied to interpret a variety of unexplained phenomena in nature.

As is now known, the experimental data obtained from the Large Hadron Collider and the James Webb Space Telescope did not confirm a number of modern models in theoretical physics and cosmology. That is why the author hopes that the new physical concepts presented in the monograph will be of interest to readers.

The monograph is essential reading for anyone interested in nonlinear field theories and their applications to unsolved problems of contemporary physics, astrophysics and cosmology: from graduate students and advanced undergraduates to physicists and astrophysicists interested in explanation of mysterious phenomena in nature.

It consists of an introduction, which describes many unexplained natural phenomena, and seven chapters.

In the first chapter, a new nonlinear generalization of the Maxwell theory of electromagnetic fields is considered which is based on the Yang–Mills field theory. The nonlinear differential equations of the considered theory are studied and several new classes of their stationary and nonstationary solutions are found. The obtained solutions are applied to investigate physical fields generated by sources with big charges and currents. The obtained results are used to explain puzzling properties of ball lightning and streak lightning, the reversals of the magnetic poles of the Earth and Sun, the mysterious appearance of crop circles in fields sowed with cereals, anomalous processes in exploding conductors and some other intriguing natural phenomena.

In the second chapter, Yang–Mills fields generated by quantum sources are considered and their general properties are studied. The obtained results are applied to describe a number of peculiar phenomena which could take place in electroweak interactions, in neutrino physics and in living cells.

In the third chapter, the problem of description of progressive waves in Yang–Mills fields is studied. In it, a number of new classes of wave solutions to the Yang–Mills equations are found. They correspond to different types of nonlinear waves and can be applied to describe radiations from cosmic sources of Yang–Mills fields.

In the fourth chapter, a new nonlinear generalization of the Yukawa meson theory of nuclear fields is proposed. The Yukawa theory well describes properties of pions which are carriers of nuclear forces. However, it cannot describe nonlinear effects of nuclear fields. The proposed nonlinear generalization of the Yukawa equation is studied and used to give theoretical formulas for the binding energies and radii of medium and heavy atomic nuclei and to investigate cooled massive neutron stars.

In the fifth chapter, a new generalization of the Dirac quantum equation for relativistic electrons is proposed to give a description of relativistic nucleons corresponding to their quark structure. This generalization results in formulas for the anomalous magnetic moments of protons and neutrons which accord with experimental data. The proposed equation for nucleons is developed to give a quantum description of light atomic nuclei.

In the sixth chapter, an original approach to describe noninertial frames of reference in general relativity is studied and a system of nonlinear differential equations for them is proposed. For the equations of general relativity with these additional equations, several important cases are examined and the solutions corresponding to them are found. In particular, solutions for propagating gravitational waves are obtained. They are of special interest since the description of gravitational waves is essentially dependent on the choice of equations for the frames of reference used by astronomers.

In the seventh chapter, a new generalization of the Einstein gravitational and Maxwell electromagnetic field equations is studied which are based on Weyl's principle of scale invariance and contain Weyl's vector field with four components. The Weyl field is given a new interpretation. It is regarded as a weak one which is caused by the physical vacuum and gives small corrections to the Einstein and Maxwell field equations but could play an important role in cosmological processes. The proposed equations lead to a new cosmology with no singularity. It is shown that this cosmology accords with observational data, gives a natural solution to the problem of dark matter and explains the evolutionary properties of spiral galaxies.

I would like to thank my friend Dr. Vladimir A. Saytanov for useful discussions of my scientific results.

About the Author

 **Alexander S. Rabinowitch** is a Professor at Moscow Institute of Electronics and Mathematics, HSE University, Russia, where he received the degree of Doctor of Sciences in Physics and Mathematics. He has published many articles in leading international journals and a book (Nova Science Publishers, New York, 2009) on unsolved problems of the Yang–Mills theory, nuclear physics, general relativity and cosmology, nonlinear generalizations in classical and quantum physics and their applications to unexplained physical phenomena. Prof. Rabinowitch graduated from Lomonosov Moscow State University, where he received his PhD in Physics and Mathematics.

Contents

Introduction

There is a paradoxical situation in modern physics. On the one hand, it has achieved amazing success and the technical devices based on modern physics make it possible what recently seemed like a miracle. Unique microelectronics, Internet, mobile telecommunication systems and many other inventions made at the end of 20th century look like something fantastic. On the other hand, by now there are a large number of natural phenomena that cannot be explained with the help of conventional physical science, which indicates its serious incompleteness. There are also many fundamental questions of physics that remain unsolved despite numerous attempts to find clues to them. All this indicates that theoretical physics needs new approaches and new theories that could shed light on the secrets of nature.

One of the ways to renew physical science is to study nonlinear generalizations of classical theories. This topic is central to the present monograph. It begins with an analysis of the famous Yang–Mills theory proposed in 1954. At first, this theory did not arouse much interest, since it was focused on the description of hypothetical particles with isospin about which the existing experimental data were silent. But after it played an important role in creating a model of electroweak interactions and correctly predicting intermediate Z and W bosons involved in them, interest in the Yang–Mills theory became very large.

However, this model of electroweak interactions was based not only on the Yang–Mills field but also on other concepts. In particular, on the Higgs field. As a result, there was no great clarity in the question of what the Yang–Mills field itself is. It should be noted that this field has always been treated as a purely quantum phenomenon, applicable only in the microcosm. However, little attention has been paid to an important feature of the Yang–Mills field equations: they are a reasonable nonlinear generalization of Maxwell's equations for the electromagnetic field. At the same time, in relation to Maxwell's theory, which is undoubtedly a reliable foundation for describing a very wide range of electromagnetic phenomena, the question sometimes arose as follows: Does it cover the entire set of electromagnetic phenomena? The reason for the emerging doubts is the linearity of Maxwell's equations at arbitrarily high intensities of electromagnetic fields. That is why such outstanding scientists as Einstein, Born, Infeld and some others believed that the true theory of the electromagnetic field should be nonlinear.

One of the most famous attempts to give a nonlinear generalization of Maxwell's equations was the Born–Infeld approach proposed in 1934. There is still a continuing interest in these nonlinear equations. But for almost nine decades of their existence, they have not received any experimental confirmation. At the same time, in 1954 a worthy competitor appeared for the Born–Infeld theory: the Yang–Mills theory. However, no special attention was paid to this, since the Yang–Mills theory was firmly assigned the label of a purely quantum theory, applicably to the description of only physical processes on a microscopic scale.

It should be said that the nonlinear equations proposed by C. N. Yang and R. Mills in 1954 are undoubtedly beautiful and have a three-dimensional Lie group of internal symmetry. They describe three vector fields, while Maxwell's equations describe only one such field. As for the energy–momentum tensors of the Yang–Mills fields, they are expressed by their field strengths in the same way as in Maxwell's theory. Besides, if the second and third Yang–Mills fields are absent, the first field is described by Maxwell's linear equations. Therefore, it makes sense to regard the Yang–Mills theory as a reasonable nonlinear generalization of Maxwell's electromagnetic theory.

For this reason, the present monograph pays great attention to the equations of nonlinear electrodynamics based on the Yang–Mills theory. To them, a number of exact solutions in the cases of spherically symmetric and axially symmetric field sources and a number of classes of exact wave solutions are found in the monograph. The results obtained are used to explain a variety of phenomena that are mysterious within the framework of Maxwell's electrodynamics. This is the mysterious properties of ball lightning, the unusual features of streak lightning, the striking phenomenon of reversals of the magnetic poles of the Sun and the Earth, the unrevealed puzzle of amazing circles on cereal-planted fields and a number of other intriguing phenomena.

Another group of questions raised in the monograph concerns the old problem of nuclear forces that bind protons and neutrons in atomic nuclei. As is known, the first successful attempt to describe them was the Yukawa meson theory proposed by him in 1935. His idea made it possible to explain the short-range character of nuclear forces and to give a correct prediction of the mass of their carriers — pions and the zero spin of these particles. However, subsequent experiments have shown that this theory, in which the nuclear potential is described by a linear equation of the Klein–Gordon type, can be valid only for relatively small values of this potential. If its value becomes large enough, then there are essentially nonlinear effects that are not described by the linear Yukawa equation. One of them is the interesting effect of saturation of nuclear forces. It manifests itself in changing the sign of nuclear forces at sufficiently large potentials, that is, when passing from the usual property of attraction, like the forces of gravity, to the phenomenon of repulsion inside nuclear matter. Unexplained in Yukawa's theory are the properties of the specific binding energies of atomic nuclei, which for some mysterious reason have a maximum at the iron nucleus, and much more. For these reasons, Yukawa's theory, after a short period of great success and bright hopes, faced a crisis. It became clear that the true equation for the nuclear potential must be essentially nonlinear. Later, a number of nonlinear models were proposed. A very interesting attempt was the now almost forgotten nonlinear meson Schiff model. However, in these models, it was not possible to achieve quantitative agreement with experiments. New hopes for a description of nuclear forces arose with the emergence of quantum chromodynamics, which is now considered the most likely candidate for the role of a correct theory of strong interactions. But it also did not lead to a quantitative theory of nuclear forces due to the impossibility of applying its main approach, perturbation theory, to atomic nuclei.

Thus, the problem of describing nuclear forces remained unresolved. At the same time, it is difficult to completely abandon the
Yukawa model, which was successful in the case of relatively small
values of nuclear potentials. In view of this, in the monograph, a
new attempt to give a nonlinear generalization of Yukawa's theory
is undertaken. It is based on the idea of the dependence of the mass
of nuclear particles on the potential of nuclear forces, which is just
not taken into account in the Yukawa model. As our studies have
shown, this dependence can be determined from the general principles of relativistic dynamics. The nonlinear generalization of the
Yukawa equation found in this way is applied to theoretically determine the binding energies and radii of medium and heavy atomic
nuclei, for which the classical approximation can be valid. The calculations performed for such nuclei showed good agreement between
the proposed model and the known experimental data. Using this
model, the dynamics of relativistic nucleons and antinucleons near
heavy atomic nuclei is studied. The obtained results can be applied
to the problem of the formation of quasi-nuclei, in which protons
or antiprotons revolve around heavy atomic nuclei. In addition, a
detailed analysis of the effect of the saturation of nuclear forces on
the equilibrium of cooled massive neutron stars is carried out. It is
shown that this effect which manifests itself in the appearance of
nuclear repulsive forces can compensate for the gravitational com-
pression of the neutron stars under consideration.

Serious attention in the monograph is also given to the problem
of the quantum description of nucleons. As is known, the proton
consists of two u-quarks having $+2/3$ of its charge and one d-quark
having $-1/3$ of its charge, and the neutron consists of one u-quark
and two d-quarks. In addition, nucleons have significant anomalous
magnetic moments, positive for the proton and negative for the neutron. In view of these properties, it is impossible to directly apply
the famous Dirac equation to the quantum description of nucleons.
This equation, which is valid for the relativistic electron, does not
give anomalous values of the magnetic moments and quark structure
of particles. It should be noted that for the description of nucleons, there is one well-known generalization of the Dirac equation. It
assumes an additional, nonminimal interaction of nucleons with an
electromagnetic field. But in it, the values of the anomalous magnetic

moments of nucleons are not determined theoretically. Their experimental values are simply inserted into the equation as coefficients. Moreover, this equation does not reflect the quark structure of nucleons. That is why the monograph poses the problem of finding a new generalization of the Dirac equation, which would not have these two defects. The idea of such a generalization is based on the multiplication of charge and mass in the Dirac equation by special matrices of the third order, composed of the quark numbers $+2/3$ and $-1/3$. It turns out that these matrices can be chosen so that the resulting generalization of the Dirac equation corresponds to the basic principles of quantum mechanics and to the quark structure of nucleons. Besides, the anomalous magnetic moments of nucleons determined on the basis of this equation agree with experimental data. Further, a more complete generalization of the Dirac equation is considered, which is used to describe the quantum properties of systems of closely spaced nucleons and light atomic nuclei. This is achieved by replacing the Dirac matrices with generalizing matrices of the Clifford algebra. A detailed study of the proposed quantum equation for systems of nucleons and light atomic nuclei is carried out to investigate the properties of their quark currents.

The last two chapters of the monograph are devoted to unsolved problems of Einstein's general theory of relativity and the cosmology based on it, as well as to a number of mysterious astronomical phenomena unexplained within its framework.

The first group of these questions is connected with the description within the framework of Einstein's general theory of relativity of a rather wide class of frames of reference: noninertial elastically deformed frames. The problem is the undoubted merit of the equations of the gravitational field of general relativity — their validity in an arbitrary coordinate system also has a negative side. Namely, the solutions of Einstein's equations contain four arbitrary functions of space–time coordinates, due to the arbitrariness of their choice, which makes it impossible to describe specific noninertial frames of reference. This circumstance leads to the emergence of significant difficulties in the study of gravitational waves since their behavior directly depends on the properties of the chosen frames of reference.

Before proceeding to the study of the class of elastically deformed frames of reference, a simpler class of frames is examined in which their own elastic deformations can be neglected. Such frames of reference are called perfect by us. Based on Einstein's principle of equivalence and general requirements for frames of reference, a system of four differential equations for the components of the metric tensor is proposed in the monograph to describe the perfect frames of reference under examination. These systems of equations are analyzed by us and a number of their exact solutions are found which correspond to several specific states of perfect frames. Further, the difference between the metric tensors in an elastically deformed frame and in a perfect frame comoving with it is determined through the strain tensor. The obtained relations make it possible to find metric tensors in elastically deformed frames from known solutions in perfect frames. Of particular interest is the form of solutions found in the monograph which describes gravitational waves relative to extended perfect frames of reference. As it turns out from their analysis, situations of significant amplification of gravitational waves are possible. Such anomalous waves can be regarded by an observer as some others, nongravitational waves. Therefore, the new results concerning gravitational waves may be important for the correct identification of these waves when observing them.

Another circle of issues studied in the monograph is associated with the need to suggest a new cosmological theory since standard cosmology cannot explain the origin of the cosmological singularity, the nature of dark matter, the existence of orderly disk galaxies more than 13.4 billion years ago, as discovered by the James Webb Space Telescope, and a number of other cosmological phenomena.

For this purpose, a generalization of the equations of general relativity is proposed in which the influence of physical vacuum on physical processes is taken into account. Physical vacuum is a special extremely rarefied state of matter, the influence of which on moving bodies is very small. However, over very long cosmological time intervals, this small effect can gradually accumulate and lead to quite noticeable results. This point of view is developed in the monograph. To describe small vacuum corrections to the equations of general relativity, the Weyl conformal geometry is applied, which generalizes the Riemann geometry used in Einstein's theory. A remarkable property of Weyl's geometry is the equivalence of expressions for the space–time intervals which differ in the scale factor. This is achieved by introducing an additional field with four potentials to the Einstein field with the ten gravitational potentials, which are components of the metric tensor. Weyl interpreted his additional field as electromagnetic, trying to create a unified theory of gravitational and electromagnetic forces. But this point of view turned out to be wrong. At the same time, it is possible to give the Weyl field a different interpretation as a very weak field caused by the physical vacuum. This approach turned out to be fruitful, allowing us to explain a number of cosmological phenomena that are difficult to explain in standard cosmology. Its development leads to generalizations of Einstein's gravitational and Maxwell's electromagnetic theories, which differ significantly from them only on cosmological space–time scales. The proposed system of field equations is applied to a homogeneous and isotropic vacuum and a new nonsingular cosmological solution to them is obtained. This solution is used to describe the influence of the Weyl field on propagating electromagnetic waves and moving free particles in vacuum. As a result, we arrive at a new cosmological theory, which is a real alternative to standard cosmology. A detailed analysis shows that it is in agreement with the known observational data, including the latest data obtained from the James Webb Space Telescope. In addition, the proposed cosmology explains the nature of dark matter and gives a simple explanation of the spiral structure of most galaxies and a number of their evolutionary properties, which have not found a satisfactory interpretation within the framework of previous theories.

Chapter 1

Yang–Mills Theory and Anomalous Physical Phenomena

One of the greatest physical theories is the Maxwell electrodynamics. It describes the most diverse electromagnetic phenomena in nature and is very well confirmed by a large number of experiments. It is the basis for numerous technical inventions: from the creation of the radio in the early 20th century to modern mobile communications. At the same time, there exist a number of anomalous electromagnetic phenomena in nature which remain still unexplained despite many attempts to interpret them within the framework of the Maxwell electrodynamics. They include mysterious features of ball and streak lightning, reversals of the magnetic poles of planets and stars and some other unexplained phenomena. Failure to understand them indicates the need to give a nonlinear generalization of the Maxwell theory which is described by linear differential equations. For this purpose, several attempts to propose nonlinear electromagnetic equations were made. However, they have got no experimental confirmations. The most popular of them was proposed by M. Born and L. Infeld in 1934. This nonlinear theory was aimed to describe electromagnetic forces at very short distances. However, experiments show that the Maxwell theory is true even in such cases and the hope of Born and Infeld did not come true.

In this chapter, we consider and study another approach to give a nonlinear generalization of the Maxwell theory. It is based on the

Yang–Mills theory [1–3] which was proposed in 1954. This nonlinear generalization will be applied to describe electromagnetic fields generated by very big electric charges and currents and to explain a number of mysterious phenomena in nature. The results obtained in our papers [4–7] are used in this chapter.

1.1.   Unsolved Problems of Classical Electrodynamics

One of the most striking achievements of physics in the 19th century is the theory of the electromagnetic field by Michael Faraday and James Maxwell. Mathematically, this theory is represented by Maxwell's equations, which describe changes in space and time of the vectors of electric and magnetic fields depending on the distribution of electric charges and currents.

These equations are truly great. They are not only amazingly beautiful, but they also describe numerous electromagnetic phenomena.

However, there is some embarrassing circumstance: the equations of the Maxwell electromagnetic field, in contrast to the equations of other physical fields, are linear at arbitrarily high intensities of the field sources. This means that with an increase in charges and currents by any number of times, the electric and magnetic fields will also increase by the same number of times.

As for such fields as nuclear and gravitational ones, they are described by linear equations only in the cases when the intensities of their sources are not too high. Sufficiently powerful sources can create highly nonlinear fields.

What if the same goes for electromagnetic fields? After all, they can also become nonlinear at very high charges and currents. Moreover, at such charges and currents that can be unattainable in laboratory conditions.

In other words, are Maxwell's equations always true? Maybe their fate will be similar to the fate of the laws of Newtonian mechanics, which gave way to Einstein's theory at speeds close to light and quantum mechanics on atomic scales.

Let us turn to experimental data that are inexplicable from the standpoint of the classical Maxwell theory. A number of mysteries arise here.

1.1.1. *Mystery of the Earth magnetism*

It is striking that there is still no clarity in the question of the nature of the Earth's magnetism. The first known attempt to explain it was undertaken by eminent English naturalist William Gilbert at the end of the 16th century. Having made a small model of the Earth from a permanent magnet, he became convinced of the similarity of its properties with the original and came to the conclusion that the Earth is a huge spherical magnet. This point of view dominated for 300 years, until it was refuted at the end of the 19th century by Pierre Curie. He experimentally proved that permanent magnets (iron and nickel) are demagnetized when a sufficiently high temperature is reached, called the Curie point. As for the interior of the Earth, the temperature in them is much higher than the Curie points for iron and nickel, and therefore Gilbert's idea had to be abandoned.

In the future, a number of other attempts were made to resolve this issue. The most popular now is the dynamo theory, in which the Earth is considered not as a permanent magnet but as an electromagnet. But for its operation, it is necessary, at a minimum, that an electromotive force is constantly maintained in the Earth, which generates electric currents and a constant geomagnetic field for a geologically long time. However, it is completely unclear how this can be done [8].

1.1.2. *Phenomenon of ball lightning*

The ball lightning is one of the most striking phenomena. Appearing sometimes during a thunderstorm, it presents a small elastic ball. As a rule, its radius does not exceed 50 cm and its lifetime can vary from seconds to several minutes [9]. The ability of the ball lightning to maintain its shape is very impressive. Indeed, it looks absolutely incomprehensible what forces hold the ball lightning back from quick disintegration.

If the ball lightning had had no electric charges, then such forces could not appear. But the classical theory is also useless if the ball lightning is a charged object. Actually, if the ball lightning had had charges of the same sign, then it would fall to pieces since such charges repulse. On the other hand, if charged layers of different signs had alternated in it, then it could not maintain its shape since neighboring layers would attract in this case. Besides, the existence of the ball lightning contradicts the classical theorem proven by British physicist and mathematician Samuel Earnshaw in 1842. According to it, no stable static configuration of electric charges can exist [10]. That is why the phenomenon of ball lightning is a direct challenge to the classical linear theory of electromagnetism which demonstrates its incompleteness.

1.1.3. *Problems of atmospheric electricity*

As follows from measurements, when the weather is fair, the electric field at the Earth's surface is about $100\,\mathrm{V/m}$. At the same time, at the height of about 50 km above this surface, the electric field is absent [11]. In order to explain this field property, it was supposed that there could be a concentric-sphere capacitor with a negatively charged part at the Earth's surface and a positively charged part at a height of about 50 km above it. However, such a capacitor should be quickly discharged because of the attractions between positive and negative charges. To save this hypothesis, its supporters suppose that the compensation of charge losses in the capacitor could be carried out by thunderstorms charging it. But modern experimental data do not confirm this point of view. Besides, as explorations of the Earth's atmosphere by rockets show, the following surprising phenomenon is observed. At the heights from 10 km to 80 km in the Earth's atmosphere, a change in the sign of the electric field repeatedly occurs [12]. Within the framework of classical physics, this phenomenon could be explained only by supposing the alternation of negatively and positively charged layers in the Earth's atmosphere. But this supposition cannot be true. Indeed, in this case, neighboring layers should attract. That is why such an atmospheric structure could not be stable: it should quickly disappear.

1.1.4. *Mysteries of atmospheric layers*

The Earth's atmosphere consists of a number of layers. These are the troposphere, stratosphere, ionosphere, protonosphere and some smaller layers. In the uppermost layer, protonosphere, temperatures are very high: their maximum value is about $1300°\mathrm{C}$. Because of such high temperatures of the layer, the Earth's gravitational field is unable to hold back its protons. That is why this layer should have rapidly dissipated. However, nothing of this sort has happened to it.

The neighboring ionosphere and mesosphere layers have sharply contrasting temperatures: In the ionosphere, they are quite high and reach $1100°$C, whereas in the mesosphere, they are very low and reach $-130°$C. It is very surprising that there is no equalization of their temperatures. The nature of this phenomenon is unknown and inexplicable within the framework of classical physics.

There are a number of other mysteries. For example, why are the north and south poles of the Sun subject to reversal every 11 years on average? It is also unknown why the reversal of the Earth's magnetic poles takes place several times every million years.

All these puzzles lead us to the following idea: The classical linear electrodynamics is inapplicable to some class of electromagnetic phenomena and its nonlinear generalization is required to describe them. But the questions arise: What way should be chosen to come to such a new theory? What is the basis for its construction?

However, such a basis has been in existence for almost 70 years and appeared in 1954. This is the famous theory proposed by C. N. Yang and R. Mills. It just presents a reasonable nonlinear generalization of the Maxwell theory. Besides, its nonlinear equations have richer symmetries as compared with the Maxwell equations.

For several years, the Yang–Mills theory was regarded as a very interesting but unfruitful one since no physical applications were seen. However, after that, the ideas suggested by Yang and Mills went through a new stage. They successfully entered into the unified theory of electroweak interactions. Besides, in 1974, a new striking event happened. Namely, on the basis of the Yang–Mills theory, a new particle with only one magnetic pole was predicted. This was the 't Hooft–Polyakov monopole. To detect such an exotic object, many experiments have been carried out. But up to now, they have not been successful. Nevertheless, the search for monopoles goes on. While on the atomic scale the existence of monopoles is an open question, on the cosmic scale, objects of the monopole type really exist. Apparently, many stars become monopoles during some periods of their lifetime. At any rate, the Sun is a clear example of that. As noted above, every 11 years on average, its north and south magnetic poles exchange their places. However, this happens quite peculiarly. In periods of the reversal of the magnetic poles, the magnetic fields in the north and south parts of the Sun change their signs not simultaneously. For several months, and sometimes for a year, the Sun becomes a magnetic monopole.

Let us now return to the Yang–Mills equations. They are intensively studied in all leading physics departments of the world. However, up to now, their use has been mainly restricted to applications to microscopic objects. This point of view has been motivated by the opinion that the only acceptable way to describe macroscopic objects is given by the well-known physics and, in particular, the Maxwell theory. But is it really true? And could the Yang–Mills theory be able to throw light upon the problems noted above that are still unsolved within the framework of the Maxwell theory?

Let us try to understand this. The nonlinear theory proposed in 1954 by Yang and Mills describes three physical fields and each of them is characterized by four potentials. At the same time, the Maxwell theory describes only one such physical field. Besides, the expressions for the energy and momentum of the Yang–Mills field are quite similar to those of the Maxwell field. Moreover, if the second and third fields of the Yang–Mills theory are absent, then its first field is described by the Maxwell linear equations. That is why the Yang–Mills nonlinear theory presents a quite reasonable generalization of the Maxwell linear theory.

These considerations have led us to the idea to use the Yang–Mills theory, instead of the Maxwell theory, in order to describe nonlinear electromagnetic fields generated by sufficiently large electric charges and currents. As will be shown later on, this idea can shed light on the puzzling natural phenomena mentioned above.

First, let us characterize the main properties of the classical Maxwell equations and then study the classical equations proposed by Yang and Mills.

1.2.   Classical Maxwell Equations

Consider the electromagnetic field generated by a classical microscopic source j^ν ($\nu = 0, 1, 2, 3$) which represents the four-dimensional vector of the form $j^\nu = (c\theta, \mathbf{j})$, where θ is the density of electric charges, c is the speed of light and $\mathbf{j}$ is the three-dimensional vector of the densities of electric currents. Then, the classical microscopic equations of the Maxwell electrodynamics can be written as [13]

$$\partial_\mu F^{\mu\nu} = (4\pi/c)j^\nu, \tag{1.2.1}$$

$$F^{\mu\nu} = \partial^\mu A^\nu - \partial^\nu A^\mu. \tag{1.2.2}$$

Here and in the following, we use the Gaussian system of units. The components A^ν and $F^{\mu\nu}$ are the potentials and strengths of the electromagnetic field, respectively, $\mu, \nu = 0, 1, 2, 3$, x^μ are coordinates of the Minkowski space–time, $x^0 = ct$, $x^1 = x$, $x^2 = y$, $x^3 = z$, t is time and x, y, z are spatial rectangular coordinates.

Hereinafter, the summation over indices met twice is implied. In inertial frames of reference, $\partial^0 = \partial_0$, $\partial^l = -\partial_l$, $l = 1, 2, 3$.

The components $F^{\mu\nu}$ have the following physical sense:

$$(F^{\mu\nu}) = \begin{pmatrix} 0 & -E_1 & -E_2 & -E_3 \\ E_1 & 0 & -H_3 & H_2 \\ E_2 & H_3 & 0 & -H_1 \\ E_3 & -H_2 & H_1 & 0 \end{pmatrix}, \tag{1.2.3}$$

where $\mathbf{E} = (E_1, E_2, E_3)$ and $\mathbf{H} = (H_1, H_2, H_3)$ are the vectors of electric and magnetic strengths, respectively.

The strengths $F_{\mu\nu}$ satisfy the equations

$$\partial_\gamma F_{\mu\nu} + \partial_\nu F_{\gamma\mu} + \partial_\mu F_{\nu\gamma} = 0. \tag{1.2.4}$$

Since the components $F^{\mu\nu}$ are antisymmetric, Eq. (1.2.1) gives the following differential equation of charge conservation:

$$\partial_\nu j^\nu = 0. \tag{1.2.5}$$

It should be noted that the Maxwell equations are covariant under the gauge transformation

$$A^\nu \to A^\nu + \partial^\nu \phi, \tag{1.2.6}$$

where ϕ is an arbitrary differentiable function of space–time coordinates.

James Clerk Maxwell
(1831–1879)

There is no doubt that Maxwell's equations describe a variety of electromagnetic fields, the quanta of which are photons. However, when the sources of these fields are powerful enough, they can also generate massive Z and W bosons. In such cases, Maxwell's equations may turn out to be incorrect, since they are applicable to fields for which only photons are their carriers.

It should be noted that the linearity of Maxwell's equations also does not speak in favor of the idea of their universal applicability. When charges and currents are very large, they could be replaced by essentially nonlinear equations.

To find a nonlinear generalization of Maxwell's equations, we turn to the Yang–Mills equations, applicable to the description of electroweak interactions whose quanta are not only photons but also Z and W bosons.

1.3.   Classical Yang–Mills Equations

The nonlinear field equations proposed by Yang and Mills in 1954 can be represented as

$$\partial_\mu F^{k,\mu\nu} + g\varepsilon_{klm}A^l_\mu F^{m,\mu\nu} = (4\pi/c)J^{k,\nu}, \qquad (1.3.1)$$

$$F^{k,\mu\nu} = \partial^\mu A^{k,\nu} - \partial^\nu A^{k,\mu} + g\varepsilon_{klm}A^{l,\mu}A^{m,\nu}, \qquad (1.3.2)$$

where $\mu, \nu = 0, 1, 2, 3$; $k, l, m = 1, 2, 3$, $A^{k,\nu}$, $F^{k,\mu\nu}$ are the potentials and strengths of a Yang–Mills field, respectively, ε_{klm} is the antisymmetric tensor, $\varepsilon_{123} = 1$, $J^{k,\nu}$ are three 4-vectors of current densities, and g is the constant of electroweak interactions.

The Yang–Mills equations (1.3.1)–(1.3.2) are covariant under the following infinitesimal gauge transformations characterizing $SO(3)$ or $SU(2)$ symmetry [1–3]:

$$\begin{aligned} J^{k,\nu} &\to J^{k,\nu} + \varepsilon_{klm}J^{l,\nu}\phi^m, \\ A^{k,\nu} &\to A^{k,\nu} + \varepsilon_{klm}A^{l,\nu}\phi^m + (1/g)\partial^\nu\phi^k, \qquad (1.3.3) \\ F^{k,\mu\nu} &\to F^{k,\mu\nu} + \varepsilon_{klm}F^{l,\mu\nu}\phi^m, \end{aligned}$$

Chen Ning Yang (1922)   **Robert Mills (1927–1999)**

where ϕ^m is a small angle, depending on the coordinates x^μ, of the rotation of the three-dimensional vectors $\mathbf{J}^\nu = (J^{1,\nu}, J^{2,\nu}, J^{3,\nu})$ about mth axis in the gauge space.

The Yang–Mills equations (1.3.1) can be represented in the form

$$D_\mu F^{k,\mu\nu} = (4\pi/c) J^{k,\nu}, \qquad (1.3.4)$$

where D_μ is the Yang–Mills covariant derivative which is defined for any differentiable vector function U^k as

$$D_\mu U^k = \partial_\mu U^k + g\varepsilon_{klm} A_\mu^l U^m. \qquad (1.3.5)$$

It should be noted that the covariant derivative D_μ satisfies the relation [1–3]

$$(D_\mu D_\nu - D_\nu D_\mu) U^k = g\varepsilon_{klm} F_{\mu\nu}^l U^m. \qquad (1.3.6)$$

Replacing U^k by $F^{k,\mu\nu}$, from this formula, we find

$$2D_\mu D_\nu F^{k,\mu\nu} = g\varepsilon_{klm} F_{\mu\nu}^l F^{m,\mu\nu}. \qquad (1.3.7)$$

Since ε_{klm} are antisymmetric, formula (1.3.7) gives the identity

$$D_\mu D_\nu F^{k,\mu\nu} = 0. \qquad (1.3.8)$$

From Eqs. (1.3.4) and (1.3.8), we obtain

$$D_\nu J^{k,\nu} = 0. \qquad (1.3.9)$$

Using (1.3.5), we can represent this formula as

$$D_\nu J^{k,\nu} \equiv \partial_\nu J^{k,\nu} + g\varepsilon_{klm} A_\nu^l J^{m,\nu} = 0. \qquad (1.3.10)$$

1.4. Nonlinear Theory of Strong Electromagnetic Fields

As was said above, there exist a number of natural phenomena that remain still unexplained in which electromagnetic fields are very strong. To find a way to explain such phenomena, let us apply the considered Yang–Mills theory.

We will here study the Yang–Mills equations (1.3.1) and (1.3.2) when their field sources have the following form:

$$J^{1,\nu} = J^{\nu}, \quad J^{2,\nu} = J^{3,\nu} = 0, \tag{1.4.1}$$

where J^{ν} is the classical 4-vector of current densities.

Then, the Yang–Mills equations are covariant under the gauge transformations

$$J^{1,\nu} \to J^{1,\nu}, \quad A^{1,\nu} \to A^{1,\nu} + (1/g)\partial^{\nu}\phi, \quad F^{1,\mu\nu} \to F^{1,\mu\nu},$$
$$A^{2,\nu} \to A^{2,\nu}\cos\phi + A^{3,\nu}\sin\phi, \quad A^{3,\nu} \to A^{3,\nu}\cos\phi - A^{2,\nu}\sin\phi,$$
$$F^{2,\mu\nu} \to F^{2,\mu\nu}\cos\phi + F^{3,\mu\nu}\sin\phi, \quad F^{3,\mu\nu} \to F^{3,\mu\nu}\cos\phi - F^{2,\mu\nu}\sin\phi, \tag{1.4.2}$$

where the angle ϕ is an arbitrary differentiable function of the coordinates x^{μ}.

Formulas (1.4.2) present a generalization for the gauge transformation of Maxwell's equations.

Substituting (1.4.1) into Eq. (1.3.10), we come to the following equations:

$$\partial_{\nu}J^{1,\nu} = 0, \tag{1.4.3}$$

$$A^{k}_{\nu}J^{1,\nu} = 0, \quad k = 2,3, \tag{1.4.4}$$

where the first equation presents the differential equation of charge conservation.

Since the considered Yang–Mills equations are covariant under the gauge transformations (1.4.2), we can choose the angle ϕ in them so as to also fulfill Eq. (1.4.4) when $k = 1$. Therefore, we can put

$$A^{k}_{\nu}J^{1,\nu} = 0, \quad k = 1,2,3. \tag{1.4.5}$$

It should be noted that the Yang–Mills equations with the field sources (1.4.1) have the class of trivial solutions of the form

$$A^{2,\nu} = A^{3,\nu} = 0, \quad F^{2,\mu\nu} = F^{3,\mu\nu} = 0, \tag{1.4.6}$$

where the components $A^{1,\nu}$ and $F^{1,\mu\nu}$ satisfy the Maxwell equations

$$\partial_\mu F^{1,\mu\nu} = (4\pi/c)J^{1,\nu},$$
$$F^{1,\mu\nu} = \partial^\mu A^{1,\nu} - \partial^\nu A^{1,\mu}. \tag{1.4.7}$$

As will be shown later on, the Yang–Mills equations with the field sources (1.4.1) have classes of nontrivial solutions in which the components $F^{2,\mu\nu}$ and $F^{3,\mu\nu}$ are nonzero.

That is why these equations can be regarded as a reasonable nonlinear generalization of the Maxwell equations.

To explain the existence of nontrivial solutions to the Yang–Mills equations with the field sources (1.4.1), let us consider the following equality which follows from Eq. (1.3.4):

$$D_\nu[D_\mu F^{k,\mu\nu} - (4\pi/c)J^{k,\nu}] = 0. \tag{1.4.8}$$

From Eqs. (1.3.10), (1.4.1) and the differential equation (1.4.3) of charge conservation, we find

$$D_\nu J^{k,\nu} \equiv 0 \text{ when } k = 1. \tag{1.4.9}$$

Using identities (1.3.8) and (1.4.9), we come to the following identity:

$$D_\nu[D_\mu F^{k,\mu\mu} - (4\pi/c)J^{k,\nu}] \equiv 0 \text{ when } k = 1. \tag{1.4.10}$$

Hence, equality (1.4.8) is an identity when $k = 1$.

This means that the Yang–Mills equations with the considered field sources of form (1.4.1) are not independent. Therefore, to uniquely determine their solutions, we need some additional equation.

For this purpose, let us represent the Yang–Mills equation (1.3.1) in the form

$$\partial_\mu F^{k,\mu\nu} = (4\pi/c)\hat{J}^{k,\nu}, \tag{1.4.11}$$

where $\hat{J}^{k,\nu}$ is defined as

$$\hat{J}^{k,\nu} = J^{k,\nu} - (cg/4\pi)\varepsilon_{klm}A_\mu^l F^{m,\mu\nu}. \qquad (1.4.12)$$

Since $\partial_\nu\partial_\mu F^{k,\mu\nu} = 0$, from (1.4.11), we derive the differential equation of charge conservation

$$\partial_\nu\hat{J}^{k,\nu} = 0. \qquad (1.4.13)$$

Taking into account Eqs. (1.4.12) and (1.4.13), we can interpret the functions $\hat{J}^{k,\nu}$ as full current densities which contain not only the source current densities $J^{k,\nu}$ but also the current densities of the quanta of Yang–Mills fields.

Let us now find a relation between the densities $J^{k,\nu}$ and $\hat{J}^{k,\nu}$ of field source currents and full currents, respectively. For this purpose, consider a small part of a field source and let Δq^k and $\Delta\hat{q}^k$ ($k = 1, 2, 3$) be its own charges and its full charges, which also include the charges of the Yang–Mills quanta created inside it, respectively.

As is known, the intrinsic electric energy of a homogeneous body with charge q is proportional to q^2. That is why let us require, taking into account the law of energy conservation, the invariance of the value $\sum_{k=1}^3 (\Delta\hat{q}^k)^2$ which is proportional to the electric energy of the considered small part of the field source. Then, since the value $\sum_{k=1}^3 (\Delta\hat{q}^k)^2$ should be the same in both cases when $\Delta\hat{q}^k = \Delta q^k$ and $\Delta\hat{q}^k \neq \Delta q^k$, we come to the following relation:

$$\sum_{k=1}^3 (\Delta\hat{q}^k)^2 = \sum_{k=1}^3 (\Delta q^k)^2. \qquad (1.4.14)$$

Using the components $J^{k,\nu}$ and $\hat{J}^{k,\nu}$ of the densities of the own currents of a field source and its full currents, respectively, relation (1.4.14) can be represented as an equality of the following relativistic invariants:

$$\sum_{k=1}^3 \hat{J}^{k,\nu}\hat{J}_\nu^k = \sum_{k=1}^3 J^{k,\nu}J_\nu^k. \qquad (1.4.15)$$

Equality (1.4.15) which has been derived from the law of energy conservation just presents the sought supplementary equation that should be added to the considered Yang–Mills equations.

We will further consider the nonlinear generalization of the Maxwell equations based on the Yang–Mills equations (1.3.1)–(1.3.2) with field sources of form (1.4.1) and the additional equation (1.4.15).

As follows from (1.4.11), Eq. (1.4.15) can be represented in the form

$$\sum_{k=1}^{3} \partial_\alpha F^{k,\alpha\nu} \partial_\beta F^{k,\beta}_{\ \ \nu} = (4\pi/c)^2 \sum_{k=1}^{3} J^{k,\nu} J^k_\nu. \qquad (1.4.16)$$

Since we consider field sources of form (1.4.1) for which $J^{2,\nu} = J^{3,\nu} = 0$, let us choose the gauge

$$F^{2,\mu\nu} F^2_{\mu\nu} = F^{3,\mu\nu} F^3_{\mu\nu}, \qquad (1.4.17)$$

which implies the equivalence of the second and third axes in the gauge space.

Let us now study the considered nonlinear generalization of Maxwell's equations in a number of interesting cases.

1.5. Yang–Mills Fields of Charged Spherical Bodies and Ball Lightning

Consider the Yang–Mills equations (1.3.1) and (1.3.2) when their sources are of the form

$$J^{1,0} = c\theta(r), \quad J^{1,l} = 0, \quad l = 1, 2, 3, \quad J^{2,\nu} = J^{3,\nu} = 0, \qquad (1.5.1)$$

where $r = \sqrt{(x^1)^2 + (x^2)^2 + (x^3)^2}$.

Then, from Eq. (1.4.5), we find

$$A^{k,0} = 0, \quad k = 1, 2, 3. \qquad (1.5.2)$$

Let us seek the potentials $A^{k,l}$ with $l \neq 0$ in the spherically symmetric case under consideration in the form

$$A^{k,l} = x^l [x^0 a^k(r) + b^k(r)], \quad l = 1, 2, 3, \qquad (1.5.3)$$

where $a^k(r)$ and $b^k(r)$ are some differentiable functions.

Substituting formulas (1.5.2) and (1.5.3) into Eq. (1.3.2), we find

$$F^{k,0l} = x^l a^k, \quad F^{k,lm} = 0, \quad l, m = 1, 2, 3. \quad (1.5.4)$$

Let us now substitute formulas (1.5.1)–(1.5.4) into Eq. (1.3.1). Then, when $\nu = 1, 2, 3$, we obtain zeros and when $\nu = 0$, we come to the following system of equations:

$$r(a^k)' + 3a^k - g\varepsilon_{klm}b^l a^m = -4\pi\theta\delta^k, \quad (1.5.5)$$

where

$$\delta^1 = 1, \quad \delta^2 = \delta^3 = 0, \quad (1.5.6)$$

and we have used the identity $\varepsilon_{klm}a^l a^m = 0$ since ε_{klm} are antisymmetric.

Let us multiply Eq. (1.5.5) by a^k and sum it over k. Then, since $\varepsilon_{klm}a^k a^m = 0$, we obtain

$$ra' + 3a = -4\pi\theta a^1/a, \quad (1.5.7)$$

where

$$a = \sqrt{(a^1)^2 + (a^2)^2 + (a^3)^2}. \quad (1.5.8)$$

Equation (1.5.5) with $k = 2, 3$ gives

$$\begin{aligned}
b^2 &= -(ga^1)^{-1}[r(a^3)' + 3a^3 - gb^1 a^2], \\
b^3 &= (ga^1)^{-1}[r(a^2)') + 3a^2 + gb^1 a^3].
\end{aligned} \quad (1.5.9)$$

Taking into account (1.5.8), let us put

$$a^1 = -a\cos\xi, \quad a^2 = -a\sin\xi\cos\eta, \quad a^3 = -a\sin\xi\sin\eta, \quad (1.5.10)$$

where ξ and η are some functions of r.

Then, Eq. (1.5.7) acquires the form

$$ra' + 3a = 4\pi\theta\cos\xi. \quad (1.5.11)$$

In order to satisfy condition (1.4.17), we put

$$\eta = \pi/4. \tag{1.5.12}$$

Let us turn to Eq. (1.4.16). Using (1.5.1) and (1.5.4), we can represent this equation as

$$\sum_{k=1}^{3} [r(a^k)' + 3a^k]^2 = (4\pi\theta)^2. \tag{1.5.13}$$

Using (1.5.10) and (1.5.12), we can rewrite Eq. (1.5.13) in the form

$$(ra' + 3a)^2 + (ra\xi')^2 = (4\pi\theta)^2. \tag{1.5.14}$$

Let us put

$$a = u(r)/r^3. \tag{1.5.15}$$

Then, Eqs. (1.5.11) and (1.5.14) acquire the form

$$du/dr = 4\pi r^2\theta \cos\xi, \tag{1.5.16}$$
$$(du/dr)^2 + (ud\xi/dr)^2 = (4\pi r^2\theta)^2. \tag{1.5.17}$$

Let us now choose the following variable q instead of r:

$$q = 4\pi \int_0^r r^2\theta(r)dr, \tag{1.5.18}$$

which is the charge of the source in the spherical region of radius r with the center at the zero point. Then, from Eqs. (1.5.16) and (1.5.17), we derive

$$du/dq = \cos\xi, \tag{1.5.19}$$
$$(du/dq)^2 + (ud\xi/dq)^2 = 1. \tag{1.5.20}$$

Substituting formula (1.5.19) into Eq. (1.5.20), we obtain

$$ud\xi/dq = \pm\sin\xi. \tag{1.5.21}$$

Dividing Eq. (1.5.19) by Eq. (1.5.21), we find

$$du/u = \pm \cot \xi. \tag{1.5.22}$$

In order to have a nonsingular solution, we choose the sign "+" in Eq. (1.5.22). Then, from this equation, we obtain

$$u = K_0 \sin \xi, \quad K_0 = \text{const.} \tag{1.5.23}$$

From Eqs. (1.5.21) and (1.5.23), we find, taking into account that the sign "+" has been chosen,

$$d\xi/dq = 1/K_0. \tag{1.5.24}$$

Therefore,

$$\xi = (q + q_0)/K_0, \quad q_0 = \text{const.} \tag{1.5.25}$$

Substituting (1.5.25) into formula (1.5.23) and using formula (1.5.15), we obtain

$$a = (K_0/r^3) \sin \left((q + q_0)/K_0 \right), \tag{1.5.26}$$

where q is defined by formula (1.5.18).

To have no singularity in this formula at $r = 0$, we should put

$$q_0/K_0 = \pi n, \tag{1.5.27}$$

where n is an integer.

Formulas (1.5.10), (1.5.12) and (1.5.25)–(1.5.27) give

$$a^1 = -\frac{K_0}{2r^3} \sin(2q/K_0), \quad a^2 = a^3 = -2^{-3/2} \frac{K_0}{r^3} [1 - \cos(2q/K_0)]. \tag{1.5.28}$$

Let us put

$$K = K_0/2 = \text{const.} \tag{1.5.29}$$

Then, from formulas (1.5.4) and (1.5.28), we obtain that the field strengths $F^{k,\mu\nu}$ are as follows:

$$F^{1,0l} = -(x^l/r^3) K \sin(q/K),$$

$$F^{2,0l} = F^{3,0l} = -2^{-1/2}(x^l/r^3) K [1 - \cos(q/K)], \tag{1.5.30}$$

$$F^{k,lm} = 0, \quad l, m = 1, 2, 3,$$

where q is determined by formula (1.5.18).

From (1.4.11), we have

$$\partial_\mu F^{1,\mu 0} = (4\pi/c)\hat{J}^{1,0}. \tag{1.5.31}$$

Formulas (1.5.30) and (1.5.31) give

$$\frac{1}{r^2}\frac{d}{dr}\left(K\sin(q/K)\right) = (4\pi/c)\hat{J}^{1,0}. \tag{1.5.32}$$

Therefore,

$$(4\pi/c)\int_0^r r^2 \hat{J}^{1,0} dr = K\sin(q/K). \tag{1.5.33}$$

The left-hand side of formula (1.5.33) presents the full charge for the index $k = 1$ in the gauge space inside the circle of radius r having its center at the zero point. The full charge includes not only the source charge q but also the charge of the Yang–Mills field quanta generated by it.

Let us denote this full charge by $\hat{q}$. Then, from (1.5.33), we obtain

$$\hat{q} = K\sin(q/K). \tag{1.5.34}$$

Formula (1.5.34) relates the full charge $\hat{q}$ and the charge q of a field source.

Using (1.5.30) and (1.5.34), we find

$$F^{1,l0} = \hat{q}x^l/r^3, \quad l = 1,2,3. \tag{1.5.35}$$

It should be noted that when $|q/K| \ll 1$, the full charge $\hat{q}$ practically coincides with the source charge q and we have the Maxwell field expressions for the strength components $F^{1,l0}$.

Let us assume that the value K in (1.5.34) is a sufficiently large positive constant. Then, formulas (1.5.34) and (1.5.35) can be regarded as a nonlinear generalization of the corresponding Maxwell field expressions for the strengths $F^{1,l0}$ when the source charge q is sufficiently large. The constant K will be estimated in the following.

1.5.1. *Earth's magnetism and ball lightning*

Consider a spherical body with a classical charge q. One of the interesting properties of the full charge $\hat{q}$ of this body is its sinusoidal dependence on body's charge q. If the charge q monotonically increases and reaches the value πK, then the value of the external electric field of the body passes by zero and after that changes its sign. This property can be called **the effect of saturation** of electric forces. It is analogues to the well-known phenomenon of saturation in nuclear interactions.

It should be stressed that the property of electric field saturation implies the existence of stable spherical bodies, which contain charged particles with the same sign and have the total classical charge $q = \pm 2\pi K$.

Indeed, as follows from formulas (1.5.34) and (1.5.35), under the surface of such a body, attractive forces should act on its charged particles, which should prevent its disintegration. As to the forces acting at its surface, they are equal to zero.

It is most likely that to such bodies, both ball lightning and spherical UFOs can be attributed, which are great puzzles for conventional physics. It is worth noting that up to now, all attempts to explain the stability of these bodies have been unsuccessful within the framework of the classical linear theory of electric fields.

At the same time, the suggested nonlinear theory of electric fields just allows one to explain the appearance of stable configurations in the form of ball lightning and spherical UFOs.

Let us also note the following. Because of formula (1.5.34), spherically symmetric bodies with classical charges which differ by $2\pi nK$, where n is an arbitrary integer, should have the same full charge $\hat{q}$ and hence the same external electric field. That is why the knowledge of the external electric field of a sufficiently big body and hence of its full charge $\hat{q}$ does not allow one to uniquely determine its classical charge q. Besides, in the case of sufficiently large values of the number n, the classical charge of the body can substantially exceed its full charge.

As an application of this property of formula (1.5.34) for the full charge, let us turn to the question of the nature of the Earth's magnetic field.

It is known that the origin of the Earth's magnetic field remains puzzling despite many attempts to explain it [8]. As an illustration of emerging difficulties, let us first note that the iron–nickel core of the Earth cannot be ferromagnetic since its temperature is considerably higher than the Curie temperature, after which ferromagnetic properties vanish. Second, since there are no reasons for appearing a constantly acting electromotive force in the Earth, any explanations of its magnetic field that are based on hypothetical currents arising in it run into serious obstacles [8].

That is why consider this question within the framework of the suggested nonlinear generalization of the Maxwell equations based on the Yang–Mills equations.

As is well known, in fair weather, the electric field near the Earth's surface is directed downwards and its value is about $100\,\mathrm{V/m}$ [11]. This means that the full charge of the Earth, which determines its external electric field, is negative and for the absolute value $|\hat{q}_{(E)}|$ of its full charge, the following estimate can be made:

$$|\hat{q}_{(E)}| \sim 5 \times 10^5 \text{ coul.} \tag{1.5.36}$$

Let us denote the classical charge of the Earth by $q_{(E)}$. Then, from (1.5.34) and (1.5.36), we obtain

$$K \sim 5 \times 10^5 / |\sin(q_{(E)}/K)| \text{ coul.} \tag{1.5.37}$$

As stated above, the full charge $\hat{q}$ of a sufficiently big spherically symmetric body, which can be determined by measuring its external field, does not allow one to uniquely determine its classical charge. This conclusion is a consequence of the fact that the function $\hat{q} = K\sin(q/K)$ is periodic and hence there exist an infinite number of values of the classical charge q corresponding to a given value $\hat{q}$.

Because of this circumstance and since the value $\hat{q}_{(E)}$ of Earth's full charge is comparatively small for such a big body as our planet, it can be supposed that the absolute value $|q_{(E)}|$ of Earth's classical charge is much larger than the value $|\hat{q}_{(E)}|$:

$$|q_{(E)}| \gg |\hat{q}_{(E)}| \sim 5 \times 10^5 \text{ coul.} \tag{1.5.38}$$

Then, it becomes possible to explain the nature of the Earth's magnetic field. Namely, owing to (1.5.38), this magnetic field can be interpreted as the result of the axial rotation of Earth's charges.

It should be noted that the charge $\hat{q}_{(E)}$ itself considered as Earth's charge within the framework of the classical linear theory of electricity is too small for such an explanation of its magnetic field.

As for the suggested nonlinear generalization of the Maxwell theory, where $q_{(E)}$ can be very large, it gives a reasonable explanation for the observed value of Earth's magnetic field.

It is worth noting that during a sufficiently large period of time, Earth's classical charge can substantially change because of both volcanic eruptions and falls of meteorites on the Earth. As a result, Earth's full charge, which is related to its classical charge by formula (1.5.34), can change its sigh and then a reversal of Earth's magnetic field becomes possible. As stated above, the phenomenon of reversals of Earth's magnetic field really takes place and happens several times per million years. However, it has not been given any satisfactory explanation within the framework of the Maxwell theory.

Let us dwell on the question of the nature of ball lightning, which up to now has not also been resolved within the framework of the classical linear electrodynamics [9].

Consider a ball lightning appearing during a thunderstorm and consisting of free electrons, neutral molecules of the atmosphere and negative ions of ozone which are formed from oxygen owing to thunderstorm electric discharges [14].

As stated above, the existence of such a stable charged ball can be explained by using formula (1.5.34). Consider this question in more detail.

With this aim, examine a ball consisting of free electrons, neutral molecules of the atmosphere and negatively charged ions of ozone and having radius $r^{(b)}$ and the classical charge $q^{(b)} = -2\pi K$.

Consider a part of this ball for which radius r lies in the range $r_1^{(b)} < r < r^{(b)}$ and which has the classical charge $q^{(b)}/2 = -\pi K$. As follows from formulas (1.5.34) and (1.5.35) for the field strengths $F^{1,l0}$, in the region $r_1^{(b)} < r < r^{(b)}$ of the examined ball, attractive forces act on negative ions of ozone and electrons which produce pressure in the ball, whereas repulsive forces act on negative charges surrounding the ball.

Therefore, the sphere $r = r^{(b)}$ can be exactly the surface bounding a stable ball lightning with the following classical charge $q^{(b)}$:

$$q^{(b)} = -2\pi K. \qquad (1.5.39)$$

Let us now estimate the number of molecules N_{mol} in a ball lightning with radius $r^{(b)}$. Let r_{mol} denote the average radius of a molecule in the atmosphere. Then, the number N_{mol} can be estimated by the following manner assuming that the molecules in the ball lightning are closely packed:

$$N_{\mathrm{mol}} \sim (r^{(b)}/r_{\mathrm{mol}})^3. \tag{1.5.40}$$

On the other hand, the number of molecules in the ball lightning can be expressed by means of the evident identity

$$N_{\mathrm{mol}} = |q^{(b)}|/[(n_{\mathrm{oz}} + n_e)e_0], \quad e_0 = 1.602 \times 10^{-19} \text{ coul.,}$$
$$n_{\mathrm{oz}} = N_{\mathrm{oz}}/N_{\mathrm{mol}}, \quad n_e = N_e/N_{\mathrm{mol}}, \tag{1.5.41}$$

where $q^{(b)}$ is the classical charge of the ball lightning, N_{oz} and N_e are the numbers of the negative ions of ozone and electrons in it, respectively, and e_0 is the elementary charge.

From (1.5.39)–(1.5.41), we find

$$r^{(b)} \sim r_{\mathrm{mol}}\{2\pi K/[(n_{\mathrm{oz}} + n_e)e_0]\}^{1/3}. \tag{1.5.42}$$

The fraction n_{oz} of ozone, which can be formed in the atmosphere from oxygen owing to thunderstorm discharges, is $\sim 1\%$ [14] and the average radius of a molecule in the atmosphere is $r_{\mathrm{mol}} \sim 1.5 \times 10^{-8}$ cm.

Thus, we have

$$n_{\mathrm{oz}} \sim 10^{-2}, \quad r_{\mathrm{mol}} \sim 1.5 \times 10^{-8} \text{ cm}, \quad e_0 = 1.602 \times 10^{-19} \text{ coul.} \tag{1.5.43}$$

Let us estimate now the maximum radius $r_{\mathrm{max}}^{(b)}$ of ball lightning.

As follows from (1.5.42) and (1.5.43), the maximum radius $r_{\mathrm{max}}^{(b)}$ is reached when $n_e \ll n_{\mathrm{oz}}$. Therefore, formula (1.5.42) gives

$$r_{\mathrm{max}}^{(b)} \sim r_{\mathrm{mol}}[2\pi K/(n_{\mathrm{oz}}e_0)]^{1/3}. \tag{1.5.44}$$

As observational data show, the maximum radius of ball lightning is $r_{\mathrm{max}}^{(b)} \sim 50$ cm [9]. Using this value and data (1.5.43), from formula (1.5.44), we obtain the following estimate for the value K:

$$K \sim 10^7 \text{ coul.,} \tag{1.5.45}$$

which refines estimate (1.5.37).

It should be stressed again that only a nonlinear theory of electric fields can describe such a phenomenon as ball lightning. This conclusion is based on the following fact: The classical linear theory of electricity is inapplicable to ball lightning in principle since Earnshaw's theorem, mentioned above in Section 1.1, which is valid in the linear theory, forbids stable existence of objects of a ball lightning kind.

Consider now a charged body which is not spherical. Its electric field at the sphere of a sufficiently big radius r depends on its charge q and is independent of its form. Therefore, the strengths $F^{1,l0}$ at $r \gg r_m$, where r_m is the maximum dimension of the body, should be described by formulas (1.5.34)–(1.5.35):

$$F^{1,l0} = K \sin(q/K) x^l/r^3, \quad r \gg r_m, \quad l = 1, 2, 3. \tag{1.5.46}$$

Let us use Eq. (1.4.11). It gives

$$\partial_l F^{1,l0} = (4\pi/c)\hat{J}^{1,0}. \tag{1.5.47}$$

Let us integrate this equation over the spherical region with a radius $r \gg r_m$. Then, using Gauss's theorem and formula (1.5.46), we obtain

$$K \sin(q/K) = \hat{q}, \tag{1.5.48}$$

where q and $\hat{q}$ are the classical charge and full charge of the considered body, respectively.

Therefore, formula (1.5.48) should be valid for a charged body with an arbitrary form.

As a result, we come to the following principle:

Principle 1.5.1. *Consider a charged body with an arbitrary form which is under the action of only its own electric field. Then, its classical charge q and its full charge $\hat{q}$ are related by formula (1.5.48), where the constant K is estimated by formula (1.5.45).*

1.6.   A Class of Solutions to the Yang–Mills Equations with Nonstationary Spherically Symmetric Sources

Consider the Yang–Mills equations with the following nonstationary spherically symmetric field sources:

$$
\begin{aligned}
J^{1,0} &= j^0(\tau, r), \quad J^{1,l} = x^l j(\tau, r), \quad \tau = x^0, \\
r &= \sqrt{(x^1)^2 + (x^2)^2 + (x^3)^2}, \quad J^{2,\nu} = J^{3,\nu} = 0,
\end{aligned}
\tag{1.6.1}
$$

where $l = 1, 2, 3$ and j^0 and j are some functions of the time coordinates x^0 and the radial coordinate r.

Let us seek nonstationary spherically symmetric solutions to the Yang–Mills equations (1.3.1) and (1.3.2) with the field sources (1.6.1) in the form

$$A^{k,0} = a^k(\tau, r), \quad A^{k,l} = x^l b^k(\tau, r), \tag{1.6.2}$$

where a^k and b^k are some differentiable functions of τ and r.

Substituting expressions (1.6.2) into formula (1.3.2) for the field strengths $F^{k,\mu\nu}$, we find

$$F^{k,0l} = x^l u^k(\tau, r), \quad F^{k,ml} = 0, \quad k, m, l = 1, 2, 3, \tag{1.6.3}$$

$$u^k = b_\tau^k + (1/r)a_r^k + g\varepsilon_{klm}a^l b^m, \tag{1.6.4}$$

where $a_r^k \equiv \partial a^k/\partial r$, $b_\tau^k \equiv \partial b^k/\partial \tau$.

Let us now substitute expressions (1.6.1)–(1.6.3) for the field sources, potentials and strengths under examination into the Yang–Mills field equation (1.3.1). Then, we come to the following system of equations:

$$ru_r^k + 3u^k - gr^2\varepsilon_{klm}b^l u^m = -(4\pi/c)j^0\delta^k, \tag{1.6.5}$$

$$u_\tau^k + g\varepsilon_{klm}a^l u^m = (4\pi/c)j\delta^k, \tag{1.6.6}$$

where

$$\delta^1 = 1, \quad \delta^2 = \delta^3 = 0. \tag{1.6.7}$$

Thus, we should study the system of the nonlinear partial differential equations (1.6.4)–(1.6.6).

Further, we will seek and find a class of nontrivial solutions to these equations for arbitrary functions j^0 and j of the arguments τ and r that accord with the differential equation (1.4.3) of charge conservation.

Let us turn to the differential equation (1.4.3) and relation (1.4.5). Substituting expressions (1.6.1) and (1.6.2) into them, we obtain

$$j_\tau^0 + rj_r + 3j = 0, \tag{1.6.8}$$

$$a^k j^0 - r^2 b^k j = 0. \tag{1.6.9}$$

Let us now multiply Eq. (1.6.5) by j and Eq. (1.6.6) by j^0 and take their sum. Then, we find

$$j^0 u^k_\tau + j(r u^k_r + 3 u^k) + g\varepsilon_{klm}(j^0 a^l - r^2 j b^l)u^m = 0. \qquad (1.6.10)$$

Using (1.6.9), from Eq. (1.6.10), we derive

$$j^0 u^k_\tau + j(r u^k_r + 3 u^k) = 0. \qquad (1.6.11)$$

This is a linear differential equation of the first order for the functions $u^k(\tau, r)$, where the functions $j^0(\tau, r)$ and $j(\tau, r)$ satisfy equation (1.6.8).

Thus, we have proved the following proposition:

Proposition 1.6.1. *One of the consequences of the nonlinear differential equations* (1.6.5) *and* (1.6.6) *and relation* (1.6.9) *is the linear differential equation* (1.6.11).

Putting

$$u^k = h^k/r^3, \qquad (1.6.12)$$

where $h^k = h^k(\tau, r)$ are some differentiable functions, from (1.6.11), we derive

$$j^0 h^k_\tau + r j h^k_r = 0. \qquad (1.6.13)$$

Let us now find the general solution to Eq. (1.6.13). For this purpose, consider the function

$$q = (4\pi/c) \int_0^r r^2 j^0(\tau, r)dr. \qquad (1.6.14)$$

As follows from (1.6.1), the function $q = q(\tau, r)$ is the charge in the spherical region of radius r having its center at the zero point. From (1.6.14), we derive

$$q_\tau = (4\pi/c) \int_0^r r^2 j^0_\tau dr, \quad q_r = (4\pi/c)r^2 j^0. \qquad (1.6.15)$$

Using (1.6.8) and (1.6.15), we find

$$q_\tau = -(4\pi/c) \int_0^r r^2 (r j_r + 3j) dr$$

$$= -(4\pi/c) \int_0^r (r^3 j)_r dr = -(4\pi/c) r^3 j. \tag{1.6.16}$$

Formulas (1.6.15) and (1.6.16) give

$$j^0 q_\tau + r j q_r = 0. \tag{1.6.17}$$

Therefore, the function q is a particular solution to Eq. (1.6.13). Using (1.6.17), we obtain that the general solution to Eq. (1.6.13) has the form

$$h^k = \Phi^k(q), \tag{1.6.18}$$

where Φ^k are arbitrary differentiable functions of the argument q. Indeed,

$$h_\tau^k = \left(d\Phi^k/dq\right) q_\tau, \quad h_r^k = \left(d\Phi^k/dq\right) q_r. \tag{1.6.19}$$

That is why, using (1.6.17), we find that the functions (1.6.18) satisfy Eq. (1.6.13).

Besides, formula (1.6.18) contains the arbitrary differentiable functions Φ^k. Therefore, it gives the general solution to Eq. (1.6.13) since this partial differential equation is of the first order.

From (1.6.12) and (1.6.18), we obtain

$$u^k = (1/r^3) \Phi^k(q). \tag{1.6.20}$$

Thus, we have proved the following proposition:

Proposition 1.6.2. *The general solution to the partial differential equation (1.6.11), in which the functions j^0 and j satisfy Eq. (1.6.8), is determined by formula (1.6.20), where $\Phi^k(q)$ are arbitrary differentiable functions and q is defined by formula (1.6.14).*

In order to have no singularity for the functions u^k at $r = 0$, we require

$$\Phi^k(0) = 0. \tag{1.6.21}$$

From formulas (1.6.9), (1.6.15), (1.6.16) and (1.6.20), we obtain

$$a^k = \frac{r^2 j}{j^0} b^k, \quad u_\tau^k = -\frac{4\pi j}{c} \frac{d\Phi^k}{dq}, \quad r u_r^k + 3 u^k = \frac{4\pi j^0}{c} \frac{d\Phi^k}{dq}. \quad (1.6.22)$$

Let us substitute formulas (1.6.20) and (1.6.22) into the studied equations (1.6.5) and (1.6.6). Then, the two equations give the same differential equation of the form

$$\frac{d\Phi^k}{dq} - \frac{gc}{4\pi r j^0} \varepsilon_{klm} b^l \Phi^m = -\delta^k. \quad (1.6.23)$$

Let us now multiply Eq. (1.6.23) by Φ^k and sum it over k. Then, using (1.6.7) and the identity $\varepsilon_{klm} \Phi^k \Phi^m = 0$ since ε_{klm} are antisymmetric, we find

$$\sum_{k=1}^{3} \Phi^k d\Phi^k / dq = -\Phi^1. \quad (1.6.24)$$

Thus, the components u^k of the considered field strengths are given by formula (1.6.20), where the functions $\Phi^k(q)$ should satisfy (1.6.21) and Eq. (1.6.24).

In addition to Eq. (1.6.24), from Eq. (1.6.23) by putting $k = 3, 2$, we obtain the following two relations for the functions b^k:

$$b^2 = -\frac{1}{\Phi^1} \left(\frac{4\pi r j^0}{gc} \frac{d\Phi^3}{dq} - b^1 \Phi^2 \right),$$

$$b^3 = \frac{1}{\Phi^1} \left(\frac{4\pi r j^0}{gc} \frac{d\Phi^2}{dq} + b^1 \Phi^3 \right). \quad (1.6.25)$$

1.6.1.   *Investigation of spherically symmetric field potentials for the considered sources*

Consider Eq. (1.6.9). It gives

$$a^k = \gamma b^k, \quad \gamma = r^2 j / j^0. \quad (1.6.26)$$

Here, the functions j^0 and j satisfy Eq. (1.6.8).

Let us turn to Eq. (1.6.4). Substituting (1.6.26) into it and using the identity $\varepsilon_{klm}b^l b^m = 0$ since ε_{klm} is antisymmetric, from this equation, we derive

$$u^k = b_\tau^k + (1/r)(\gamma b^k)_r. \tag{1.6.27}$$

As stated above, the functions u^k are given by formulas (1.6.20). Thus, we have Eqs. (1.6.25) and (1.6.27) for the functions a^k and b^k.

Let us prove the following proposition:

Proposition 1.6.3. *Equation* (1.6.27) *with* $k = 2, 3$ *is a consequence of formulas* (1.6.20), (1.6.25) *and this equation with* $k = 1$.

Proof. Let us use formula (1.6.20) and the third equality in (1.6.22). Then, Eqs. (1.6.25) can be represented as

$$\begin{aligned}
b^2 &= (gr^3u^1)^{-1}(gr^3u^2b^1 - r^2u_r^3 - 3ru^3), \\
b^3 &= (gr^3u^1)^{-1}(gr^3u^3b^1 + r^2u_r^2 + 3ru^2).
\end{aligned} \tag{1.6.28}$$

Let us introduce the complex functions

$$w = b^2 + ib^3, \tag{1.6.29}$$

$$p = u^2 + iu^3. \tag{1.6.30}$$

Then, Eqs. (1.6.28) can be represented as

$$w = (gu^1)^{-1}(gpb^1 + ir^{-1}p_r + 3ir^{-2}p). \tag{1.6.31}$$

As follows from Proposition 1.6.2, formula (1.6.20) for u^k satisfies Eq. (1.6.11). From this equation, formula (1.6.30) and the expression for γ in (1.6.26), we find

$$ru_\tau^1 + \gamma(u_r^1 + 3u^1/r) = 0, \quad rp_\tau + \gamma(p_r + 3p/r) = 0. \tag{1.6.32}$$

Using (1.6.31) and the second formula in (1.6.32), we obtain

$$\gamma w = (gu^1)^{-1}(g\gamma pb^1 - ip_\tau). \tag{1.6.33}$$

Differentiating Eq. (1.6.31) with respect to τ and Eq. (1.6.33) with respect to r, we derive

$$w_\tau = -(gu^1)^{-1}[gu^1_\tau w - g(p_\tau b^1 + pb^1_\tau) - ir^{-2}(rp_{\tau r} + 3p_\tau)],$$
(1.6.34)

$$(\gamma w)_r = -(gu^1)^{-1}[gu^1_r\gamma w - g(\gamma b^1)_r p - gp_r\gamma b^1 + ip_{\tau r}].$$
(1.6.35)

This gives

$$w_\tau + (1/r)(\gamma w)_r = -(gru^1)^{-1}\{(ru^1_\tau + \gamma u^1_r)gw - gp[rb^1_\tau + (\gamma b^1)_r]$$
$$-g(rp_\tau + \gamma p_r)b^1 - 3ir^{-1}p_\tau\}.$$
(1.6.36)

From Eq. (1.6.27) with $k = 1$ and Eq. (1.6.32), we find

$$rb^1_\tau + (\gamma b^1)_r = ru^1, \quad ru^1_\tau + \gamma u_r = -3\gamma u^1/r, \quad rp_\tau + \gamma p_r = -3\gamma p/r.$$
(1.6.37)

Substituting formulas (1.6.37) into Eq. (1.6.36), we get

$$w_\tau + (1/r)(\gamma w)_r = p + 3(gr^2 u^1)^{-1}(g\gamma wu^1 - g\gamma pb^1 + ip_\tau).$$ (1.6.38)

As follows from (1.6.33),

$$g\gamma wu^1 - g\gamma pb^1 + ip_\tau = 0.$$
(1.6.39)

Therefore, from Eq. (1.6.38), we obtain

$$w_\tau + (1/r)(\gamma w)_r = p.$$
(1.6.40)

From this equality and formulas (1.6.29) and (1.6.30), we derive that Eq. (1.6.27) is fulfilled for $k = 2, 3$.

Thus, we have proved Proposition 1.6.3.

Let us now turn to Eq. (1.6.27) with $k = 1$. Using (1.6.20), we can represent it as

$$rb^1_\tau + \gamma b^1_r + \gamma_r b^1 = (1/r^2)\Phi^1(q).$$
(1.6.41)

In any region where the sign of the function $j^0(\tau, r)$ does not change, we can introduce the inverse function

$$r = r(\tau, q)$$
(1.6.42)

for the function $q = q(\tau, r)$ of form (1.6.14).

Let us apply this inverse function to solve Eq. (1.6.41). Using (1.6.42), we can put

$$b^1 = B(\tau, q), \qquad (1.6.43)$$

Then, using formulas (1.6.15) and (1.6.16), we find

$$b^1_\tau = B_\tau - (4\pi/c)r^3 j B_q, \quad b^1_r = (4\pi/c)r^2 j^0 B_q. \qquad (1.6.44)$$

Substituting formulas (1.6.43) and (1.6.44) into Eq. (1.6.41), we obtain

$$rB_\tau + (4\pi/c)r^2(\gamma j^0 - r^2 j)B_q + \gamma_r B = (1/r^2)\Phi^1(q). \qquad (1.6.45)$$

As follows from (1.6.26), $\gamma j^0 - r^2 j = 0$. Therefore, Eq. (1.6.45) acquires the form

$$B_\tau + (\gamma_r/r)B = (1/r^3)\Phi^1(q). \qquad (1.6.46)$$

Using (1.6.42), we can put

$$\gamma_r/r = P(\tau, q), \quad 1/r^3 = Q(\tau, q), \qquad (1.6.47)$$

where P and Q are some functions of τ and q.

From (1.6.46) and (1.6.47), we obtain the following ordinary differential equation of the first order for the function $B = B(\tau, q)$:

$$B_\tau + P(\tau, q)B = Q(\tau, q)\Phi^1(q). \qquad (1.6.48)$$

From this linear differential equation for the function B, we readily find

$$B = \Phi^1(q)\exp\left(-\int_a^\tau P d\tau\right)\left[\int_a^\tau Q \exp\left(\int_a^\tau P d\tau\right)d\tau + C(q)\right],$$
$$a = \text{const}, \qquad (1.6.49)$$

where $C(q)$ is an arbitrary function.

As follows from (1.6.43), formula (1.6.49) determines the function b^1.

Thus, formulas (1.6.25), (1.6.26) and (1.6.49) give expressions for the field potentials $A^{k,\nu}$ and formulas (1.6.20), (1.6.21) and the differential relation (1.6.24) give expressions for the field strengths $F^{k,\mu\nu}$.

It should be noted that the functions $\Phi^k(q)$ in (1.6.20) can be arbitrary differentiable functions satisfying (1.6.21) and relation (1.6.24).

The obtained class of solutions to the Yang–Mills equations (1.3.1) and (1.3.2) with the field source components of form (1.6.1) can be extended by using the gauge transformations (1.4.2).

1.6.2. *Investigation of nonstationary spherically symmetric solutions to the considered Yang–Mills equations with additional physical conditions*

Let us now study nonstationary spherically symmetric solutions satisfying not only the Yang–Mills equations (1.3.1) and (1.3.2) but also the additional differential condition (1.4.16). For this purpose, let us turn to expressions (1.6.3) and (1.6.20) for the field strengths $F^{k,\mu\nu}$.

From (1.6.3) and (1.6.22), we find

$$\partial_\mu F^{k,\mu 0} = -\frac{4\pi j^0}{c}\frac{d\Phi^k}{dq}, \quad \partial_\mu F^{k,\mu l} = \frac{4\pi j x^l}{c}\frac{d\Phi^k}{dq}, \quad l = 1,2,3.$$

$$(1.6.50)$$

Substituting expressions (1.6.1) and (1.6.50) into Eq. (1.4.16), we come to the following equation:

$$\sum_{k=1}^{3}\left(d\Phi^k/dq\right)^2 = 1. \tag{1.6.51}$$

This equation is additional to Eq. (1.6.24) for the functions $\Phi^k(q)$. Let us put

$$\Phi^1 = -\Phi\cos\xi, \quad \Phi^2 = -\Phi\sin\xi\cos\eta, \quad \Phi^3 = -\Phi\sin\xi\sin\eta,$$

$$(1.6.52)$$

where $\Phi = \Phi(q)$, $\xi = \xi(q)$, $\eta = \eta(q)$ are some differentiable functions.

In order to satisfy condition (1.4.17), let us choose the following gauge:

$$\eta = \pi/4. \tag{1.6.53}$$

Then, from (1.6.24), (1.6.51) and (1.6.52), we derive

$$d\Phi/dq = \cos\xi, \tag{1.6.54}$$

$$(d\Phi/dq)^2 + (\Phi d\xi/dq)^2 = 1. \tag{1.6.55}$$

Substituting (1.6.54) into (1.6.55), we obtain

$$\Phi d\xi/dq = \pm \sin \xi. \tag{1.6.56}$$

Dividing Eq. (1.6.54) by Eq. (1.6.56), we find

$$d\Phi/\Phi = \pm \cot \xi \, d\xi. \tag{1.6.57}$$

In order to have no singularity, let us choose the sign '+' in (1.6.57) and hence in (1.6.56). Then, integrating Eq. (1.6.57), we obtain

$$\Phi = K_0 \sin \xi, \quad K_0 = \text{const.} \tag{1.6.58}$$

Let us substitute (1.6.58) into (1.6.56). Then, taking into account that the sign '+' is chosen in (1.6.56) and using (1.6.21), we get

$$\xi = q/K_0 + \pi n, \tag{1.6.59}$$

where n is an integer.

Formulas (1.6.52), (1.6.53), (1.6.58) and (1.6.59) give

$$\Phi^1 = -(K_0/2)\sin(2q/K_0),$$
$$\Phi^2 = \Phi^3 = -2^{-3/2} K_0 [1 - \cos(2q/K_0)]. \tag{1.6.60}$$

Formulas (1.6.60) can be represented in the form

$$\Phi^1 = -K \sin(q/K), \quad \Phi^2 = \Phi^3 = -2^{-1/2} K [1 - \cos(q/K)], \tag{1.6.61}$$

where

$$K = K_0/2 = \text{const.} \tag{1.6.62}$$

From formulas (1.6.3) and (1.6.20), we obtain that the field strengths $F^{k,\mu\nu}$ are as follows:

$$F^{k,0l} = (x^l/r^3)\Phi^k, \quad F^{k,lm} = 0, \quad l, m = 1, 2, 3, \tag{1.6.63}$$

where the functions Φ^k are determined by expressions (1.6.61).

Formulas (1.6.63) give

$$\partial_\mu F^{k,\mu 0} = -(1/r^2)\Phi_r^k. \tag{1.6.64}$$

Let us substitute formula (1.6.64) into Eq. (1.4.11) with $\nu = 0$. Then, we get

$$\Phi_r^k = -(4\pi/c)r^2 \hat{J}^{k,0}. \tag{1.6.65}$$

Taking into account (1.6.21), from (1.6.65), we find

$$\Phi^k = -(4\pi/c) \int_0^r r^2 \hat{J}^{k,0} dr. \tag{1.6.66}$$

Formulas (1.6.61) and (1.6.66) with $k = 1$ give

$$(4\pi/c) \int_0^r r^2 \hat{J}^{1,0} dr = K \sin(q/K). \tag{1.6.67}$$

The left-hand side of formula (1.6.67) presents the full charge for the index $k = 1$ in the gauge space inside the circle of radius r having its center at the zero point. The full current includes not only the classical charge q but also the charge of the Yang–Mills field quanta generated by it.

Let us denote this full charge by $\hat{q}$. Then, from (1.6.67), we obtain

$$\hat{q} = K \sin(q/K). \tag{1.6.68}$$

Formula (1.6.68) relates the full charge $\hat{q}$ and the classical charge q of a field source. It generalizes formula (1.5.34) obtained in the stationary case.

Using (1.6.61), (1.6.63) and (1.6.68), we find

$$F^{1,l0} = \hat{q}x^l/r^3, \quad l = 1, 2, 3. \tag{1.6.69}$$

It should be noted that when $|q/K| \ll 1$, the full charge $\hat{q}$ practically coincides with the classical charge q and we have the Maxwell field expressions for the strength components $F^{1,l0}$.

Formulas (1.6.68) and (1.6.69) can be regarded as a nonlinear generalization of the corresponding Maxwell field expressions for the strengths $F^{1,l0}$ when the source charge q is sufficiently large.

1.7.   A Class of One-Dimensional Solutions to the Yang–Mills Equations

Consider the Yang–Mills equations (1.3.1) and (1.3.2) when their sources are of the form

$$J^{1,0} = c\theta(z), \quad J^{1,l} = 0, \; l = 1, 2, 3, \quad J^{2,\nu} = J^{3,o} = 0, \quad (x, y) \in S,$$
$$(1.7.1)$$

where $x = x^1$, $y = x^2$, $z = x^3$ and S is some region in the plane (x, y).

Then, from Eq. (1.4.5), we find

$$A^{k,0} = 0, \quad k = 1, 2, 3. \tag{1.7.2}$$

Let us seek the potentials $A^{k,l}$ with $l \neq 0$ in the form

$$A^{k,1} = A^{k,2} = 0, \quad A^{k,3} = x^0 a^k(z) + b^k(z), \tag{1.7.3}$$

where $a^k(z)$ and $b^k(z)$ are some differentiable functions.

Substituting formulas (1.7.2) and (1.7.3) into Eq. (1.3.2), we find

$$F^{k,01} = F^{k,02} = 0, \quad F^{k,03} = a^k(z), \quad F^{k,lm} = 0, \quad l, m = 1, 2, 3. \tag{1.7.4}$$

Let us now substitute formulas (1.7.1)–(1.7.4) into Eq. (1.3.1). Then, when $\nu = 1, 2, 3$, we obtain zeros and when $\nu = 0$, we come to the following system of equations:

$$(a^k)' - g\varepsilon_{klm}b^l a^m = -4\pi\theta\delta^k, \tag{1.7.5}$$

where

$$\delta^1 = 1, \quad \delta^2 = \delta^3 = 0, \tag{1.7.6}$$

and we have used the identity $\varepsilon_{klm}a^l a^m = 0$ since ε_{klm} are antisymmetric.

Let us multiply Eq. (1.7.5) by a^k and sum it over k. Then, since $\varepsilon_{klm}a^k a^m = 0$, we obtain

$$a' = -4\pi\theta a^1/a, \qquad (1.7.7)$$

where

$$a = \sqrt{(a^1)^2 + (a^2)^2 + (a^3)^2}. \qquad (1.7.8)$$

Equation (1.7.5) with $k = 2, 3$ gives

$$\begin{aligned}
b^2 &= -(ga^1)^{-1}[(a^3)' - gb^1 a^2], \\
b^3 &= (ga^1)^{-1}[(a^2)') + gb^1 a^3].
\end{aligned} \qquad (1.7.9)$$

Taking into account (1.7.8), let us put

$$a^1 = -a\cos\xi, \quad a^2 = -a\sin\xi\cos\eta, \quad a^3 = -a\sin\xi\sin\eta, \qquad (1.7.10)$$

where ξ and η are some functions of z.

Then, Eq. (1.7.7) acquires the form

$$a' = 4\pi\theta\cos\xi. \qquad (1.7.11)$$

In order to satisfy condition (1.4.17), we put

$$\eta = \pi/4. \qquad (1.7.12)$$

Let us turn to Eq. (1.4.16). Using (1.7.1) and (1.7.4), we can represent this equation as

$$\sum_{k=1}^{3}[(a^k)']^2 = (4\pi\theta)^2. \qquad (1.7.13)$$

Using (1.7.10) and (1.7.12), we can rewrite Eq. (1.7.13) in the form

$$(a')^2 + (a\xi')^2 = (4\pi\theta)^2. \qquad (1.7.14)$$

Let us now choose the following variable q instead of z:

$$q = S\int_0^z \theta(z)dz, \qquad (1.7.15)$$

which is the charge of the field source in the region $(x^1, x^2) \in S$, $0 \le x^3 \le z$.

Then, from Eqs. (1.7.11) and (1.7.14), we derive

$$da/dq = (4\pi/S)\cos\xi, \tag{1.7.16}$$

$$(da/dq)^2 + (ad\xi/dq)^2 = (4\pi/S)^2. \tag{1.7.17}$$

Substituting formula (1.7.16) into Eq. (1.7.17), we obtain

$$ad\xi/dq = \pm(4\pi/S)\sin\xi. \tag{1.7.18}$$

Dividing Eq. (1.7.16) by Eq. (1.7.18), we find

$$da/a = \pm\cot\xi. \tag{1.7.19}$$

In order to have a nonsingular solution, we choose the sign "+" in Eq. (1.7.19). Then, from this equation, we obtain

$$a = K_0\sin\xi, \quad K_0 = \text{const.} \tag{1.7.20}$$

From (1.7.18) and (1.7.20), we find, taking into account that the sign "+" has been chosen,

$$d\xi/dq = 4\pi/(K_0 S). \tag{1.7.21}$$

Therefore,

$$\xi = 4\pi(q + q_0)/(K_0 S) \quad q_0 = \text{const.} \tag{1.7.22}$$

From formulas (1.7.10), (1.7.12) and (1.7.20), we find

$$a^1 = -(K_0/2)\sin(2\xi), \quad a^2 = a^3 = -2^{-3/2}K_0[1 - \cos(2\xi)], \tag{1.7.23}$$

which give the nonzero strengths $F^{k,03}$.

From formulas (1.7.4) and (1.4.11), we obtain

$$\partial_\mu F^{1,\mu 0} = -da^1/dz = (4\pi/c)\hat{J}^{1,0}. \tag{1.7.24}$$

Therefore,

$$a^1 = -(4\pi/c)\int_0^z \hat{J}^{1,0}dz. \tag{1.7.25}$$

From formulas (1.7.22), (1.7.23) and (1.7.25), we find

$$S \int_0^z (\hat{J}^{1,0}/c)dz = \frac{K_0 S}{8\pi} \sin\left(\frac{8\pi q}{K_0 S}\right). \qquad (1.7.26)$$

Let us put

$$K = \frac{K_0 S}{8\pi}. \qquad (1.7.27)$$

Then, formula (1.7.26) acquires the form

$$S \int_0^z (\hat{J}^{1,0}/c)dz = K \sin(q/K). \qquad (1.7.28)$$

The left-hand side of formula (1.7.28) is the full charge for the index $k = 1$ in the gauge space in the region $(x^1, x^2) \in S$, $0 \le x^3 \le z$. The full charge includes not only the classical charge q but also the charge of the Yang–Mills field quanta generated by it.

Let us denote this full charge by $\hat{q}$. Then, from (1.7.28), we obtain

$$\hat{q} = K \sin(q/K). \qquad (1.7.29)$$

Formula (1.7.29) relates the full charge $\hat{q}$ and the classical charge q of a field source and accords with Principle 1.5.1.

Using (1.7.4), (1.7.22), (1.7.23) and (1.7.27), we find

$$F^{1,30} = \frac{4\pi K}{S} \sin\left(\frac{q}{K}\right). \qquad (1.7.30)$$

It should be noted that when $|q/K| \ll 1$, the full charge $\hat{q}$ practically coincides with the classical charge q and we have the Maxwell field expression for the strength component $F^{1,30}$.

One of the interesting applications of the obtained solutions is the explanation of the mystery of the existence of living organisms and, in particular, fishes in the depths of the oceans. According to traditional ideas, the pressures arising there are so huge that no living body is able to withstand them. However, living bodies are found in certain areas of the oceans and, in particular, near the bottom of the Mariana Trench, where the pressure can reach 1100 atm. The reason for this may be large electric charges in the depths of the oceans, which arise as a result of ionization under the action of high pressure of the salts dissolved in them. Then, owing to formula (1.7.30), regions may arise in the oceans in which the nonlinear electric force compensates for gravity. In such areas, the pressure may decrease sharply, which may make it possible for living organisms to exist there.

1.8. A Class of Solutions to the Yang–Mills Equations with Stationary Axisymmetric Currents and Streak Lightning

Consider the Yang–Mills equations with the following stationary axisymmetric field sources:

$$J^{1,0} = 0, \quad J^{1,1} = xb(\rho, z), \quad J^{1,2} = yb(\rho, z),$$

$$J^{1,3} = j(\rho, z), \quad \rho = \sqrt{x^2 + y^2}, \quad J^{2,\nu} = J^{3,\nu} = 0, \quad (1.8.1)$$

where $x = x^1$, $y = x^2$, $z = x^3$, b and j are some functions of the coordinates ρ and z.

Let us study axisymmetric stationary solutions to the Yang–Mills equations (1.3.1) and (1.3.2) when the field sources are given by expressions (1.8.1).

We will seek the field potentials in the considered case in the form

$$A^{k,0} = 0, \quad A^{k,1} = xu^k(\rho, z), \quad A^{k,2} = yu^k(\rho, z), \quad A^{k,3} = v^k(\rho, z),$$

$$(1.8.2)$$

where u^k and v^k are some differentiable functions of ρ and z.

Substituting expressions (1.8.2) into formula (1.3.2) for the field strengths $F^{k,\mu\nu}$, we find

$$F^{k,0\nu} = 0, \quad F^{k,12} = 0, \quad F^{k,13} = xh^k, \quad F^{k,23} = yh^k, \quad (1.8.3)$$

$$h^k = -(1/\rho)v_\rho^k + u_z^k + g\varepsilon_{klm}u^l v^m, \quad (1.8.4)$$

where $v_\rho^k \equiv \partial v^k / \partial \rho$, $u_z^k \equiv \partial u^k / \partial z$. Further, we will use such notation for partial derivatives.

Let us now substitute expressions (1.8.1)–(1.8.3) for the field sources, potentials and strengths under examination into the Yang–Mills field equation (1.3.1). Then, we come to the following system of equations:

$$h^k_z - g\varepsilon_{klm}v^l h^m = -(4\pi/c)b\delta^k, \qquad (1.8.5)$$

$$\rho h^k_\rho + 2h^k - g\rho^2\varepsilon_{klm}u^l h^m = (4\pi/c)j\delta^k, \qquad (1.8.6)$$

where

$$\delta^1 = 1, \quad \delta^2 = \delta^3 = 0. \qquad (1.8.7)$$

Thus, we should study the system of nonlinear partial differential equations (1.8.4)–(1.8.7).

Further, we will seek and find a class of nontrivial solutions to these equations for arbitrary functions b and j of the arguments ρ and z that accord with the differential equation (1.4.3) of charge conservation.

1.8.1. *Investigation of axisymmetric stationary field strengths for the considered sources*

Let us turn to the differential equation (1.4.3) and relation (1.4.5). Substituting expressions (1.8.1) and (1.8.2) into them, we obtain

$$\rho b_\rho + 2b + j_z = 0, \qquad (1.8.8)$$

$$\rho^2 bu^k + jv^k = 0. \qquad (1.8.9)$$

Let us now multiply Eq. (1.8.5) by j and Eq. (1.8.6) by b and take their sum. Then, we obtain

$$b(\rho h^k_\rho + 2h^k) + jh^k_z - g\varepsilon_{klm}(\rho^2 bu^l + jv^l)h^m = 0. \qquad (1.8.10)$$

Using (1.8.9), from Eq. (1.8.10), we derive

$$b(\rho h^k_\rho + 2h^k) + jh^k_z = 0. \qquad (1.8.11)$$

This is a linear differential equation of the first order for the functions $h^k(\rho, z)$, where the functions $b(\rho, z)$ and $j(\rho, z)$ satisfy Eq. (1.8.8).

Thus, we have proved the following proposition:

Proposition 1.8.1. *One of the consequences of the nonlinear differential equations (1.8.5) and (1.8.6) and relation (1.8.9) is the linear differential equation (1.8.11).*

Putting

$$h^k = q^k/\rho^2, \qquad (1.8.12)$$

where $q^k = q^k(\rho, z)$ are some differentiable functions, from (1.8.11), we derive

$$\rho b q_\rho^k + j q_z^k = 0. \qquad (1.8.13)$$

Let us now find the general solution to Eq. (1.8.13). For this purpose, consider the function

$$I = 2\pi \int_0^\rho \rho j(\rho, z) d\rho. \qquad (1.8.14)$$

As follows from (1.8.1), the function $I = I(\rho, z)$ is the source classical current passing in the direction z through the circle of radius ρ with the center $(0, 0)$ in the plane (x, y).

From (1.8.14), we derive

$$I_\rho = 2\pi \rho j, \quad I_z = 2\pi \int_0^\rho \rho j_z d\rho. \qquad (1.8.15)$$

Using (1.8.8) and (1.8.15), we find

$$I_z = -2\pi \int_0^\rho (\rho^2 b_\rho + 2\rho b) d\rho = -2\pi \int_0^\rho (\rho^2 b)_\rho d\rho = -2\pi \rho^2 b. \qquad (1.8.16)$$

Formulas (1.8.15) and (1.8.16) give

$$\rho b I_\rho + j I_z = 0. \qquad (1.8.17)$$

Therefore, the function I is a particular solution to Eq. (1.8.13).

Using (1.8.17), we obtain that the general solution to Eq. (1.8.13) has the form

$$q^k = \Phi^k(I), \qquad (1.8.18)$$

where Φ^k are arbitrary differentiable functions of the argument I.

Indeed,

$$q_\rho^k = (d\Phi^k/dI) I_\rho, \quad q_z^k = (d\Phi^k/dI) I_z. \qquad (1.8.19)$$

That is why, using (1.8.17), we find that the functions (1.8.18) satisfy Eq. (1.8.13).

Besides, formula (1.8.18) contains the arbitrary differentiable functions Φ^k. Therefore, it gives the general solution to Eq. (1.8.13) since this partial differential equation is of the first order.

From (1.8.12) and (1.8.18), we obtain

$$h^k = (1/\rho^2)\Phi^k(I). \tag{1.8.20}$$

Thus, we have proved the following proposition:

Proposition 1.8.2. *The general solution to the partial differential equation* (1.8.11), *in which the functions b and j satisfy Eq.* (1.8.8), *is determined by formula* (1.8.20), *where $\Phi^h(I)$ are arbitrary differentiable functions and I is defined by formula* (1.8.14).

In order to have no singularity for the functions h^k at $\rho = 0$, we require

$$\Phi^k(0) = 0. \tag{1.8.21}$$

From formulas (1.8.9), (1.8.15), (1.8.16) and (1.8.20), we obtain

$$v^k = -\frac{\rho^2 b}{j}u^k, \quad \rho h_\rho^k + 2h^k = 2\pi j\frac{d\Phi^k}{dI}, \quad h_z^k = -2\pi b\frac{d\Phi^k}{dI}. \tag{1.8.22}$$

Let us substitute formulas (1.8.20) and (1.8.22) into the studied equations (1.8.5) and (1.8.6). Then, the two equations give the same differential equation of the form

$$\frac{d\Phi^k}{dI} - \frac{g}{2\pi j}\varepsilon_{klm}u^l\Phi^m = \frac{2}{c}\delta^k. \tag{1.8.23}$$

Let us now multiply Eq. (1.8.23) by Φ^k and sum it over k. Then, using (1.8.7) and the identity $\varepsilon_{klm}\Phi^k\Phi^m = 0$ since ε_{klm} are antisymmetric, we find

$$\sum_{k=1}^{3}\Phi^k d\Phi^k/dI = (2/c)\Phi^1. \tag{1.8.24}$$

Thus, the components h^k of the considered field strengths are given by formula (1.8.20), where the functions $\Phi^k(I)$ should satisfy (1.8.21) and Eq. (1.8.24).

In addition to Eq. (1.8.24), from Eq. (1.8.23) by putting $k = 3, 2$, we obtain the following two relations for the functions u^k:

$$u^2 = (g\Phi^1)^{-1}(g\Phi^2 u^1 - 2\pi j d\Phi^3/dI),$$
$$u^3 = (g\Phi^1)^{-1}(g\Phi^3 u^1 + 2\pi j d\Phi^2/dI). \tag{1.8.25}$$

1.8.2. *Investigation of axisymmetric stationary field potentials for the considered sources*

Consider Eq. (1.8.9). It gives

$$v^k = -\gamma u^k, \quad \gamma = \rho^2 b/j. \tag{1.8.26}$$

Here, the functions b and j satisfy Eq. (1.8.8).

Let us turn to Eq. (1.8.4). Substituting (1.8.26) into it and using the identity $\varepsilon_{klm} u^l u^m = 0$ since ε_{klm} are antisymmetric, from this equation, we derive

$$h^k = (1/\rho)(\gamma u^k)_\rho + u_z^k. \tag{1.8.27}$$

As stated above, the functions h^k are given by formulas (1.8.20).

Thus, we have Eqs. (1.8.25)–(1.8.27) for the functions u^k and v^k. Let us prove the following proposition:

Proposition 1.8.3. *Equation (1.8.27) with $k = 2, 3$ is a consequence of formulas (1.8.20), (1.8.25) and this equation with $k = 1$.*

Proof. Let us use formula (1.8.20) and the second equality in (1.8.22). Then, Eqs. (1.8.25) can be represented as

$$u^2 = (g\rho^2 h^1)^{-1}(g\rho^2 h^2 u^1 - \rho h_\rho^3 - 2h^3),$$
$$u^3 = (g\rho^2 h^1)^{-1}(g\rho^2 h^3 u^1 + \rho h_\rho^2 + 2h^2). \tag{1.8.28}$$

Let us introduce the complex functions

$$w = u^2 + iu^3, \tag{1.8.29}$$

$$p = h^2 + ih^3. \tag{1.8.30}$$

Then, Eqs. (1.8.28) can be represented as

$$w = (gh^1)^{-1}(gpu^1 + i\rho^{-1}p_\rho + 2i\rho^{-2}p). \tag{1.8.31}$$

As follows from Proposition 1.8.2, formula (1.8.20) for h^k satisfies Eq. (1.8.11). From this equation, formula (1.8.30) and the expression for γ in (1.8.26), we find

$$\gamma(h_\rho^1 + 2h^1/\rho) + \rho h_z^1 = 0, \quad \gamma(p_\rho + 2p/\rho) + \rho p_z = 0. \tag{1.8.32}$$

Using (1.8.31) and the second formula in (1.8.32), we obtain

$$\gamma w = (gh^1)^{-1}(g\gamma p u^1 - ip_z). \tag{1.8.33}$$

Differentiating Eq. (1.8.31) with respect to z and Eq. (1.8.33) with respect to ρ, we derive

$$w_z = -(gh^1)^{-1}[gh_z^1 w - g(p_z u^1 + pu_z^1) - i\rho^{-2}(\rho p_{\rho z} + 2p_z)], \tag{1.8.34}$$

$$(\gamma w)_\rho = -(gh^1)^{-1}[gh_\rho^1 \gamma w - g(\gamma u^1)_\rho p - g\gamma p_\rho u^1 + ip_{\rho z}]. \tag{1.8.35}$$

This gives

$$(1/\rho)(\gamma w)_\rho + w_z = -(g\rho h^1)^{-1}\{(\gamma h_\rho^1 + \rho h_z^1)gw - gp[(\gamma u^1)_\rho + pu_z^1]$$
$$-g(\gamma p_\rho + \rho p_z)u^1 - 2i\rho^{-1}p_z\}. \tag{1.8.36}$$

From Eq. (1.8.27) with $k = 1$ and Eqs. (1.8.32), we find

$$(\gamma u^1)_\rho + \rho u_z^1 = \rho h^1, \quad \gamma h_\rho^1 + \rho h_z^1 = -2\gamma h^1/\rho,$$
$$\gamma p_\rho + \rho p_z = -2\gamma p/\rho. \tag{1.8.37}$$

Substituting formulas (1.8.37) into Eq. (1.8.36), we get

$$(1/\rho)(\gamma w)_\rho + w_z = p + 2(g\rho^2 h^1)^{-1}(g\gamma w h^1 - g\gamma p u^1 + ip_z). \tag{1.8.38}$$

As follows from (1.8.33),

$$g\gamma w h^1 - g\gamma p u^1 + ip_z = 0. \tag{1.8.39}$$

Therefore, from Eq. (1.8.38), we obtain

$$(1/\rho)(\gamma w)_\rho + w_z = p. \tag{1.8.40}$$

From this equality and formulas (1.8.29) and (1.8.30), we derive that Eq. (1.8.27) is fulfilled for $k = 2, 3$.

Thus, we have proved Proposition 1.8.3.

Let us now turn to Eq. (1.8.27) with $k = 1$. Using (1.8.20), we can represent it as

$$\gamma u_\rho^1 + \gamma_\rho u^1 + \rho u_z^1 = (1/\rho)\Phi^1(I). \tag{1.8.41}$$

In any region where the sign of the function $j(\rho, z)$ does not change, we can introduce the inverse function

$$\rho = \rho(I, z) \tag{1.8.42}$$

for the function $I = I(\rho, z)$ of form (1.8.14).

Let us apply this inverse function to solve Eq. (1.8.41).

Using (1.8.42), we can put

$$u^1 = U(I, z), \tag{1.8.43}$$

where $U(I, z)$ is some differentiable function.

Then, using formulas (1.8.15) and (1.8.16), we find

$$u_\rho^1 = 2\pi\rho j U_I, \quad u_z^1 = U_z - 2\pi\rho^2 b U_I. \tag{1.8.44}$$

Substituting formulas (1.8.43) and (1.8.44) into Eq. (1.8.41), we obtain

$$2\pi\rho(\gamma j - \rho^2 b)U_I + \rho U_z + \gamma_\rho U = (1/\rho)\Phi^1(I). \tag{1.8.45}$$

As follows from (1.8.26), $\gamma j - \rho^2 b = 0$. Therefore, Eq. (1.8.45) acquires the form

$$U_z + (\gamma_\rho/\rho)U = (1/\rho^2)\Phi^1(I). \tag{1.8.46}$$

Using (1.8.42), we can put

$$\gamma_\rho/\rho = P(I, z), \quad 1/\rho^2 = Q(I, z), \tag{1.8.47}$$

where P and Q are some functions of I and z.

From (1.8.46) and (1.8.47), we obtain the following ordinary differential equation of the first order for the function $U = U(I, z)$:

$$U_z + P(I, z)U = Q(I, z)\Phi^1(I). \qquad (1.8.48)$$

From this linear differential equation for the function U, we readily find

$$U = \Phi^1(I) \exp\left(-\int_a^z P\,dz\right) \left[\int_a^z Q \exp\left(\int_a^z P\,dz\right) dz + C(I)\right],$$

$$a = \text{const}, \qquad (1.8.49)$$

where $C(I)$ is an arbitrary function.

As follows from (1.8.43), formula (1.8.49) determines the function u^1.

Thus, formulas (1.8.25), (1.8.26) and (1.8.49) give expressions for the field potentials $A^{k,\nu}$ and formulas (1.8.20), (1.8.21) and the differential relation (1.8.24) give expressions for the field strengths $F^{k,\mu\nu}$.

It should be noted that the functions $\Phi^k(I)$ in (1.8.20) can be arbitrary differentiable functions satisfying (1.8.21) and relation (1.8.24).

The obtained class of solutions to the Yang–Mills equations (1.3.1) and (1.3.2) with the field source components of form (1.8.1) can be extended by using the gauge transformations (1.4.2).

1.8.3. *Investigation of axisymmetric stationary solutions to the considered Yang–Mills equations with additional physical conditions*

Let us now study axisymmetric stationary solutions satisfying not only the Yang–Mills equations (1.3.1) and (1.3.2) but also the additional differential condition (1.4.16). For this purpose, let us turn to expressions (1.8.3) and (1.8.20) for the field strengths $F^{k,\mu\nu}$.

From (1.8.3) and (1.8.22), we find

$$\begin{aligned}
&\partial_\mu F^{k,\mu 0} = 0, \quad \partial_\mu F^{k,\mu 1} = 2\pi x b\, d\Phi^k/dI, \\
&\partial_\mu F^{k,\mu 2} = 2\pi y b\, d\Phi^k/dI, \quad \partial_\mu F^{k,\mu 3} = 2\pi j\, d\Phi^k/dI.
\end{aligned} \qquad (1.8.50)$$

Substituting expressions (1.8.1) and (1.8.50) into Eq. (1.4.16), we come to the following equation:

$$\sum_{k=1}^{3} (d\Phi^k/dI)^2 = (2/c)^2. \tag{1.8.51}$$

This equation is additional to Eq. (1.8.24) for the functions $\Phi^k(I)$. Let us put

$$\Phi^1 = \Phi \cos \xi, \quad \Phi^2 = \Phi \sin \xi \cos \eta, \quad \Phi^3 = \Phi \sin \xi \sin \eta, \tag{1.8.52}$$

where $\Phi = \Phi(I), \xi = \xi(I), \eta = \eta(I)$ are some differentiable functions.

In order to satisfy condition (1.4.17), let us choose the following gauge:

$$\eta = \pi/4. \tag{1.8.53}$$

Then, from (1.8.24), (1.8.51) and (1.8.52), we derive

$$d\Phi/dI = (2/c) \cos \xi, \tag{1.8.54}$$

$$(d\Phi/dI)^2 + (\Phi d\xi/dI)^2 = (2/c)^2. \tag{1.8.55}$$

Substituting (1.8.54) into (1.8.55), we obtain

$$\Phi d\xi/dI = \pm(2/c) \sin \xi. \tag{1.8.56}$$

Dividing Eq. (1.8.54) by Eq. (1.8.56), we find

$$d\Phi/\Phi = \pm \cot \xi \, d\xi. \tag{1.8.57}$$

In order to have no singularity, let us choose the sign '+' in (1.8.57) and hence in (1.8.56). Then, integrating Eq. (1.8.57), we obtain

$$\Phi = D_0 \sin \xi, \quad D_0 = \text{const.} \tag{1.8.58}$$

Let us substitute (1.8.58) into (1.8.56). Then, taking into account that the sign '+' is chosen in (1.8.56) and using (1.8.21), we get

$$\xi = 2I/(cD_0) + \pi n, \tag{1.8.59}$$

where n is an integer.

Formulas (1.8.52), (1.8.53), (1.8.58) and (1.8.59) give

$$\Phi^1 = (D_0/2)\sin(4I/(cD_0)),$$

$$\Phi^2 = \Phi^3 = 2^{-3/2}D_0[1 - \cos(4I/(cD_0))]. \qquad (1.8.60)$$

Formulas (1.8.60) can be represented in the form

$$\Phi^1 = (2D/c)\sin(I/D), \quad \Phi^2 = \Phi^3 = \sqrt{2}(D/c)[1 - \cos(I/D)], \qquad (1.8.61)$$

where

$$D = cD_0/4 = \text{const.} \qquad (1.8.62)$$

From formulas (1.8.3) and (1.8.20), we obtain that the field strengths $F^{k,\mu\nu}$ are as follows:

$$F^{k,0\nu} = 0, \quad F^{k,12} = 0, \quad F^{k,13} = (x/\rho^2)\Phi^k, \quad F^{k,23} = (y/\rho^2)\Phi^k, \qquad (1.8.63)$$

where the functions Φ^k are determined by expressions (1.8.61).
Formulas (1.8.63) give

$$\partial_\mu F^{k,\mu3} = (1/\rho)\Phi_\rho^k. \qquad (1.8.64)$$

Let us substitute formula (1.8.64) into Eq. (1.4.11) with $\nu = 3$. Then, we get

$$\Phi_\rho^k = (4\pi/c)\rho\hat{J}^{k,3}. \qquad (1.8.65)$$

Taking into account (1.8.21), from (1.8.65), we find

$$\Phi^k = (4\pi/c)\int_0^\rho \rho\hat{J}^{k,3}d\rho. \qquad (1.8.66)$$

Formulas (1.8.61) and (1.8.66) with $k = 1$ give

$$2\pi\int_0^\rho \rho\hat{J}^{1,3}d\rho = D\sin(I/D). \qquad (1.8.67)$$

The left-hand side of formula (1.8.67) is the full current for the index $k = 1$ in the gauge space that passes in the direction z through the circle of radius ρ with the center $(0, 0)$ in the plane (x, y). The full current includes not only the classical current I but also the current of Yang–Mills field quanta generated by it.

Let us denote this full current by $\hat{I}$. Then, from (1.8.67), we obtain

$$\hat{I} = D \sin(I/D). \qquad (1.8.68)$$

Formula (1.8.68) relates the full current $\hat{I}$ and the classical current I of a field source.

Using (1.8.61), (1.8.63) and (1.8.68), we find

$$F^{1,13} = (2x/\rho^2)\hat{I}/c, \quad F^{1,23} = (2y/\rho^2)\hat{I}/c. \qquad (1.8.69)$$

It should be noted that when $|I/D| \ll 1$, the full current $\hat{I}$ practically coincides with the classical current I and we have the Maxwell field expressions for the strength components $F^{1,13}$ and $F^{1,23}$.

Let us assume that the value D in (1.8.68) is a sufficiently large positive constant. Then, formulas (1.8.68) and (1.8.69) can be regarded as a nonlinear generalization of the corresponding Maxwell field expressions for the strengths $F^{1,13}$ and $F^{1,23}$ when the source classical current I is sufficiently large.

As follows from (1.8.68), the constant D is a maximum value of the full current $\hat{I}$. In order to estimate this constant, let us turn to streak lightning and its following property: As is known from observations, the maximum value of lightning currents is about $200\,\text{kA}$ [15].

That is why the constant D could be estimated as $\sim 200\,\text{kA}$.

As a result, we come to the following principle:

Principle 1.8.2. *Consider a body with the flow of a classical electric current when there are no external fields. Then, the classical current I in the body and its full current $\hat{I}$ are related by formula (1.8.68), where the constant $D \sim 200\,kA$.*

Formulas (1.8.68) and (1.8.69) describe a nonlinear effect of field saturation. Namely, let the absolute value of the source classical current I increase starting from zero. Then, when it reaches the value πD, the strengths $F^{1,13}$ and $F^{1,23}$ become equal to zero and after that, they change their signs.

This property could be applied to give a new interpretation for the unusual phenomenon of bipolar lightning that actually changes its polarity (positive becoming negative or vice versa) [15].

It is also interesting to note that puzzling data for lightning were obtained by the Fermi Gamma-ray Space Telescope which could be explained by formulas (1.8.68) and (1.8.69). Namely, some of the lightning storms had the surprising sign of positrons, and the conclusion was made that the normal orientation for an electromagnetic field associated with a lightning storm somehow reversed [16].

To explain these data, let us note that as follows from (1.8.68), the sign of the full current $\hat{I}$ can differ from the sign of the source classical current I when the latter is sufficiently large.

1.9.　Yang–Mills Fields inside a Thin Plane Layer with Axisymmetric Stationary Sources

Let us study Yang–Mills fields when their source is inside a thin plane layer $-\frac{1}{2}d \leq z \leq \frac{1}{2}d$, where d is its small thickness, and has the following form:

$$J^{1,0} = c\theta(\rho), \quad J^{1,l} = 0, \quad l = 1,2,3, \quad J^{2,\nu} = J^{3,\nu} = 0, \quad (1.9.1)$$

where $\theta(\rho)$ is some function of $\rho = \sqrt{x^2 + y^2}$.

Let us seek the field potentials $A^{k,\nu}$ in the considered thin layer in the form

$$A^{k,0} = A^{k,3} = 0, \quad A^{k,1} = x[\tau\alpha^k(\rho) + \beta^k(\rho)],$$

$$A^{k,2} = y[\tau\alpha^k(\rho) + \beta^k(\rho)], \quad \tau = x^0, \quad -\frac{1}{2}d \leq z \leq \frac{1}{2}d, \quad (1.9.2)$$

where α^k and β^k are some differentiable functions of ρ.

Then, as follows from (1.3.2), we have the following stationary expressions for the field strengths in the thin layer:

$$F^{k,01} = x\alpha^k(\rho), \quad F^{k,02} = y\alpha^k(\rho), \quad F^{k,03} = 0,$$

$$F^{k,12} = F^{k,13} = F^{k,23} = 0. \quad (1.9.3)$$

Substituting formulas (1.9.1)–(1.9.3) into the Yang–Mills equation (1.3.1) and taking into account that ε_{klm} are antisymmetric and hence $\varepsilon_{klm}\alpha^l\alpha^m \equiv 0$, we come to the equations

$$\rho(\alpha^k)' + 2\alpha^k + g\rho^2\varepsilon_{klm}\alpha^l\beta^m = -4\pi\theta\delta^k,$$

$$k = 1,2,3, \quad \delta^1 = 1, \quad \delta^2 = \delta^3 = 0. \quad (1.9.4)$$

Therefore, we have three equations for the six functions $\alpha^k(\rho)$ and $\beta^k(\rho)$.

Multiplying Eq. (1.9.4) by α^k, summing it over k and using the identity $\varepsilon_{klm}\alpha^k\alpha^l = 0$, we find

$$\rho\alpha' + 2\alpha = -4\pi\theta\alpha^1/\alpha, \tag{1.9.5}$$

where

$$\alpha = \sqrt{(\alpha^1)^2 + (\alpha^2)^2 + (\alpha^3)^2}. \tag{1.9.6}$$

From the second and third equations in (1.9.4) $(k = 2, 3)$, we obtain

$$\begin{aligned}
\beta^2 &= -(g\rho^2\alpha^1)^{-1}[\rho(\alpha^3)' + 2\alpha^3 - g\rho^2\alpha^2\beta^1], \\
\beta^3 &= (g\rho^2\alpha^1)^{-1}[\rho(\alpha^2)' + 2\alpha^2 + g\rho^2\alpha^3].
\end{aligned} \tag{1.9.7}$$

Taking into account condition (1.4.17), let us represent the functions α^k in the form

$$\alpha^1 = -\alpha\cos\xi, \quad \alpha^2 = \alpha^3 = -2^{-1/2}\alpha\sin\xi, \tag{1.9.8}$$

where $\alpha = \alpha(\rho)$ and $\xi = \xi(\rho)$.

From (1.9.5) and (1.9.8), we find

$$\rho\alpha' + 2\alpha = 4\pi\theta\cos\xi. \tag{1.9.9}$$

Substituting expressions (1.9.1) and (1.9.3) into Eq. (1.4.16), we get

$$\sum_{k=1}^{3}[\rho(\alpha^k)' + 2\alpha^k]^2 = (4\pi\theta)^2. \tag{1.9.10}$$

Using formulas (1.9.8), we can rewrite this equation as

$$(\rho\alpha' + 2\alpha)^2 + (\rho\alpha\xi')^2 = (4\pi\theta)^2. \tag{1.9.11}$$

Taking into account (1.9.9), from Eq. (1.9.11), we obtain

$$\rho\alpha\xi' = \pm 4\pi\theta\sin\xi. \tag{1.9.12}$$

Equations (1.9.9) and (1.9.12) give

$$\rho\alpha' + 2\alpha = \pm\rho\alpha\xi'\cot\xi. \tag{1.9.13}$$

Dividing this equation by $\rho\alpha$ and then integrating it, we find

$$\int \left(\frac{\alpha'}{\alpha} + \frac{2}{\rho}\right) d\rho = \pm \int \cot\xi \, d\xi. \qquad (1.9.14)$$

From Eq. (1.9.14), we obtain

$$\ln|\rho^2\alpha| = \pm\ln|\sin\xi| + \text{const.} \qquad (1.9.15)$$

In order to have the function $\alpha(\rho)$ nonsingular, we choose the sign '+' in Eq. (1.9.15) and hence in Eq. (1.9.13). Then, from (1.9.15), we find

$$\alpha = (B_0/\rho^2)\sin\xi, \quad B_0 = \text{const.} \qquad (1.9.16)$$

Substituting this formula into Eq. (1.9.12) and taking into account that the sign '+' is chosen, we obtain

$$\xi' = 4\pi\theta\rho/B_0. \qquad (1.9.17)$$

From (1.9.16) and (1.9.17), we find

$$\alpha = B_0\frac{\sin\xi}{\rho^2}, \quad \xi = \frac{4\pi}{B_0}\int_0^\rho \theta\rho \, d\rho + \pi n, \qquad (1.9.18)$$

where n is an integer and hence $\alpha(\rho)$ is nonsingular at $\rho = 0$.

Formulas (1.9.8) and (1.9.18) give

$$\alpha^1 = -\frac{2K}{d}\sin\left(\frac{q}{K}\right)\frac{1}{\rho^2}, \quad K = \frac{B_0 d}{4},$$

$$\alpha^2 = \alpha^3 = -\frac{\sqrt{2}K}{d}\left[1 - \cos\left(\frac{q}{K}\right)\right]\frac{1}{\rho^2}, \qquad (1.9.19)$$

$$q = 2\pi d\int_0^\rho \theta\rho \, d\rho, \qquad (1.9.20)$$

where $q = q(\rho)$ is the classical charge in the cylindrical region with radius ρ and the small thickness d.

From formulas (1.9.3) and (1.9.19), we find the following expressions for the field strengths $F^{1,l0}$:

$$F^{1,10} = \frac{2\hat{q}}{d}\frac{x}{\rho^2}, \quad F^{1,20} = \frac{2\hat{q}}{d}\frac{y}{\rho^2}, \quad F^{1,30} = 0,$$

$$\hat{q} = K\sin\left(\frac{q}{K}\right), \quad q = q(\rho). \qquad (1.9.21)$$

Here, as follows from Principle 1.5.1, $\hat{q}$ is the full charge in the region $-\frac{1}{2}d \leq z \leq \frac{1}{2}d$, $x^2 + y^2 \leq \rho^2$, and the constant K was estimated in Section 1.5 and is $\sim 10^7$ coul.

When $|q| \ll K$, the full charge $\hat{q}$ practically coincides with the classical charge q, and we have the classical expressions of the Maxwell theory for the strengths $F^{1,l0}$. When the classical charge q is large, we get their nonlinear generalization.

The obtained formulas (1.9.21) describe a nonlinear effect of electric field saturation. Namely, when the absolute value of the charge q increases from zero and reaches the value πK, the strengths $F^{1,10}$ and $F^{1,20}$ become zero and after that, they change their sign.

This property can be used to interpret the mysterious circles that appear during lightning discharges in the fields sown with cereals. Indeed, let, as a result of the action of the electric field during these discharges, the ears of grain will be positively charged. Then, if their number is large enough, then their total charge can become very large and can be several times higher than the constant $2\pi K$.

In this case, we find from formulas (1.9.21) that concentric circular regions with the charge equal to $2\pi K$ should appear in such fields, on which the change in the electric field should be repeated in a similar way. Namely, taking into account the sinusoidal dependence of the full charge on the classical charge inside such circular regions, the electric field should increase from zero to a positive maximum, then decrease to zero, change sign, reach its negative minimum and then increase to zero. Therefore, in such a field, periodic lodging of charged cereal ears should be seen, with an alternating change in its direction, which is exactly what is observed in a number of cases after lightning discharges.

1.10. Yang–Mills Fields inside Thin Circular Cylinders with Nonstationary Plasma

Consider now the Yang–Mills field inside a thin circular cylinder filled with a nonstationary plasma when the field sources are of the form

$$J^{1,0} = J^0(\tau, z), \quad J^{1,1} = J^{1,2} = 0,$$
$$J^{1,3} = J(\tau, z), \quad J^{2,\nu} = J^{3,\nu} = 0, \tag{1.10.1}$$
$$0 \leq \rho \leq \rho_0, \quad 0 \leq z \leq L,$$

where J^0 and J are some functions, ρ_0 is the radius of the field source and L is its length.

The boundary $\rho = \rho_0$ of the plasma is considered free and we add the following boundary conditions:

$$F^{k,01} = F^{k,02} = F^{k,13} = F^{k,23} = 0 \quad \text{at } \rho = \rho_0. \tag{1.10.2}$$

Let us seek the potentials $A^{k,\nu}$ inside the considered thin cylinder in the form

$$A^{k,0} = \lambda^k(\tau, z), \quad A^{k,1} = A^{k,2} = 0, \quad A^{k,3} = \gamma^k(\tau, z), \quad 0 \le \rho \le \rho_0. \tag{1.10.3}$$

Then, from formula (1.3.2) for the strengths $F^{k,\mu\nu}$, we find

$$
\begin{aligned}
F^{k,01} &= F^{k,02} = 0, \quad F^{k,03} = p^k(\tau, z), \\
F^{k,12} &= F^{k,13} = F^{k,23} = 0,
\end{aligned} \tag{1.10.4}
$$

where

$$p^k = \gamma_\tau^k + \lambda_z^k + g\varepsilon_{klm}\lambda^l\gamma^m. \tag{1.10.5}$$

The obtained expressions for the strengths $F^{k,\mu\nu}$ satisfy the boundary conditions (1.10.2).

Substituting formulas (1.10.1) and (1.10.3) into Eqs. (1.4.3) and (1.4.5), we obtain

$$J_\tau^0 + J_z = 0, \tag{1.10.6}$$

$$J^0\lambda^k = J\gamma^k. \tag{1.10.7}$$

Since λ^k and γ^k are proportional, we have $\varepsilon_{klm}\lambda^l\gamma^m = 0$. Therefore, formula (1.10.5) acquires the form

$$p^k = \gamma_\tau^k + \lambda_z^k. \tag{1.10.8}$$

Substituting formulas (1.10.1), (1.10.3) and (1.10.4) into the Yang–Mills equations (1.3.1), we obtain

$$p_z^k - g\varepsilon_{klm}\gamma^l p^m = -(4\pi/c)\delta^k J^0, \tag{1.10.9}$$

$$p_\tau^k + g\varepsilon_{klm}\lambda^l p^m = (4\pi/c)\delta^k J, \tag{1.10.10}$$

$$\delta^1 = 1, \quad \delta^2 = \delta^3 = 0. \tag{1.10.11}$$

Let us multiply Eq. (1.10.9) by J and Eq. (1.10.10) by J^0 and then add them. Then, using equality (1.10.7), we obtain

$$J^0 p_\tau^k + J p_z^k = 0. \tag{1.10.12}$$

Let us introduce the function

$$q = \frac{\pi \rho_0^2}{c} \left[\int_0^z J^0(0, z) dz - \int_0^\tau J(\tau, z) d\tau \right], \tag{1.10.13}$$

which is the classical charge at the time τ in the region $0 \le x^3 \le z$ of the considered plasma.

Using (1.10.6) and (1.10.13), we have

$$q_\tau = -(\pi \rho_0^2 / c) J(\tau, z), \tag{1.10.14}$$

$$q_z = -(\pi \rho_0^2 / c) \left[\int_0^\tau J_z(\tau, z) d\tau - J^0(0, z) \right]$$

$$= (\pi \rho_0^2 / c) \left[\int_0^\tau J_\tau^0(\tau, z) d\tau + J^0(0, z) \right]$$

$$= (\pi \rho_0^2 / c) J^0(\tau, z). \tag{1.10.15}$$

Formulas (1.10.14) and (1.10.15) give

$$J^0 q_\tau + J q_z = 0. \tag{1.10.16}$$

Taking into account this equality, we come to the following solution of the partial differential equation (1.10.12) of the first order:

$$p^k = p^k(q), \tag{1.10.17}$$

where $p^k(q)$ are arbitrary differentiable functions.

Indeed, substituting (1.10.17) into Eq. (1.10.12) and using equality (1.10.16), we obtain

$$J^0 p_\tau^k + J p_z^k = (dp^k / dq)(J^0 q_\tau + J q_z) = 0. \tag{1.10.18}$$

Therefore, the functions of form (1.10.17) satisfy Eq. (1.10.12).

Thus, as follows from (1.10.4), (1.10.13) and (1.10.17), the nonzero strengths $F^{k,03}$ inside the considered source depend on the all classical charge passing through its transversal section since the emerging of the current.

Let us turn to Eq. (1.10.8) and taking into account equality (1.10.7), let us seek the functions λ^k and γ^k in the form

$$\gamma^k = (\pi\rho_0^2/c)J^0[b^k(q) + \vartheta(\tau, z)p^k(q)],$$
$$\lambda^k = (\pi\rho_0^2/c)J[b^k(q) + \vartheta(\tau, z)p^k(q)], \tag{1.10.19}$$

where $b^k(q)$ are arbitrary differentiable functions and $\vartheta(\tau, z)$ is some differentiable function which will be determined in the following.

Then, substituting (1.10.19) into Eq. (1.10.8) and using equality (1.10.6), we obtain

$$p^k = (\pi\rho_0^2/c)\{[db^k/dq + \vartheta dp^k/dq](J^0q_\tau + Jq_z) + (J^0\vartheta_\tau + J\vartheta_z)p^k\}. \tag{1.10.20}$$

Using equality (1.10.16), we come to the following equation for the function ϑ:

$$J^0\vartheta_\tau + J\vartheta_z = \frac{c}{\pi\rho_0^2}. \tag{1.10.21}$$

When $J^0 = 0$, from (1.10.6) and (1.10.21), we have $J = J(\tau)$ and $\vartheta = cz\pi^{-1}\rho_0^{-2}/J(\tau) + \vartheta_0(\tau)$, where $\vartheta_0(\tau)$ is an arbitrary function.

Consider the case $J^0 \neq 0$. Then, to solve Eq. (1.10.21), it is convenient to choose the variable q instead of z and put

$$\vartheta = \phi(\tau, q), \quad q = q(\tau, z). \tag{1.10.22}$$

Indeed, using (1.10.14) and (1.10.15), we find

$$\vartheta_\tau = \phi_\tau - (\pi\rho_0^2/c)J\phi_q, \quad \vartheta_z = (\pi\rho_0^2/c)J^0\phi_q \tag{1.10.23}$$

and substituting (1.10.23) into Eq. (1.10.21), we obtain

$$J^0\phi_\tau = \frac{c}{\pi\rho_0^2}. \tag{1.10.24}$$

Therefore, we come to the formula for ϕ

$$\phi(\tau, q) = \frac{c}{\pi\rho_0^2} \int d\tau/J^0(\tau, q), \tag{1.10.25}$$

where J^0 is regarded as a function of τ and q.

Let us now substitute formulas (1.10.17) and (1.10.19) into Eqs. (1.10.9) and (1.10.10). Then, using formulas (1.10.14) and (1.10.15) and the evident identity $\varepsilon_{klm}p^l p^m \equiv 0$, we obtain that Eqs. (1.10.9) and (1.10.10) give the same equation of the following form:

$$dp^k/dq - g\varepsilon_{klm}p^l b^m = -(4/\rho_0^2)\delta^k, \quad k = 1, 2, 3,$$
$$p^k = p^k(q), \quad b^k = b^k(q), \quad \delta^1 = 1, \quad \delta^2 = \delta^3 = 0. \tag{1.10.26}$$

Multiplying Eq. (1.10.26) by $2p^k$ and summing it over k, taking into account the antisymmetry of ε_{klm}, we obtain

$$\frac{d}{dq}\sum_{k=1}^{3}(p^k)^2 = -(8/\rho_0^2)p^1. \tag{1.10.27}$$

In addition to this equation, from the second and third equations in (1.10.26), we can find relations between the functions $b^2(q)$ and $b^3(q)$ and the functions $b^1(q)$ and $p^k(q)$.

Let us put

$$p^1 = -(4/\rho_0^2)\beta(q)\cos\zeta(q), \quad p^2 = p^3 = -(4/\rho_0^2)2^{-1/2}\beta(q)\sin\zeta(q), \tag{1.10.28}$$

in accordance with condition (1.4.17). Here, $\beta(q)$ and $\zeta(q)$ are some differentiable functions.

Substituting expressions (1.10.28) into Eq. (1.10.27), we find

$$\beta' = \cos\zeta. \tag{1.10.29}$$

Let us turn to Eq. (1.4.16). From it and formulas (1.10.1) and (1.10.4), we have

$$\sum_{k=1}^{3}[(p_z^k)^2 - (p_\tau^k)^2] = (4\pi/c)^2[(J^0)^2 - (J)^2]. \tag{1.10.30}$$

Using formulas (1.10.14), (1.10.15) and (1.10.17), from Eq. (1.10.30), we derive

$$\sum_{k=1}^{3}(dp^k/dq)^2 = (2/\rho_0)^4. \tag{1.10.31}$$

Substituting formulas (1.10.28) into this equation, we obtain

$$(\beta')^2 + (\beta\zeta')^2 = 1, \qquad (1.10.32)$$

where $\beta = \beta(q)$ and $\zeta = \zeta(q)$.

Substituting now expression (1.10.29) for β' into Eq. (1.10.32), we find

$$\beta\zeta' = \pm\sin\zeta. \qquad (1.10.33)$$

Equations (1.10.29) and (1.10.33) give

$$\beta'/\beta = \pm\cot\zeta\,\zeta'. \qquad (1.10.34)$$

Let us choose the sign '+' in this equation, in order to get its nonsingular solution. Then, integrating Eq. (1.10.34), we obtain

$$\beta = K_0\sin\zeta, \qquad (1.10.35)$$

where K_0 is some constant.

Substituting expression (1.10.35) into Eq. (1.10.33) and taking into account that the sign '+' is chosen, we find

$$\zeta' = 1/K_0, \quad \zeta = q/K_0 + \zeta_0, \qquad (1.10.36)$$

where ζ_0 is some constant.

From formulas (1.10.35) and (1.10.36), we obtain

$$\beta = K_0\sin(q/K_0 + \zeta_0), \qquad (1.10.37)$$

where q is given by formula (1.10.13).

Since p^k should be zero when $q = 0$, we put

$$\zeta_0 = \pi n, \qquad (1.10.38)$$

where n is an integer.

Using formulas (1.10.4), (1.10.13), (1.10.28) and (1.10.36)–(1.10.38), we find

$$\begin{aligned}
F^{1,03} &= (8/\rho_0^2)K\sin(q/K), \quad K = K_0/2 = \text{const}, \\
F^{2,03} &= F^{3,03} = (4\sqrt{2}/\rho_0^2)K[1 - \cos(q/K)],
\end{aligned} \qquad (1.10.39)$$

where q is defined by formula (1.10.13).

As follows from Principle 1.5.1 in Section 1.5, the value $\hat{q} = K\sin(q/K)$ is the full charge in the region $0 \leq x^3 \leq z$ of the considered plasma, where the constant K is $\sim 10^7$ coul.

When $|q/K| \ll 1$, the full charge $\hat{q}$ practically coincides with the classical charge q and the expression in (1.10.39) for $F^{1,03}$ corresponds to Maxwell's electrodynamics. When q is a very large charge, we come to a new result, substantially different from the classical electrodynamics.

Let us apply the obtained results to the mysterious phenomenon of current pause, which occurs in exploding conductors [17]. This phenomenon takes place in three stages. At the moment the electrical circuit is closed, a sufficiently large current flows through the wire, causing it to explode. Then, after a while, the current flow through the wire stops and the current pause period begins. After a while, the current pause can end and the current can resume.

The physical nature of the current pause is still poorly understood within the framework of Maxwell's electrodynamics. Therefore, we turn to its nonlinear generalization based on the Yang–Mills equations, which we are just dealing with. For this purpose, we will use formulas (1.10.39) and apply them to an exploding wire. As follows from formulas (1.10.39), in some time after the beginning of the current flow through the wire, the field strength becomes equal to zero. At this moment, the current in the wire should stop. Consequently, formulas (1.10.39) make it possible to explain the appearance of a current pause in an exploding wire. After a while, when a certain redistribution of charges occurs in the wire, this pause may end and the current flow will resume.

Chapter 2

Peculiar Interactions
in Yang–Mills Fields

In this chapter, interactions of particles with Yang–Mills fields generated by sources with three types of charges are studied. First, we consider a new model of electroweak interactions based only on the Yang–Mills theory and show that it agrees with known experimental data, as well as the standard model. Then, we study Yang–Mills fields generated by stationary spherically symmetric sources with three types of charges and apply the obtained results to explain the amazing phenomenon of nitrogen fixation in a number of types of bacteria. At the end of the chapter, the dynamic equations for a system of particles with three types of charges moving under the action of their own Yang–Mills fields are studied.

2.1. Electroweak Interactions in Yang–Mills Fields

In this section, we study electroweak interactions by using only the Yang–Mills theory considered in Chapter 1 and not introducing any additional fields, such as the Higgs field and spontaneous symmetry breaking of the standard model. As a result, we come to a new model of electroweak interactions, which agrees with the Fermi theory of weak interactions and known experimental data. The section is based on our paper [18].

2.1.1.   *Yang–Mills fields in vacuum*

Consider a Yang–Mills field described by Eqs. (1.3.1) and (1.3.2) of
Chapter 1 in a vacuum region which is far enough away from the
field sources. In this vacuum region, the fields' strengths $F^{k,\mu\nu}$ are
zero and we have the condition at infinity

$$F^{k,\mu\nu} = 0, \quad r \to \infty, \tag{2.1.1}$$

where r is the distance from the spatial zero point.

Let us find nonzero constant potentials $\overline{A}^{k,\nu}$, which correspond to
condition (2.1.1). For them, we have

$$A^{k,\nu} = \overline{A}^{k,\nu}, \quad r \to \infty, \quad \overline{A}^{k,\nu} = \text{const}, \quad \sum_{k=1}^{3} \overline{A}^{k,\nu}\overline{A}^{k}_{\nu} \neq 0. \tag{2.1.2}$$

From formulas (1.3.1), (2.1.1) and (2.1.2), we find

$$\varepsilon_{klm}\overline{A}^{l,\mu}\overline{A}^{m,\nu} = 0. \tag{2.1.3}$$

Let us choose the gauge so as to have $\overline{A}^{1,0} \neq 0$. Then, from (2.1.3)
when $\mu = 0$ and $k = 2, 3$, we find

$$\overline{A}^{1,0}\,\overline{A}^{3,\nu} - \overline{A}^{3,0}\,\overline{A}^{1,\nu} = 0, \quad \overline{A}^{1,0}\,\overline{A}^{2,\nu} - \overline{A}^{2,0}\,\overline{A}^{1,\nu} = 0,$$
$$\overline{A}^{3,\nu} = (\overline{A}^{3,0}/\overline{A}^{1,0})\overline{A}^{1,\nu}, \quad \overline{A}^{2,\nu} = (\overline{A}^{2,0}/\overline{A}^{1,0})\overline{A}^{1,\nu}. \tag{2.1.4}$$

Putting

$$\overline{\lambda}^{k} = \overline{A}^{k,0}/\overline{A}^{1,0}, \quad \overline{\beta}^{\nu} = \overline{A}^{1,\nu}, \tag{2.1.5}$$

from (2.1.4), we get the equality

$$\overline{A}^{k,\nu} = \overline{\lambda}^{k}\,\overline{\beta}^{\nu}, \tag{2.1.6}$$

which is an evident identity when $k = 1$.

As follows from (1.3.2) and (2.1.6), the constant potentials $A^{k,\nu} = \overline{A}^{k,\nu}$ satisfy condition (2.1.1) at infinity since ε_{klm} are antisymmetric
and hence $\varepsilon_{klm}\overline{\lambda}^{l}\overline{\lambda}^{m} = 0$.

Let us put

$$\lambda^k = \overline{\lambda}^k a, \quad \beta^\nu = \overline{\beta}^\nu / a, \quad a = (\overline{\beta}^\mu \overline{\beta}_\mu)^{1/2}. \tag{2.1.7}$$

Then, we come to the following condition at infinity, which accords with (2.1.1):

$$A^{k,\nu} = \lambda^k \beta^\nu, \quad r \to \infty,$$
$$(\lambda^1)^2 + (\lambda^2)^2 + (\lambda^3)^2 = \Lambda^2, \quad \beta^\nu \beta_\nu = 1, \tag{2.1.8}$$

where $\lambda^k, \beta^\nu, \Lambda$ are some real numbers.

As for the constant Λ in (2.1.8), it will be regarded as a fundamental positive constant of the physical vacuum.

It should be noted that condition (2.1.8) at infinity is covariant under the Lorentz transformations.

The Yang–Mills equations (1.3.1)–(1.3.2) supplemented by the boundary condition (2.1.8) at infinity will be used in the following to describe weak interactions.

2.1.2. *Fields of neutrinos*

Consider now weak interactions in a small spatial region and represent the field potentials $A^{k,\nu}$ corresponding to them in the form

$$A^{k,\nu} = \lambda^k \beta^\nu + u^{k,\nu}, \tag{2.1.9}$$

where, as follows from (2.1.8), $u^{k,\nu} \to 0$ as $r \to \infty$. For the considered weak interactions, let us assume that $|u^{k,\nu}| \ll \Lambda$. Then, from (1.3.1), (1.3.2) and (2.1.9), we derive the following linear equations for the small functions $u^{k,\nu}$, in which the small values of the second order are neglected:

$$\partial^\mu \partial_\mu u^{k,\nu} - \partial^\nu \partial_\mu u^{k,\mu} + g\varepsilon_{klm}(2\lambda^l \beta^\mu \partial_\mu u^{m,\nu} + \lambda^m \beta^\nu \partial_\mu u^{l,\mu}$$
$$+ \lambda^m \beta_\mu \partial^\nu u^{l,\mu}) - g^2 \lambda_m \beta_\mu (\lambda^k \beta^\nu u^{m,\mu} + \lambda^m \beta^\mu u^{k,\nu} - \lambda^k \beta^\mu u^{m,\nu}$$
$$- \lambda^m \beta^\nu u^{k,\mu}) = (4\pi/c)J^{k,\nu}, \quad \lambda_m \equiv \lambda^m. \tag{2.1.10}$$

Let us study Eq. (2.1.10) for a Z^0 boson at rest. Then, $u^{k,\nu}$ should represent its wave functions of the de Broglie form. That is why and using that this boson is neutral, we should put

$$J^{k,\nu} = 0, \quad u^{k,\nu} = a^{k,\nu}\exp(-iM_{Z^0}c^2 t/\hbar), \qquad (2.1.11)$$

where $a^{k,\nu}$ = const, t is time and M_{Z^0} is the mass at rest of the Z^0 boson.

Since we consider a boson at rest, the numbers β^ν in expression (2.1.9) for the field potentials $A^{k,\nu}$ should be independent of the choice of spatial axes. Taking this into account and using (2.1.8), we find

$$\beta^0 = \pm 1, \quad \beta^1 = \beta^2 = \beta^3 = 0. \qquad (2.1.12)$$

Let us choose the Lorentz gauge for the potentials $A^{k,\nu}$:

$$\partial_\nu A^{k,\nu} = 0. \qquad (2.1.13)$$

Then, from (2.1.9) and (2.1.11), we obtain

$$a^{k,0} = 0. \qquad (2.1.14)$$

As follows from (2.1.11), (2.1.12) and (2.1.14), Eq. (2.1.10) is an evident identity when $\nu = 0$. Putting $\nu = 1, 2, 3$ in Eq. (2.1.10), we come to the following system of linear equations for $a^{k,\nu}$:

$$\sum_{k=1}^{3} a^{k,\nu}d_{ik} = 0, \quad \nu = 1,2,3, \ i,k = 1,2,3, \qquad (2.1.15)$$

where the matrix D, consisting of the elements d_{ik}: $D = (d_{ik})$, has the form

$$D = \begin{pmatrix} \omega^2 + g^2(\lambda_2^2 + \lambda_3^2) & -2ig\gamma\lambda_3 - g^2\lambda_1\lambda_2 & 2ig\gamma\lambda_2 - g^2\lambda_1\lambda_3 \\ 2ig\gamma\lambda_3 - g^2\lambda_1\lambda_2 & \omega^2 + g^2(\lambda_1^2 + \lambda_3^2) & -2ig\gamma\lambda_1 - g^2\lambda_2\lambda_3 \\ -2ig\gamma\lambda_2 - g^2\lambda_1\lambda_3 & 2ig\gamma\lambda_1 - g^2\lambda_2\lambda_3 & \omega^2 + g^2(\lambda_1^2 + \lambda_2^2) \end{pmatrix}.$$

$$(2.1.16)$$

Here,

$$\lambda_k \equiv \lambda^k, \quad \omega = M_{Z^0}c/\hbar, \quad \gamma = \beta^0\omega, \quad \beta^0 = \pm 1. \qquad (2.1.17)$$

In order to have nonzero solutions to the system of linear equations (2.1.15), we should require

$$\det(D) = 0. \tag{2.1.18}$$

From (2.1.16)–(2.1.18), after calculating the determinant of the matrix D, we obtain the following equation for the mass M_{Z^0}:

$$\begin{aligned}
\det(D) &= \omega^6 - 2g^2(\lambda_1^2 + \lambda_2^2 + \lambda_3^2)\omega^4 \\
&\quad + g^4(\lambda_1^2 + \lambda_2^2 + \lambda_3^2)^2\omega^2 = 0.
\end{aligned} \tag{2.1.19}$$

From (2.1.8), (2.1.17) and (2.1.19), we get

$$\begin{aligned}
\omega^2(\omega^2 - g^2\Lambda^2)^2 = 0, \quad \Lambda^2 = \lambda_1^2 + \lambda_2^2 + \lambda_3^2, \\
\lambda_k \equiv \lambda^k, \quad \omega = M_{Z^0}c/\hbar
\end{aligned} \tag{2.1.20}$$

and for the mass M_{Z^0} of the Z^0 boson, we find

$$M_{Z^0} = g\Lambda\hbar/c. \tag{2.1.21}$$

In order to further find a relation between the constant Λ and the Fermi constant of weak interactions, let us consider electroweak fields described by Eq. (2.1.10).

Let us multiply Eq. (2.1.10) by λ^k and sum it over k. Then, since ε_{klm} are antisymmetric and hence $\varepsilon_{klm}\lambda^k\lambda^l = 0$, we easily obtain

$$\partial^\mu\partial_\mu y^\nu - \partial^\nu\partial_\mu y^\mu = (4\pi/c)\lambda_k J^{k,\nu}, \quad y^\nu = \lambda_k u^{k,\nu}, \quad \lambda_k \equiv \lambda^k. \tag{2.1.22}$$

Multiplying now Eq. (2.1.10) by β_ν, summing it over ν and using again the antisymmetry of ε_{klm}, we find

$$\begin{aligned}
\partial^\mu\partial_\mu z^k - \beta_\nu\partial^\nu\partial_\mu u^{k,\mu} + g\varepsilon_{klm}(\lambda^l\beta^\mu\partial_\mu z^m + \lambda^m\partial_\mu u^{l,\mu}) \\
= (4\pi/c)\beta_\nu J^{k,\nu}, \quad z^k = \beta_\nu u^{k,\nu}.
\end{aligned} \tag{2.1.23}$$

Consider an electron neutrino or a muon neutrino. Then, since they are neutral particles, let us impose the following condition on their potentials:

$$u^{k,\nu} = 0 \quad \text{outside the neutrino, where } J^{k,\nu} = 0. \tag{2.1.24}$$

From (2.1.22) and (2.1.24), we find that in the region occupied by the neutrino,

$$\lambda_k J^{k,\nu} = 0, \tag{2.1.25}$$

since if (2.1.25) had not been satisfied, then from (2.1.22), we would obtain that outside the neutrino, $\lambda_k u^{k,\nu} = y^\nu \neq 0$, which contradicts (2.1.24).

From (2.1.22) and (2.1.25), we find

$$\lambda_k u^{k,\nu} = y^\nu = 0. \tag{2.1.26}$$

Since the neutrino mass is very small, as compared with the mass M_{Z^0} of the Z^0 boson, and taking into account (2.1.21) and the de Broglie formula for the wavelength, we have the following relation for the radius r_ν of the region occupied by the neutrino:

$$1/r_\nu \ll M_{Z^0} c/\hbar = g\Lambda. \tag{2.1.27}$$

As follows from (2.1.27) and (2.1.8), we can neglect the linear terms with respect to λ^l in the left-hand side of Eq. (2.1.10), as compared with its quadratic terms. Therefore, from Eq. (2.1.10), we derive the approximate equation

$$g^2 \lambda_m \beta_\mu (\lambda^k \beta^\nu u^{m,\mu} + \lambda^m \beta^\mu u^{k,\nu} \\ - \lambda^k \beta^\mu u^{m,\nu} - \lambda^m \beta^\nu u^{k,\mu}) = -\frac{4\pi}{c} J^{k,\nu}. \tag{2.1.28}$$

From (2.1.26), (2.1.28) and (2.1.8), we have

$$u^{k,\nu} - \beta^\nu z^k = -4\pi J^{k,\nu}/(cg^2\Lambda^2), \quad z^k = \beta_\nu u^{k,\nu}. \tag{2.1.29}$$

Let us use the differential equations of charge conservation

$$\partial_\nu J^{k,\nu} = 0. \tag{2.1.30}$$

From (2.1.29) and (2.1.30), we find

$$\partial_\nu u^{k,\nu} = \beta^\nu \partial_\nu z^k. \tag{2.1.31}$$

Substituting (2.1.31) for $\partial_\mu u^{k,\mu}$ and $\partial_\mu u^{l,\mu}$ into (2.1.23) and using the antisymmetry of ε_{klm}, we obtain

$$\partial^\mu \partial_\mu z^k - \beta^\mu \beta^\nu \partial_\mu \partial_\nu z^k = (4\pi/c)\beta_\nu J^{k,\nu}. \tag{2.1.32}$$

From (2.1.32) and (2.1.24), we find that in the region occupied by the neutrino,

$$\beta_\nu J^{k,\nu} = 0. \tag{2.1.33}$$

Indeed, if (2.1.33) had not been satisfied, then from (2.1.32), we would obtain that outside the neutrino, $\beta_\nu u^{k,\nu} = z^k \neq 0$, which contradicts (2.1.24).

From (2.1.32) and (2.1.33), we have

$$\beta_\nu u^{k,\nu} = z^k = 0 \tag{2.1.34}$$

and from (2.1.29) and (2.1.34), we find the potentials of weak interaction:

$$u^{k,\nu} = -4\pi J^{k,\nu}/(cg^2\Lambda^2). \tag{2.1.35}$$

It follows from (2.1.25) and (2.1.33) that the found potentials (2.1.35) satisfy relations (2.1.26) and (2.1.34).

2.1.3. *Electroweak interactions*

Let us introduce three charges q_k for a particle taking part in electroweak interactions and represent the densities of its currents $J^{k,\nu}$ in the form

$$J^{k,\nu}/c = q_k I^\nu, \tag{2.1.36}$$

where I^ν is the 4-vector proportional to the particle velocity.

We will assume that for the proton and electron, the charge q_1 coincides with their classical electric charge and $q_2 = q_3 = 0$. The neutrino charges will be considered later on.

Let us seek a relation for the three charges of a charged elementary particle. This relation should satisfy the following two properties:

(1) The relation should be covariant under the gauge rotations by constant angles preserving the Yang–Mills equations and the differential equations of charge conservation (2.1.30).
(2) The charges $q_1 = \pm e_p$, $q_2 = q_3 = 0$ corresponding to the proton and electron, where e_p is the proton charge, should satisfy the relation.

Using these properties, we come to the equality

$$q_1^2 + q_2^2 + q_3^2 = e_p^2. \tag{2.1.37}$$

Consider an electron neutrino and a muon neutrino. Since they are neutral particles having weak interactins with electrons, we assume that they have two charges q_1, q_2 or q_1, q_3 neutralizing each other, namely,

$$q_1 = -q_2, \quad q_3 = 0 \quad \text{or} \quad q_1 = -q_3, \quad q_2 = 0. \tag{2.1.38}$$

As follows from (2.1.25) and (2.1.36), formulas (2.1.38) imply that $\lambda_1 = \lambda_2$ or $\lambda_1 = \lambda_3$ inside the considered neutrinos.

Formulas (2.1.37) and (2.1.38) give

$$q_1 = \pm 2^{-1/2} e_p, \quad q_2 = -q_1, \quad q_3 = 0 \quad \text{or} \quad q_2 = 0, \quad q_3 = -q_1. \tag{2.1.39}$$

Let us now turn to a generalization of the classical Dirac equation [1] for the relativistic electron to describe a fermion with three charges q_1, q_2, q_3. This generalization can be represented in the form

$$i\hbar \partial \Psi / \partial t = [c\alpha_l (i\hbar \partial^l - q_k A^{k,l}/c) + \gamma^0 mc^2 + q_k A^{k,0}]\Psi,$$
$$\alpha_l = \gamma^0 \gamma^l, \quad l = 1, 2, 3, \tag{2.1.40}$$

where Ψ is a bispinor consisting of four wave functions of the considered fermion, γ^ν are the Dirac matrices, m is the rest mass of the fermion and $A^{k,\nu}$ are the potentials of a Yang–Mills field.

Consider infinitesimal gauge transformations (1.3.3) that do not change the fermion current densities $J^{k,\nu}$. Since $J^{k,\nu}$ are proportional to q_k, as follows from (2.1.36), and $\varepsilon_{klm} q^l q^m = 0$, where $q^l \equiv q_l$, in the region occupied by the fermion, these transformations can be represented as

$$\phi^l = \eta q_l, \quad J^{k,\nu} \to J^{k,\nu},$$
$$A^{k,\nu} \to A^{k,\nu} - \varepsilon_{klm} \eta q^l A^{m,\nu} + (1/g) q^k \partial^\nu \eta, \quad q^l \equiv q_l, \tag{2.1.41}$$

where η is a small function.

Taking into account that $\varepsilon_{klm} q^k q^l = 0$, from (2.1.40) and (2.1.41), we have the following infinitesimal gauge transformations that do not change the fermion state:

$$q_k A^{k,\nu} \to q_k A^{k,\nu} + (1/g) q_k \partial^\nu \phi^k,$$
$$\Psi \to \Psi \exp(-iq_k \phi^k/(\hbar cg)), \quad \phi^k = \eta q_k. \tag{2.1.42}$$

Consider the Hamiltonian of interaction H_{int} with a weak field of a fermion satisfying Eq. (2.1.40).

For the potentials $A^{k,\nu}$, we can use formula (2.1.9). As for its constant term $\lambda^k \beta^\nu$ in this formula, it can be removed by the gauge transformation

$$\Psi \to \Psi \exp(iq_k \lambda^k \beta_\nu x^\nu/(c\hbar)), \tag{2.1.43}$$

where x^ν are space–time coordinates in (2.1.40).

Using this, from (2.1.9) and (2.1.36), we find that the Hamiltonian of interaction H_{int} [1] has the form

$$H_{\text{int}} = q_k I_\nu u^{k,\nu}, \quad I^\nu = \overline{\Psi} \gamma^\nu \Psi, \quad \overline{\Psi} = \Psi^+ \gamma^0, \tag{2.1.44}$$

where $J^{k,\nu} = cq_k I^\nu$ are current densities and Ψ^+ is the Hermitian conjugate of Ψ.

For the electron charges, we have $q_1 = -e_p, q_2 = q_3 = 0$. Consider an electron moving in the weak field of an electron antineutrino. Since the energy of their interaction is positive, from (2.1.39), (2.1.44), (2.1.35) and (2.1.36), we obtain that for the electron antineutrino,

$$q_1 = -q_2 = 2^{-1/2} e_p, \quad q_3 = 0. \tag{2.1.45}$$

Therefore, from (2.1.35), (2.1.36) and (2.1.44), we find that the Hamiltonian of the interaction of an electron with an antineutrino has the form

$$H_{\text{int}} = 2^{-1/2} 4\pi (g\Lambda)^{-2} e_p^2 I_{(e)}^\mu I_{\mu(\overleftarrow{\nu})}, \tag{2.1.46}$$

where the indices $(e), (\overleftarrow{\nu})$ correspond to the electron and the electron antineutrino, respectively.

In the considered case, the Hamiltonian of interaction can be represented as [1]

$$H_{\text{int}} = 2^{-1/2} G_F I_{(e)}^\mu I_{\mu(\bar{\nu})}, \quad G_F = 10^{-5} \hbar^3 (cm_p^2)^{-1}, \tag{2.1.47}$$

where G_F is the Fermi constant and m_p is the proton mass at rest.

Comparing formulas (2.1.46) and (2.1.47), we come to the equality

$$G_F = 10^{-5}\hbar^3/(cm_p^2) = 4\pi e_p^2/(g\Lambda)^2. \tag{2.1.48}$$

From (2.1.48) and (2.1.21), we find

$$10^{-5}\hbar^3/(cm_p^2) = 4\pi[e_p\hbar/(M_{Z^0}c)]^2. \tag{2.1.49}$$

This equality gives the following value of the mass M_{Z^0} of the Z^0 boson at rest:

$$M_{Z^0} = 200m_p[10\pi e_p^2/(\hbar c)]^{1/2} = 89.8\,\text{GeV}. \tag{2.1.50}$$

As is seen from (2.1.50), the obtained theoretical value of the rest mass M_{Z^0} of the Z^0 boson well accords with its experimental value, which is also approximately equal to 90 GeV [1].

As is well known, the masses M_W of the $W^\pm$ bosons are approximately equal to 80 GeV [1]. The difference between the masses of the Z^0 and $W^\pm$ bosons can be explained by the existence of electric fields of the latter.

Let us now find a relation between the constants g and e_p. For this purpose, consider an electron described by Eq. (2.1.40) when the electroweak field potentials $A^{k,\nu}$ are subject to the gauge transformations (2.1.41).

Since for the electron $q_1 = -e_p, q_2 = q_3 = 0$, from (2.1.41), we obtain

$$A^{1,\nu} \to A^{1,\nu} + (1/g)\partial^\nu\phi^1, \quad A^{2,\nu} \to A^{2,\nu} + A^{3,\nu}\phi^1,$$
$$A^{3,\nu} \to A^{3,\nu} - A^{2,\nu}\phi^1, \tag{2.1.51}$$

where ϕ^1 is a small angle of rotation of the two-dimensional vector $(A^{2,\nu}, A^{3,\nu})$ about its initial point.

Making such a rotation N times by the small angle ϕ^1/N, where N is a sufficiently large number, we readily generalize the gauge transformations (2.1.51) for an arbitrary angle ϕ^1 of rotation of the two-dimensional vector $(A^{2,\nu}, A^{3,\nu})$ about its initial point:

$$A^{1,\nu} \to A^{1,\nu} + (1/g)\partial^\nu\phi^1,$$
$$A^{2,\nu} \to A^{2,\nu}\cos\phi^1 + A^{3,\nu}\sin\phi^1, \tag{2.1.52}$$
$$A^{3,\nu} \to A^{3,\nu}\cos\phi^1 - A^{2,\nu}\sin\phi^1.$$

In addition to this, from (2.1.42), we obtain the gauge transformation for the wave functions Ψ of the electron:

$$\Psi_{\phi^1} = \Psi_0 \exp(ie_p\phi^1/(\hbar cg)), \qquad (2.1.53)$$

where Ψ_{ϕ^1} are the wave functions Ψ corresponding to the angle ϕ^1 in (2.1.52).

Suppose that in the spatial region occupied by an electron at rest, the angle ϕ^1 is slowly changing from the constant value $\phi^1 = 0$ to the constant value $\phi^1 = 2\pi$. Then, as follows from (2.1.52), at the moments with $\phi^1 = 0$ and $\phi^1 = 2\pi$, the potentials $A^{k,\nu}$ are the same. Since these potentials determine the wave functions of the considered electron at rest, we come to the conclusion that at the moments with $\phi^1 = 0$ and $\phi^1 = 2\pi$, the wave functions Ψ_0 and $\Psi_{2\pi}$ are also the same. Therefore, taking into account (2.1.53), we find

$$\Psi_{2\pi} = \Psi_0 \exp(2\pi ie_p/(\hbar cg)) = \Psi_0. \qquad (2.1.54)$$

From (2.1.54), we come to the following formula for the elementary charge e_p:

$$e_p = \hbar cg. \qquad (2.1.55)$$

Formula (2.1.55) explains the existence of the elementary charge e_p since in it, this charge is determined by the fundamental constants $\hbar, c, g$ and it gives the value $e_p/(\hbar c)$ of the constant g.

From (2.1.48) and (2.1.55), we find the value of the constant Λ:

$$\begin{aligned}
\Lambda &= 2\hbar c(\pi/G_F)^{1/2} = 1121\, m_p(c^3/\hbar)^{1/2} \\
&= 2.997 \times 10^8\, \mathrm{g}^{1/2}\mathrm{cm}^{1/2}/\mathrm{sec}.
\end{aligned} \qquad (2.1.56)$$

2.2. Stationary Yang–Mills Fields Generated by Spherical Sources with Three Charges

Consider the classical Yang–Mills fields in a more general case when their sources have three charges. Then, they can be represented in the form

$$\partial_\mu F^{k,\mu\nu} + g\varepsilon_{klm}A_\nu^l F^{m,\mu\nu} = 4\pi\theta^k\, dx^\nu/ds, \qquad (2.2.1)$$

$$F^{k,\mu\nu} = \partial^\mu A^{k,\nu} - \partial^\nu A^{k,\mu} + g\varepsilon_{klm}A^{l,\mu}A^{m,\nu}, \qquad (2.2.2)$$

where θ^k are the densities of the charges of the field source in a comoving local inertial frame of reference and dx^ν/ds is the 4-vector of velocities of its points.

The Yang–Mills equations (2.2.1)–(2.2.2) should be supplemented by condition (1.4.16) proposed in Chapter 1. It expresses the conservation of the intrinsic electric energy of a small part of the source when charged quanta of the Yang–Mills field are generated inside the source. This condition can be represented in the form

$$\sum_{k=1}^{3} \partial_\alpha F^{k,\alpha\nu} \partial_\beta F^{k,\beta}{}_\nu = (4\pi)^2 \sum_{k=1}^{3} (\theta^k)^2. \tag{2.2.3}$$

Let us turn to the case of stationary spherically symmetric sources. Then, we have

$$\begin{aligned} \theta^1 &= \theta^1(r), \quad \theta^2 = \theta^2(r), \quad \theta^3 = \theta^3(r), \\ dx^0/ds &= 1, \quad dx^l/ds = 0, \quad l = 1, 2, 3, \end{aligned} \tag{2.2.4}$$

where r is the distance from the source center.

We will seek spherically symmetric solutions to Eqs. (2.2.1)–(2.2.3) in the form

$$A^{k,0} = 0, \quad A^{k,l} = x^l[x^0 u^k(r) + v^k(r)], \quad l = 1, 2, 3, \tag{2.2.5}$$

where

$$x^0 = ct, \quad x^1 = x, \quad x^2 = y, \quad x^3 = z, \quad r = \sqrt{x^2 + y^2 + z^2}. \tag{2.2.6}$$

Here, t is time and x, y, z are rectangular coordinates.

From Eq. (2.2.2), we obtain the following field strengths:

$$F^{k,0l} = u^k(r)x^l, \quad F^{k,mn} = 0, \quad k, l, m, n = 1, 2, 3. \tag{2.2.7}$$

Substituting formulas (2.2.4), (2.2.5) and (2.2.7) into the Yang–Mills equation (2.2.1), we come to the system of equations

$$r(u^k)' + 3u^k + gr^2 \varepsilon_{klm} u^l v^m = -4\pi\theta^k. \tag{2.2.8}$$

Thus, we obtain three equations for the six unknown functions u^k and v^k.

Multiplying Eq. (2.2.8) by u^k, summing it over the index k and taking into account the antisymmetry of ε_{klm}, we find

$$ru' + 3u = -(4\pi/u) \sum_{k=1}^{3} \theta^k u^k, \quad (u)^2 = \sum_{k=1}^{3} (u^k)^2. \tag{2.2.9}$$

Using the second and third equations ($k = 2, 3$) of the system of equations (2.2.8), we obtain relations between v^2, v^3 and v^1, u^1, u^2, u^3.

Consider the case

$$\theta^1 = \theta(r)\cos\omega(r), \quad \theta^2 = \theta(r)\sin\omega(r), \quad \theta^3 = 0, \tag{2.2.10}$$

where θ and ω are some functions of the argument r.

In this case, we seek the unknown functions $u^k(r)$ in the form

$$u^1 = -R(r)\cos\xi(r)/r^3, \quad u^2 = -R(r)\sin\xi(r)/r^3, \quad u^3 = 0, \tag{2.2.11}$$

where $R(r)$ and $\xi(r)$ are some functions.

Then, using (2.2.10), (2.2.11) and Eq. (2.2.9), we find

$$R' = 4\pi\, r^2\theta\cos(\xi - \omega). \tag{2.2.12}$$

Let us turn to Eq. (2.2.3). Substituting expressions (2.2.10) and (2.2.11) into it, we obtain

$$R'^2 + R^2\xi'^2 = (4\pi\theta)^2 r^4. \tag{2.2.13}$$

Taking into account (2.2.12), from this equation, we have

$$R\xi' = \pm 4\pi r^2\theta\sin(\xi - \omega). \tag{2.2.14}$$

Consider Eqs. (2.2.12) and (2.2.14) in the region occupied by the field source. In it, these equations can be represented as

$$\frac{dR}{dq} = \cos(\xi - \omega), \quad \frac{d\xi}{dq} = \pm\frac{\sin(\xi - \omega)}{R}, \quad q = 4\pi \int_0^r r^2\theta dr, \tag{2.2.15}$$

where $q = q(r)$ is the charge distributed with the density θ in the spherical region with radius r and having its center at the zero point. Here, R and ξ are represented as functions of the argument q.

From (2.2.15), we find

$$dR/R = \pm \cot(\xi - \omega)d\xi. \qquad (2.2.16)$$

Integrating this equation and writing $d\xi = d(\xi - \omega) + d\omega$, we obtain

$$\ln|R| = \pm \ln|\sin(\xi - \omega)| \pm \int \cot(\xi - \omega)d\omega. \qquad (2.2.17)$$

Let us choose the sign '+' in (2.2.14)–(2.2.17) to have a nonsingular solution to Eq. (2.2.17). Then, we get

$$R = K_0 \sin(\xi - \omega) \exp\left(\int_0^q \cot(\xi - \omega)\frac{d\omega}{dq}dq\right), \quad K_0 = \text{const.} \qquad (2.2.18)$$

It should be noted that the principal value of the integral in (2.2.18) should be taken when $\cot(\xi - \omega)$ can be infinite.

As follows from formula (2.2.11), we should put

$$R(0) = 0. \qquad (2.2.19)$$

Since the sign '+' is chosen in Eqs. (2.2.15) and (2.2.17), from them, we obtain

$$\frac{d\xi}{dq} = \frac{1}{K_0} \exp\left(-\int_0^q \cot(\xi - \omega)\frac{d\omega}{dq}dq\right). \qquad (2.2.20)$$

Consider the case $\omega = \omega_0 = \text{const}$. Then, from Eqs. (2.2.18)–(2.2.20), we find

$$\xi = \frac{q}{K_0^*} + \alpha_0, \quad \alpha_0 = \text{const},$$
$$R = K_0^* \sin\left(\frac{q}{K_0^*}\right), \qquad (2.2.21)$$

where K_0^* is the value of the constant K_0 in the considered case.

From (2.2.7), (2.2.11) and (2.2.21), we come to the following expressions for the nonzero components $F^{1,l0}$ and $F^{2,l0}$ ($l \neq 0$) of the field strengths:

$$F^{1,l0} = K_0 \sin\left(\frac{q}{K_0}\right) \cos\left(\frac{q}{K_0} + \alpha_0\right) \frac{x^l}{r^3},$$

$$F^{2,l0} = K_0 \sin\left(\frac{q}{K_0}\right) \sin\left(\frac{q}{K_0} + \alpha_0\right) \frac{x^l}{r^3}, \quad l = 1, 2, 3.$$

(2.2.22)

In order to have the classical expressions for the field strengths when the value of q is small, let us put $\alpha_0 = \omega_0$. Then, in a small vicinity of the point $r = 0$, where the values of q are small, we obtain

$$F^{1,l0} = q_1 x^l / r^3, \quad F^{2,l0} = q_2 x^l / r^3, \quad \alpha_0 = \omega_0,$$

$$q_1 = q \cos \omega_0, \quad q_2 = q \sin \omega_0, \quad |q/K_0| \ll 1,$$

(2.2.23)

where q_1 and q_2 are two charges of the considered spherical region with radius r.

From (2.2.22), putting $\alpha_0 = \omega_0$, we find

$$F^{1,l0} = Q_1 x^l / r^3, \quad F^{2,l0} = Q_2 x^l / r^3,$$

$$Q_1 = K[\sin(q/K + \omega_0) - \sin \omega_0],$$

$$Q_2 = K[\cos \omega_0 - \cos(q/K + \omega_0)], \quad K = K_0/2,$$

(2.2.24)

where Q_1 and Q_2 can be regarded as the full charges in the spherical region of radius r.

In the studied case $\omega = \omega_0$, from (2.2.10), we have

$$q^2 = q_1^2 + q_2^2, \quad \tan \omega_0 = q_2/q_1,$$

(2.2.25)

where

$$q_1 = 4\pi \int_0^r r^2 \theta^1 dr, \quad q_2 = 4\pi \int_0^r r^2 \theta^2 dr.$$

(2.2.26)

The constant K in (2.2.24) characterizes the relation between the classical charges q_1 and q_2 of the field source and its full charges Q_1 and Q_2. Its value is estimated in Chapter 1, where one charge is considered, and is $\sim 10^7$ coul. If the charge q is not large: $|q/K| \ll 1$, then from (2.2.24), we have the classical expressions (2.2.23) for the field strengths $F^{1,l0}$ and $F^{2,l0}$.

Let us assume that a stationary field source occupies a region bounded by the radius r_* and $\bar{q}_1$ and $\bar{q}_2$ are its classical charges. Consider the field generated by it at distances r far enough from it: $r \gg r_*$. Let us require that at such distances, its field should depend only on its charges $\bar{q}_1$ and $\bar{q}_2$ and be independent of their distribution. Then, using (2.2.24) and (2.2.25), we come to the following conditions at infinity for the field strengths $F^{1,l0}$ and $F^{2,l0}$:

$$F^{1,l0} = \bar{Q}_1 x^l / r^3, \quad F^{2,l0} = \bar{Q}_2 x^l / r^3, \quad r \gg r_*,$$

$$\bar{Q}_1 = K[\sin(\bar{q}/K + \bar{\omega}) - \sin\bar{\omega}], \quad \tan\bar{\omega} = \bar{q}_2/\bar{q}_1, \quad (2.2.27)$$

$$\bar{Q}_2 = K[\cos\bar{\omega} - \cos(\bar{q}/K + \bar{\omega})], \quad \bar{q}^2 = \bar{q}_1^2 + \bar{q}_2^2,$$

where $\bar{q}_1, \bar{q}_2$ and $\bar{Q}_1, \bar{Q}_2$ can be regarded as the classical and full charges, respectively, of the field source.

When the charge $\bar{q}$ is small as compared with the constant K, from formula (2.2.27), we have

$$F^{1,l0} = \bar{q}_1 x^l / r^3, \quad F^{2,l0} = \bar{q}_2 x^l / r^3, \quad |\bar{q}/K| \ll 1, \quad r \gg r_*.$$
$$(2.2.28)$$

Consider now the case $\omega \neq$ const. Then, as follows from (2.2.18), if the derivative $d\omega/dq$ is sufficiently large, at some point $r = r_1$, field strengths can reach large values inside the considered field source. Therefore, if the densities θ_1 and θ_2 of the source charges are both nonzero and not proportional to each other, then significant amplification of field strengths becomes possible inside the field source.

It should be noted that applying to the obtained solution a gauge transformation corresponding to a rotation of the 4-vectors of the source currents about the first axis in the gauge space, we come to the case in which all the three charge densities $\theta_1, \theta_2, \theta_3$ of the source are nonzero.

Let us turn now to the following case:

$$\omega = 0, \quad 0 \leq q < b,$$
$$\omega = \omega_1, \quad b < q \leq q_*. \quad (2.2.29)$$

Here, ω_1 and b are some constants and $q_* = q(r_*)$, where the spherical surface $r = r_*$ is the boundary of the field source.

Then, from (2.2.18), (2.2.20) and (2.2.29), we find

$$\xi = q/K_0, \quad R = K_0 \sin(q/K_0), \quad 0 \le q < b. \tag{2.2.30}$$

Consider now the region $b < q \le q_*$. Then, from (2.2.20) and (2.2.29), we obtain

$$\frac{d\xi}{dq} = \frac{1}{K_0} \exp\left(-\int_0^q \cot(\xi(b) - w)dw\right) = \frac{1}{K_0} \frac{\sin(\xi(b) - \omega_1)}{\sin\xi(b)}. \tag{2.2.31}$$

Therefore,

$$\xi = \frac{q}{K_1} + \xi_1, \quad K_1 = K_0 \frac{\sin\xi(b)}{\sin(\xi(b) - \omega_1)}, \quad b < q \le q_*, \tag{2.2.32}$$

where ξ_1 is a constant.

From (2.2.30) and (2.2.32), we get

$$\xi(b) = b/K_0 = b/K_1 + \xi_1, \quad \xi_1 = b(1/K_0 - 1/K_1). \tag{2.2.33}$$

Formulas (2.2.18) and (2.2.29) give

$$R = K_0 \sin(\xi - \omega_1) \exp\left(\int_0^q \cot(\xi(b) - w)dw\right), \quad b < q \le q_*. \tag{2.2.34}$$

From this formula, we find

$$R = K_0 \sin(\xi - \omega_1) \frac{\sin\xi(b)}{\sin(\xi(b) - \omega_1)}. \tag{2.2.35}$$

Using (2.2.33), formula (2.2.35) can be represented as

$$R = K_0 \sin(\xi - \omega_1) \frac{\sin(b/K_0)}{\sin(b/K_0 - \omega_1)}, \quad b < q \le q_*. \tag{2.2.36}$$

From (2.2.7), (2.2.11) and (2.2.30), we obtain when $0 \leq q < b$:

$$F^{1,l0} = K \sin(q/K)x^l/r^3, \quad F^{2,l0} = K[1 - \cos(q/K)]x^l/r^3,$$

$$K = K_0/2, \tag{2.2.37}$$

where $l = 1, 2, 3$.

When $b < q \leq q_*$, from (2.2.7), (2.2.11) and (2.2.36), we find

$$
\begin{aligned}
F^{1,l0} &= K[\sin(2\xi - \omega_1) - \sin\omega_1]\frac{\sin\left(\frac{1}{2}b/K\right)}{\sin\left(\frac{1}{2}b/K - \omega_1\right)}\frac{x^l}{r^3}, \\
F^{2,l0} &= K[\cos\omega_1 - \cos(2\xi - \omega_1)]\frac{\sin\left(\frac{1}{2}b/K\right)}{\sin\left(\frac{1}{2}b/K - \omega_1\right)}\frac{x^l}{r^3},
\end{aligned}
\tag{2.2.38}
$$

where ξ is determined by formula (2.2.32).

As follows from formulas (2.2.38), the field strengths $F^{1,l0}$ and $F^{2,l0}$ can acquire very large values when $\frac{1}{2}b/K - \omega_1 \approx \pi n$, where n is an integer.

Let us use this result to explain the amazing phenomenon of nitrogen fixation in a number of types of bacteria. Nitrogen fixation is known to be a process in which nitrogen is taken from inert molecular forms in the atmosphere and converted to nitrogen compounds, such as ammonia, nitrates and nitrogen dioxide. Such processes require very high energies, and the only nonbiological way of their implementation, found in nature, is lightning discharges.

To explain their implementation in bacteria, it can be assumed that living cells have not one but two or three charges, leading to the appearance of Yang–Mills fields in them. Then, we can apply the result obtained above, which is that some sources of Yang–Mills fields can create significant field strengths inside themselves. Therefore, the above assumption just makes it possible to explain why nitrogen fixation, which requires very high energies, is carried out in certain types of bacteria.

It should be noted that the question of the fundamental differences between living matter and inanimate matter is very intriguing and one of the most important for the development of science. After all, a living cell consists of the same atoms and molecules as inanimate formations, but its vital activity looks truly miraculous and incomprehensible from the standpoint of existing physical theories. At the same time, the theory of nonlinear Yang–Mills fields, which turned out to be useful for explaining the mysterious phenomenon of nitrogen fixation in a number of bacteria, can play an important role in the study of the features of physical processes in living matter.

2.3. General Dynamic Equations for a System of Particles with Three Charges

First, consider the classical problem of the movement of a dust-like matter consisting of a system of particles with electric charges under the action of their own Maxwell electromagnetic field. In an inertial frame of reference, it is described by the equation [13]

$$c^2 \rho_0 \frac{d^2 x^\nu}{ds^2} = \theta F^\nu{}_\alpha \frac{dx^\alpha}{ds}, \tag{2.3.1}$$

where x^ν are space–time coordinates of a particle, $\nu = 0, 1, 2, 3$, ds is the four-dimensional interval, $ds^2 = (dx^0)^2 - (dx^1)^2 - (dx^2)^2 - (dx^3)^2$, ρ_0 is the mass density at rest of the matter, θ is the charge density at rest of the matter and $F_{\mu\nu}$ is the tensor of strengths of the electromagnetic field.

The energy–momentum tensor $T^{\mu\nu}$ of the considered dust-like matter and its electromagnetic field has the form [13]

$$T^{\mu\nu} = T_m^{\mu\nu} + T_e^{\mu\nu}, \tag{2.3.2}$$

$$T_m^{\mu\nu} = c^2 \rho_0 \frac{dx^\mu}{ds} \frac{dx^\nu}{ds}, \quad T_e^{\mu\nu} = \frac{1}{4\pi} \left(-F^{\mu\alpha} F^\nu{}_\alpha + \frac{1}{4} g^{\mu\nu} F_{\alpha\beta} F^{\alpha\beta} \right), \tag{2.3.3}$$

where $T_m^{\mu\nu}$ is the energy–momentum tensor of the matter, $T_e^{\mu\nu}$ is the energy–momentum tensor of the electromagnetic field and $g^{\mu\nu}$ are the components of the Minkowski metric tensor.

The tensor $T^{\mu\nu}$ satisfies the differential equation of energy and momentum conservation [13]

$$\partial_\mu T^{\mu\nu} = 0. \tag{2.3.4}$$

It should be stressed that the application of formulas (2.3.2)–(2.3.4) and the differential equation of mass conservation

$$\partial_\mu(\rho_0 dx^\mu/ds) = 0 \tag{2.3.5}$$

leads to the dynamic equations (2.3.1) [13].

Let us turn now to the movement of a dust-like matter consisting of particles with three charges under the action of their own Yang–Mills field. Then, formulas (2.3.2) and (2.3.3) for the energy–momentum tensor of the considered system of particles can be generalized as follows:

$$T^{\mu\nu} = T_m^{\mu\nu} + T_{YM}^{\mu\nu}, \tag{2.3.6}$$

$$T_m^{\mu\nu} = c^2 \rho_0 \frac{dx^\mu}{ds}\frac{dx^\nu}{ds}, \quad T_{YM}^{\mu\nu} = \frac{1}{4\pi}\left(-F^{k,\mu\alpha}F^{k,\nu}{}_\alpha + \frac{1}{4}g^{\mu\nu}F_{\alpha\beta}^k F^{k,\alpha\beta}\right), \tag{2.3.7}$$

where $T_{YM}^{\mu\nu}$ is the energy–momentum tensor of the Yang–Mills field. Here, the summation over the index $k = 1, 2, 3$ is implied, as well as over the indices $\alpha, \beta = 0, 1, 2, 3$.

Let us apply the differential equations of energy–momentum conservation (2.3.4) to the energy–momentum tensor of form (2.3.6)–(2.3.7) to generalize the dynamic equations (2.3.1).

First, let us calculate $\partial_\mu T_m^{\mu\nu}$, using (2.3.5) and (2.3.7). For this expression, we find

$$\partial_\mu T_m^{\mu\nu} = \partial_\mu(c^2 \rho_0 dx^\mu/ds\, dx^\nu/ds)$$
$$= c^2 \partial_\mu(\rho_0 dx^\mu/ds)dx^\nu/ds + c^2 \rho_0 \partial_\mu(dx^\nu/ds)dx^\mu/ds$$
$$= c^2 \rho_0 d^2 x^\nu/ds^2. \tag{2.3.8}$$

Consider now $\partial_\mu T_{YM}^{\mu\nu}$. For its calculation, let us use the Yang–Mills equations (2.2.1) and the following well-known identities for the tensor of field strengths $F_{\mu\nu}^k$ [1]:

$$D_\gamma F_{\mu\nu}^k + D_\mu F_{\nu\gamma}^k + D_\nu F_{\gamma\mu}^k = 0, \tag{2.3.9}$$

where D_μ is the Yang–Mills covariant derivative, which is defined for an arbitrary differentiable vector function U^k as

$$D_\mu U^k = \partial_\mu U^k + g\varepsilon_{klm}A_\mu^l U^m. \tag{2.3.10}$$

Using it, we can represent the Yang–Mills equation (2.2.1) in the form

$$D_\mu F^{k,\mu\nu} = 4\pi\theta^k dx^\nu/ds, \qquad (2.3.11)$$

where θ^k are the densities of the three charges of the considered particles, $k = 1, 2, 3$.

Using (2.3.7), we find

$$\partial_\mu T^{\mu\nu}_{YM} = \frac{1}{4\pi}\left(-\partial_\mu F^{k,\mu\alpha} F^{k,\nu}{}_\alpha - F^{k,\mu\alpha}\partial_\mu F^{k,\nu}{}_\alpha + \frac{1}{2}F^{k,\alpha\beta}\partial^\nu F^k_{\alpha\beta}\right). \qquad (2.3.12)$$

From (2.3.10)–(2.3.12), we obtain

$$\begin{aligned}
\partial_\mu T^{\mu\nu}_{YM} = {}& -\frac{1}{4\pi}F^{k,\nu}{}_\alpha\left(4\pi\theta^k dx^\alpha/ds + g\varepsilon_{klm}F^{l,\mu\alpha}A^m_\mu\right)\\
& -\frac{1}{4\pi}F^{k,\mu\alpha}\left(D_\mu F^{k,\nu}{}_\alpha + g\varepsilon_{klm}F^{l,\nu}{}_\alpha A^m_\mu\right) \qquad (2.3.13)\\
& +\frac{1}{8\pi}F^{k,\alpha\beta}\left(D^\nu F^k_{\alpha\beta} + g\varepsilon_{klm}F^l_{\alpha\beta}A^{m,\nu}\right).
\end{aligned}$$

Taking into account identities (2.3.9), we have the following equality:

$$\begin{aligned}
F^{k,\alpha\beta}D_\nu F^k_{\alpha\beta} &= -F^{k,\alpha\beta}(D_\alpha F^k_{\beta\nu} + D_\beta F^k_{\nu\alpha})\\
&= -2F^{k,\alpha\beta}D_\alpha F^k_{\beta\nu} = -2F^{k,\mu\alpha}D_\mu F^k_{\alpha\nu}. \qquad (2.3.14)
\end{aligned}$$

This gives

$$F^{k,\alpha\beta}D^\nu F^k_{\alpha\beta} = 2F^{k,\mu\alpha}D_\mu F^{k,\nu}{}_\alpha. \qquad (2.3.15)$$

Besides, we have the following evident equalities:

$$\varepsilon_{klm}F^{k,\nu}{}_\alpha F^{l,\mu\alpha} = -\varepsilon_{klm}F^{k,\mu\alpha}F^{l,\nu}{}_\alpha, \quad \varepsilon_{klm}F^{k,\alpha\beta}F^l_{\alpha\beta} = 0. \quad (2.3.16)$$

Using the obtained equalities (2.3.15) and (2.3.16) in formula (2.3.13), we come to the relation

$$\partial_\mu T^{\mu\nu}_{YM} = -\theta^k F^{k,\nu}{}_\alpha dx^\alpha/ds. \qquad (2.3.17)$$

Substituting formulas (2.3.6), (2.3.8) and (2.3.17) into Eq. (2.3.4), we find

$$c^2 \rho_0 \frac{d^2 x^\nu}{ds^2} = \theta^k F^{k,\nu}{}_\alpha \frac{dx^\alpha}{ds}. \tag{2.3.18}$$

The obtained dynamic equations (2.3.18) for the system of particles with three charges moving under the action of their own Yang–Mills field present a reasonable generalization of the dynamic equations (2.3.1) for the system of particles with one classical charge moving under the action of their own Maxwell field.

Chapter 3

Nonlinear Waves in Yang–Mills Fields

The Yang–Mills equations were studied in many works [19, 20] and a part of them was devoted to their solutions describing nonabelian plane waves [21–27]. In this chapter, we find new classes of nonlinear wave solutions to the Yang–Mills equations by using the results obtained in our papers [28–32].

3.1. Spherical Waves in Yang–Mills Fields

Consider the Yang–Mills equations (1.3.1) and (1.3.2) in the region outside their sources. Then, they acquire the form

$$\partial_\mu F^{k,\mu\nu} + g\varepsilon_{klm}A_\mu^l F^{m,\mu\nu} = 0, \tag{3.1.1}$$

$$F^{k,\mu\nu} = \partial^\mu A^{k,\nu} - \partial^\nu A^{k,\mu} + g\varepsilon_{klm}A^{l,\mu}A^{m,\nu}. \tag{3.1.2}$$

Let us turn to the case of a nonstationary spherically symmetric wave radiated from a source located at the zero point of a rectangular coordinate system at the moment $t = 0$ and generating the field potentials of the form

$$A^{k,0} = 0, \ A^{k,l} = (\sigma/r)\varepsilon_{klm}x^m, \ \sigma = \sigma(\tau, r),$$

$$k, l, m = 1, 2, 3, \ \tau \equiv x^0, \ r \equiv [(x^1)^2 + (x^2)^2 + (x^3)^2]^{1/2}.$$

$$\tag{3.1.3}$$

It should be noted that in the stationary case $\sigma = \sigma(r)$, this form of potential was used in the Wu–Yang solution to the Yang–Mills equations [19].

Let us substitute expressions (3.1.3) into Eq. (3.1.2). Then, we obtain the following expressions for the field strengths $F^{k,\mu\nu}$:

$$
\begin{aligned}
& F^{k,0l} = \sigma_\tau \varepsilon_{klm} x^m / r, \ F^{1,21} = \phi \, x^1 x^3, \\
& F^{1,13} = \phi x^1 x^2, \ \phi = (\sigma_r - \sigma/r - g\sigma^2)/r^2, \\
& F^{1,23} = \sigma_r + \sigma/r - \phi(x^1)^2, \ F^{2,21} = \phi \, x^2 x^3, \qquad (3.1.4) \\
& F^{2,31} = \sigma_r + \sigma/r - \phi(x^2)^2, \\
& F^{2,32} = \phi \, x^1 x^2, \ F^{3,12} = \sigma_r + \sigma/r - \phi(x^3)^2, \\
& F^{3,13} = \phi \, x^2 x^3, \ F^{3,32} = \phi \, x^1 x^3,
\end{aligned}
$$

where $\sigma_\tau \equiv \partial\sigma/\partial\tau$ and $\sigma_r \equiv \partial\sigma/\partial r$.

Let us denote the left-hand side of Eq. (3.1.1) by $\bar{J}^{k,\nu}$,

$$
\bar{J}^{k,\nu} \equiv \partial_\mu F^{k,\mu\nu} + g\varepsilon_{klm} A_\mu^l F^{m,\mu\nu}. \qquad (3.1.5)
$$

Then, substituting expressions (3.1.3) and (3.1.4) for $A^{k,\nu}$ and $F^{k,\mu\nu}$ into formula (3.1.5), we find after calculations:

$$
\begin{aligned}
& \bar{J}^{k,0} = 0, \ \bar{J}^{1,1} = \bar{J}^{2,2} = \bar{J}^{3,3} = 0, \ \bar{J}^{1,2} = -\bar{J}^{2,1} = jx^3/r, \\
& \bar{J}^{2,3} = -\bar{J}^{3,2} = jx^1/r, \ \bar{J}^{3,1} = -\bar{J}^{1,3} = jx^2/r, \qquad (3.1.6) \\
& j = \sigma_{\tau\tau} - \sigma_{rr} - 2\sigma_r/r + 2\sigma/r^2 + 3g\sigma^2/r + g^2\sigma^3,
\end{aligned}
$$

where $\sigma_{\tau\tau} \equiv \partial^2\sigma/\partial\tau^2$ and $\sigma_{rr} \equiv \partial^2\sigma/\partial r^2$.

From (3.1.1), we have $\bar{J}^{k,\nu} = 0$. Therefore, formulas (3.1.6) give the following nonlinear partial differential equation for the function $\sigma(\tau, r)$:

$$
\sigma_{\tau\tau} - \sigma_{rr} - 2\sigma_r/r + 2\sigma/r^2 + 3g\sigma^2/r + g^2\sigma^3 = 0. \qquad (3.1.7)
$$

Putting

$$
\sigma = (gr)^{-1}(\psi - 1), \qquad (3.1.8)
$$

from (3.1.7), we come to the following nonlinear wave equation for the function $\psi(\tau, r)$:

$$
\psi_{\tau\tau} - \psi_{rr} + \psi(\psi^2 - 1)/r^2 = 0. \qquad (3.1.9)
$$

It should be noted that Eq. (3.1.9) has the evident particular solutions

$$\psi = 0, \ \psi = \pm 1. \tag{3.1.10}$$

For large values of r, this equation has the wave solution

$$\psi = f_1(\tau - r) + f_2(\tau + r), \ r \to \infty, \tag{3.1.11}$$

where f_1 and f_2 are differentiable functions, $\tau = ct$ and t is the time.

Consider now solutions to Eq. (3.1.9) corresponding to the waves radiated at the moment $t = 0$ from a source located at the point $r = 0$.

Let us introduce the dimensionless variable

$$\xi = r/\tau, \ \xi \geq 0, \tag{3.1.12}$$

and seek such solutions in the form

$$\psi = 0 \text{ when } \xi > 1(\tau - r < 0),$$
$$\psi = f(\xi) \text{ when } \xi \leq 1(\tau - r \geq 0), \tag{3.1.13}$$

where $f(\xi)$ is some dimensionless function.

When $\xi \leq 1$, from (3.1.12) and (3.1.13), we find

$$\psi_\tau = -\xi^2 f'/r, \ \psi_{\tau\tau} = \xi^3(\xi f'' + 2f')/r^2,$$
$$\psi_r = \xi f'/r, \ \psi_{rr} = \xi^2 f''/r^2, \tag{3.1.14}$$

where $f' \equiv df/d\xi, \ f'' \equiv d^2 f/d\xi^2$.

Substituting (3.1.12)–(3.1.14) into Eq. (3.1.9), we obtain when $\xi \leq 1$

$$\xi^2(\xi^2 - 1)f'' + 2\xi^3 f' + f(f^2 - 1) = 0, \ f = f(\xi). \tag{3.1.15}$$

We seek solutions to Eq. (3.1.15) in the form of the power series

$$f(\xi) = 1 + d_1\xi^2 + d_2\xi^4 + \cdots + d_n\xi^{2n} + \cdots, \tag{3.1.16}$$

where d_n are some numbers, and put $d_0 = 1$.

From (3.1.16), we find

$$f^2 - 1 = \sum_{k=1}^{\infty} \xi^{2k} \sum_{i=0}^{k} d_i d_{k-i}, \ d_0 = 1,$$

$$f(f^2 - 1) = \sum_{n=1}^{\infty} \xi^{2n} \sum_{k=1}^{n} \sum_{i=0}^{k} d_i d_{k-i} d_{n-k}. \tag{3.1.17}$$

Substituting (3.1.16) and (3.1.17) into Eq. (3.1.15) and dividing it by ξ^2, we obtain

$$2(\xi^2 - 1) \sum_{n=0}^{\infty} (2n+1)(n+1)d_{n+1}\xi^{2n} + 4\xi^2 \sum_{n=0}^{\infty} (n+1)d_{n+1}\xi^{2n}$$

$$+ \sum_{n=1}^{\infty} \xi^{2(n-1)} \sum_{k=1}^{n} \sum_{i=0}^{k} d_i d_{k-i} d_{n-k} = 0. \tag{3.1.18}$$

Equating the coefficients for ξ^{2n}, $n = 0, 1, 2, \ldots$, in the left-hand side of (3.1.18) to zero, we find

$$-2(2n+1)(n+1)d_{n+1} + 2n(2n+1)d_n$$

$$+ \sum_{k=1}^{n+1} \sum_{i=0}^{k} d_i d_{k-i} d_{n+1-k} = 0, \ d_0 = 1, \ n \geq 0. \tag{3.1.19}$$

It can be easily shown that when $n = 0$, equality (3.1.19) is identically satisfied.

When $n \geq 1$, from (3.1.19), we obtain the following recurrence relation:

$$d_{n+1} = \frac{1}{2n(2n+3)} \left[2n(2n+1)d_n + \sum_{k=1}^{n} \sum_{i=0}^{k} d_i d_{k-i} d_{n+1-k} \right.$$

$$\left. + \sum_{i=1}^{n} d_i d_{n+1-i} \right], \ d_0 = 1, \ n \geq 1,$$

$$\tag{3.1.20}$$

where the number d_1 can be arbitrarily chosen.

Using (3.1.20), it is easy to prove by induction that the values $|d_{n+1}|$, where $n \geq 1$, with a negative number d_1 are less than the positive values d_{n+1} when this number is positive.

When $d_1 = 0$ from (3.1.20), we find that $d_{n+1} = 0$, $n = 1, 2, 3, \ldots$, which correspond to the particular solution $f = 1$ to Eq. (3.1.15).

Consider the case $d_1 = 1/2$. Then, from (3.1.20), we derive

$$d_0 = 1, \ d_1 = 1/2, \ d_2 = 3/8, \ d_3 = 5/16, \ldots. \tag{3.1.21}$$

Let us show that this case corresponds to an exact particular solution to Eq. (3.1.15). With this aim, we represent the function $f(\xi)$ in the form

$$f(\xi) = (1 - \xi^2)^{-1/2} u(\xi). \tag{3.1.22}$$

Then, for the derivatives $f'(\xi)$ and $f''(\xi)$, we have

$$f' = (1 - \xi^2)^{-3/2}[(1 - \xi^2)u' + \xi u],$$
$$f'' = (1 - \xi^2)^{-5/2}[(1 - \xi^2)^2 u'' + 2\xi(1 - \xi^2)u' + (1 + 2\xi^2)u].$$
$$\tag{3.1.23}$$

Substituting (3.1.22) and (3.1.23) into Eq. (3.1.15), we come to the following equation:

$$\xi^2(1 - \xi^2)^2 u'' + u(1 - u^2) = 0, \ u = u(\xi). \tag{3.1.24}$$

This equation has the particular solution

$$u = 1. \tag{3.1.25}$$

From (3.1.22) and (3.1.25), we find the following particular exact solution to Eq. (3.1.15):

$$f(\xi) = (1 - \xi^2)^{-1/2}. \tag{3.1.26}$$

It should be noted that this solution exactly corresponds to the case $d_1 = 1/2$ and has the coefficients (3.1.21) of its expansion in Maclaurin's series.

Consider the obtained equation (3.1.24) when $\xi \to 1$. From it, we have that $u(1)$ cannot be infinite since in such a case, the term u^3 of Eq. (3.1.24) would tend to an infinite value as $\xi \to 1$ faster than its other terms. As for finite values of $u(1)$, from (3.1.24), we find that they can be as follows:

$$u(1) = 0 \text{ or } u(1) = \pm 1. \tag{3.1.27}$$

Let us turn now to the recurrence relation (3.1.20) for the coefficients d_n when $0 \leq d_1 \leq 1/2$ and let $\bar{d}_n$ denote the values d_n when $d_1 = 1/2$. Then, taking into account that the terms in (3.1.20) are nonnegative, we find, easily proved by induction, that the following inequalities hold:

$$0 \leq d_n < \bar{d}_n \text{ when } 0 \leq d_1 < 1/2 \text{ and } n \geq 1, \tag{3.1.28}$$

$$d_n > \bar{d}_n \text{ when } d_1 > 1/2 \text{ and } n \geq 1. \tag{3.1.29}$$

In the case $d_1 = 1/2$, when $d_n = \bar{d}_n$, the power series (3.1.16) is convergent when $|\xi| < 1$ since it corresponds to the function of form (3.1.26):

$$(1 - \xi^2)^{-1/2} = \sum_{n=0}^{\infty} \bar{d}_n \xi^{2n}, \quad \bar{d}_0 = 1, \quad |\xi| < 1. \tag{3.1.30}$$

Therefore, taking into account (3.1.28), we find that the power series (3.1.16) is convergent when $|\xi| < 1$ and $0 \leq d_1 \leq 1/2$.

Consider properties of the power series (3.1.16) when $0 \leq |\xi| < 1$ and $0 < d_1 < 1/2$. Using formulas (3.1.16), (3.1.22) and (3.1.30), we can represent the function $u(\xi)$ in the form

$$u(\xi) = \sum_{n=0}^{\infty} d_n \xi^{2n} \bigg/ \sum_{n=0}^{\infty} \bar{d}_n \xi^{2n}, \quad d_0 = \bar{d}_0 = 1. \tag{3.1.31}$$

From (3.1.28) and (3.1.31), we find

$$u(0) = 1, \quad u'(0) = 0, \quad 0 < u(\xi) < 1, \quad 0 < \xi < 1,$$
$$0 < d_1 < 1/2. \tag{3.1.32}$$

Using Eq. (3.1.24) and the inequalities for $u(\xi)$ in (3.1.32), we obtain

$$u''(\xi) < 0, \quad 0 < \xi < 1. \tag{3.1.33}$$

Therefore, the function $u'(\xi)$ decreases in the region $0 \leq \xi \leq 1$ when $0 < d_1 < 1/2$. That is why, taking into account the equality

$u'(0) = 0$ in (3.1.32), we get

$$u'(\xi) < 0, \ 0 < \xi < 1. \tag{3.1.34}$$

From formulas (3.1.32) and (3.1.34), we find

$$u(0) = 1, \ u'(\xi) < 0, \ 0 < \xi < 1, \qquad 0 < d_1 < 1/2. \tag{3.1.35}$$

These formulas show that $u(\xi)$ decreases in the region $0 \le \xi \le 1$ and, therefore, we have the inequality $u(1) < 1$. Thus, taking into account (3.1.27), we obtain

$$u(1) = 0, \ 0 < d_1 < 1/2. \tag{3.1.36}$$

In this case, let us consider the behavior of $u(\xi)$ when $\xi \to 1$. Because of (3.1.36), near the point $\xi = 1$, Eq. (3.1.24) can be approximated as

$$4(1 - \xi)^2 u'' + u = 0, \ u = u(\xi), \ \xi = 1 - \delta, \tag{3.1.37}$$

where δ takes an arbitrarily small positive value.

As can be readily verified, Eq. (3.1.37) has the solution

$$u(\xi) = [a + b\ln(1 - \xi)](1 - \xi)^{1/2}, \ \xi = 1 - \delta, \ a, b = \text{const.} \tag{3.1.38}$$

Therefore, in the considered case (3.1.36), the function $f(\xi) = (1 - \xi^2)^{-1/2} u(\xi)$ has a logarithmic singularity near the point $\xi = 1$.

Let us show now that the obtained exact solution to Eq. (3.1.15) of form (3.1.26), for which $d_1 = 1/2$, is limiting for the physically possible solutions to this equation. For this purpose, let us consider the power series (3.1.16) in the case $d_1 > 1/2$ and show that its sum can become infinite for some value ξ inside the region $|\xi| < 1$.

Actually, suppose that this is not true and the power series (3.1.16) is convergent when $|\xi| < 1$ and $d_1 > 1/2$. Then, from (3.1.29) and (3.1.31), we obtain

$$u(0) = 1, \ u'(0) = 0, \ u(\xi) > 1, \ 0 < \xi < 1, \ d_1 > 1/2. \tag{3.1.39}$$

From (3.1.24) and the inequality for $u(\xi)$ in (3.1.39), we find

$$u'' > 0, \ 0 < \xi < 1. \tag{3.1.40}$$

Therefore, the derivative $u'(\xi)$ increases in the region $0 \le \xi \le 1$. That is why, taking into account the equality $u'(0) = 0$ in (3.1.39), we have

$$u'(\xi) > 0, \ 0 < \xi < 1. \tag{3.1.41}$$

From (3.1.39) and (3.1.41), we get $u(1) > 1$, which contradicts the possible values (3.1.27) for $u(1)$. This contradiction shows that in the case $d_1 > 1/2$, the sum of the power series (3.1.16) becomes infinite at some point ξ in the region $|\xi| < 1$, which is physically impossible. Thus, the value $d_1 = 1/2$ is maximum for physically possible solutions to Eq. (3.1.15). Therefore, the exact solution (3.1.26) to Eq. (3.1.15), for which $d_1 = 1/2$, should be regarded as limiting.

Using formulas (3.1.8) and (3.1.12), we find the function $\sigma(\tau, r)$ corresponding to the exact solution (3.1.26):

$$\sigma(\tau, r) = (gr)^{-1}[(1 - \xi^2)^{-1/2} - 1], \ \xi = r/\tau. \tag{3.1.42}$$

Consider a field quantum radiated from the point $r = 0$ at the moment $\tau = 0$ and moving along a straight line at some speed v. Then, taking into account (3.1.12) and the relation $\tau = ct$, where t is time, we obtain for this quantum

$$\xi = r/\tau = v/c. \tag{3.1.43}$$

Therefore, for the exact solution (3.1.42), we find

$$\sigma(\tau, r) = (gr)^{-1}[(1 - v^2/c^2)^{-1/2} - 1]. \tag{3.1.44}$$

As follows from (3.1.3), the function $\sigma(\tau, r)$ determines the field potentials $A^{k,l}$.

Formula (3.1.44), which contains the multiplier $(1 - v^2/c^2)^{-1/2}$ of the relativistic mass of the quantum, can be interpreted as follows: The function $\sigma(\tau, r)$ and hence the potentials $A^{k,\nu}$ corresponding to it depend on the relativistic mass of the quantum, which is at the moment τ on the sphere with radius r, having its center at the zero point.

It is also interesting to note that the potentials $A^{k,\nu}$, which are determined by formula (3.1.3), satisfy the Lorentz gauge $\partial_\nu A^{k,\nu} = 0$.

Let us substitute formulas (3.1.13) and (3.1.16) into (3.1.8). Then, we obtain

$$\sigma = (gr)^{-1}\xi^2[d_1 + d_2\xi^2 + \cdots + d_n\xi^{2(n-1)} + \cdots]. \tag{3.1.45}$$

Substituting (3.1.45) into formulas (3.1.4), we find that the obtained field strengths $F^{k,\mu\nu}$ are bounded when $\tau > 0$ and $r \to 0$.

3.2. Axisymmetric Wave Solutions to the Yang–Mills Equations

Let us seek axisymmetric wave solutions to the Yang–Mills equations (3.1.1) and (3.1.2) in the form

$$
\begin{aligned}
A^{k,0} &= a^{k,0}, \ A^{k,1} = (\alpha^{k,1}x + \alpha^{k,2}y)/\rho, \\
A^{k,2} &= (\alpha^{k,1}y - \alpha^{k,2}x)/\rho, \ A^{k,3} = \alpha^{k,3}, \\
\alpha^{k,\nu} &= \alpha^{k,\nu}(\tau, \rho, z), \ \tau \equiv x^0, \ \rho \equiv \sqrt{x^2 + y^2},
\end{aligned} \tag{3.2.1}
$$

where $x^0 \equiv ct$, t is the time, $x \equiv x^1$, $y \equiv x^2$, $z \equiv x^3$ and x^1, x^2, x^3 are rectangular spatial coordinates.

It follows from formulas (3.2.1) that each of the three two-dimensional vectors $(A^{k,1}, A^{k,2})$, $k = 1, 2, 3$, is the sum of the vectors $(\alpha^{k,1}/\rho) \cdot (x, y)$ and $(\alpha^{k,2}/\rho) \cdot (y, -x)$, which are orthogonal to each other. Therefore, the considered field potentials $A^{k,\nu}$ are covariant under rotations of the coordinate system (x, y, z) about the axis z and hence are axially symmetric.

Substituting formulas (3.2.1) into expressions (3.1.2) for the field strengths $F^{k,\mu\nu}$, we obtain

$$
\begin{aligned}
F^{k,01} &= \frac{f^{k,1}x + f^{k,2}y}{\rho}, \ F^{k,12} = f^{k,4}, \ F^{k,13} = \frac{f^{k,5}x + f^{k,6}y}{\rho}, \\
F^{k,02} &= \frac{f^{k,1} - f^{k,2}x}{\rho}, \ F^{k,23} = \frac{f^{k,5}y - f^{k,6}x}{\rho}, \\
F^{k,03} &= f^{k,3}, \ f^{k,q} = f^{k,q}(\tau, \rho, z), \ q = 1, 2, \ldots, 6,
\end{aligned} \tag{3.2.2}
$$

where the functions $f^{k,q}(\tau, \rho, z)$ are given by the formulas

$$
f^{k,1} = \frac{\partial \alpha^{k,1}}{\partial \tau} + \frac{\partial \alpha^{k,0}}{\partial \rho} + g\varepsilon_{klm}\alpha^{l,0}\alpha^{m,1},
$$

$$
f^{k,2} = \frac{\partial \alpha^{k,2}}{\partial \tau} + g\varepsilon_{klm}\alpha^{l,0}\alpha^{m,2},
$$

$$f^{k,3} = \frac{\partial \alpha^{k,3}}{\partial \tau} + \frac{\partial \alpha^{k,0}}{\partial z} + g\varepsilon_{klm}\alpha^{l,0}\alpha^{m,3},$$

$$f^{k,4} = \frac{\partial \alpha^{k,2}}{\partial \rho} + \frac{\alpha^{k,2}}{\rho} - g\varepsilon_{klm}\alpha^{l,1}\alpha^{m,2},$$

$$f^{k,5} = \frac{\partial \alpha^{k,1}}{\partial z} - \frac{\partial \alpha^{k,3}}{\partial \rho} + g\varepsilon_{klm}\alpha^{l,1}\alpha^{m,3},$$

$$f^{k,6} = \frac{\partial \alpha^{k,2}}{\partial z} + g\varepsilon_{klm}\alpha^{l,2}\alpha^{m,3}. \tag{3.2.3}$$

Let us substitute expressions (3.2.1) and (3.2.2) for $A^{k,\nu}$ and $F^{k,\mu\nu}$ into the left-hand side of the Yang–Mills equation (3.1.1), which is denoted in the following by $\bar{J}^{k,\nu}$. Then, we find

$$\bar{J}^{k,\nu} \equiv \partial_\mu F^{k,\mu\nu} + g\varepsilon_{klm}A^l_\mu F^{m,\mu\nu},$$

$$\bar{J}^{k,0} = j^{k,0}, \quad \bar{J}^{k,1} = \frac{j^{k,1}x + j^{k,2}y}{\rho}, \tag{3.2.4}$$

$$\bar{J}^{k,2} = \frac{j^{k,1}y - j^{k,2}x}{\rho}, \quad \bar{J}^{k,3} = j^{k,3}, \quad j^{k,\nu} = j^{k,\nu}(\tau, \rho, z),$$

where the functions $j^{k,\nu}$ are given by the formulas

$$j^{k,0} = -\frac{\partial f^{k,1}}{\partial \rho} - \frac{f^{k,1}}{\rho} - \frac{\partial f^{k,3}}{\partial z} - g\varepsilon_{klm}(f^{l,1}\alpha^{m,1} + f^{l,2}\alpha^{m,2} + f^{l,3}\alpha^{m,3}),$$

$$j^{k,1} = \frac{\partial f^{k,1}}{\partial \tau} - \frac{\partial f^{k,5}}{\partial z} - g\varepsilon_{klm}(f^{l,1}\alpha^{m,0} - f^{l,4}\alpha^{m,2} + f^{l,5}\alpha^{m,3}),$$

$$j^{k,2} = \frac{\partial f^{k,2}}{\partial \tau} - \frac{\partial f^{k,4}}{\partial \rho} - \frac{\partial f^{k,6}}{\partial z} - g\varepsilon_{klm}(f^{l,2}\alpha^{m,0} + f^{l,4}\alpha^{m,1} + f^{l,6}\alpha^{m,3}),$$

$$j^{k,3} = \frac{\partial f^{k,3}}{\partial \tau} + \frac{\partial f^{k,5}}{\partial \rho} + \frac{f^{k,5}}{\rho} + g\varepsilon_{klm}(f^{l,5}\alpha^{m,1} + f^{l,6}\alpha^{m,2} - f^{l,3}\alpha^{m,0}).$$

$$\tag{3.2.5}$$

From the Yang–Mills equations (3.1.1) and formulas (3.2.4), we obtain the system of equations

$$j^{k,\nu}(\tau, \rho, z) = 0, \quad k = 1, 2, 3, \quad \nu = 0, 1, 2, 3, \tag{3.2.6}$$

where $j^{k,\nu}$ are defined by formulas (3.2.5) and (3.2.3).

We seek the components $\alpha^{k,\nu}$ of the field potentials that satisfy Eqs. (3.2.6) in the form

$$\alpha^{1,0} = 0, \ \alpha^{2,0} = P(\eta, \rho), \ \alpha^{3,0} = Q(\eta, \rho), \ \eta = \tau - z,$$

$$\alpha^{1,2} = \frac{\phi(\rho)}{g}, \ \alpha^{2,2} = \alpha^{3,2} = 0, \quad (3.2.7)$$

$$\alpha^{k,1} = 0, \ \alpha^{k,3} = \alpha^{k,0}, \ k = 1, 2, 3,$$

where $P(\eta, \rho)$, $Q(\eta, \rho)$ and $\phi(\rho)$ are some differentiable functions. Then, from (3.2.3), we find

$$f^{1,1} = 0, \ f^{2,1} = \frac{\partial P}{\partial \rho}, \ f^{3,1} = \frac{\partial Q}{\partial \rho},$$

$$f^{1,2} = 0, \ f^{2,2} = \phi Q, \ f^{3,2} = -\phi P,$$

$$f^{k,3} = 0, \ f^{1,4} = \frac{1}{g}\left(\frac{d\phi}{d\rho} + \frac{\phi}{\rho}\right), \ f^{2,4} = f^{3,4} = 0, \quad (3.2.8)$$

$$f^{k,5} = -f^{k,1}, \ f^{k,6} = -f^{k,2}, \ k = 1, 2, 3.$$

Substituting formulas (3.2.7) and (3.2.8) into expressions (3.2.5) for $j^{k,\nu}$, we obtain

$$j^{1,0} = 0, \ j^{2,0} = -\frac{\partial^2 P}{\partial \rho^2} - \frac{1}{\rho}\frac{\partial P}{\partial \rho} + \phi^2 P,$$

$$j^{3,0} = -\frac{\partial^2 Q}{\partial \rho^2} - \frac{1}{\rho}\frac{\partial Q}{\partial \rho} + \phi^2 Q, \ j^{1,2} = -\frac{1}{g}\frac{d}{d\rho}\left(\frac{d\phi}{d\rho} + \frac{\phi}{\rho}\right), \quad (3.2.9)$$

$$j^{2,2} = j^{3,2} = 0, \ j^{k,1} = 0, \ j^{k,3} = j^{k,0}, \ k = 1, 2, 3.$$

From (3.2.6) and (3.2.9), we find

$$\frac{\partial^2 P}{\partial \rho^2} + \frac{1}{\rho}\frac{\partial P}{\partial \rho} - \phi^2 P = 0, \ \frac{\partial^2 Q}{\partial \rho^2} + \frac{1}{\rho}\frac{\partial Q}{\partial \rho} - \phi^2 Q = 0, \quad (3.2.10)$$

$$\frac{d}{d\rho}\left(\frac{d\phi}{d\rho} + \frac{\phi}{\rho}\right) = 0. \quad (3.2.11)$$

Equation (3.2.11) has the following nonzero solution that tends to zero at infinity:

$$\phi = \frac{b}{\rho}, \ b = \text{const} \neq 0. \quad (3.2.12)$$

Substituting (3.2.12) into (3.2.10), we obtain two differential equations for the functions P and Q of the form

$$\frac{\partial^2 P}{\partial \rho^2} + \frac{1}{\rho}\frac{\partial P}{\partial \rho} - \left(\frac{b}{\rho}\right)^2 P = 0, \quad \frac{\partial^2 Q}{\partial \rho^2} + \frac{1}{\rho}\frac{\partial Q}{\partial \rho} - \left(\frac{b}{\rho}\right)^2 Q = 0.$$
(3.2.13)

It is easy to verify that the functions ρ^b and ρ^{-b} satisfy Eqs. (3.2.13). Therefore, the solutions to Eqs. (3.2.13) that tend to zero at infinity can be represented as

$$P = \frac{G(\tau - z)}{\rho^{|b|}}, \quad Q = \frac{H(\tau - z)}{\rho^{|b|}}, \quad b \neq 0,$$
(3.2.14)

where G and H are arbitrary differentiable functions.

Using formulas (3.2.1), (3.2.7), (3.2.12) and (3.2.14), we can represent the field potentials $A^{k,\nu}$ in the form

$$A^{1,0} = 0, \quad A^{2,0} = \frac{G(\tau - z)}{\rho^{|b|}}, \quad A^{3,0} = \frac{H(\tau - z)}{\rho^{|b|}},$$

$$A^{1,1} = \frac{by}{g\rho^2}, \quad A^{2,1} = A^{3,1} = 0, \quad A^{1,2} = -\frac{bx}{g\rho^2},$$
(3.2.15)

$$A^{2,2} = A^{3,2} = 0, \quad A^{k,3} = A^{k,0}, \quad k = 1,2,3.$$

Formulas (3.2.2), (3.2.8), (3.2.12) and (3.2.14) give the following expressions for the field strengths:

$$F^{1,01} = F^{1,02} = 0, \quad F^{k,03} = F^{k,12} = 0,$$

$$F^{2,01} = -\frac{|b|\,G(\tau - z)x - bH(\tau - z)y}{\rho^{2+|b|}},$$

$$F^{3,01} = \frac{-bG(\tau - z)y - |b|\,H(\tau - z)x}{\rho^{2+|b|}},$$

$$F^{2,02} = -\frac{|b|\,G(\tau - z)y + bH(\tau - z)x}{\rho^{2+|b|}},$$
(3.2.16)

$$F^{3,02} = -\frac{-bG(\tau - z)x + |b|\,H(\tau - z)y}{\rho^{2+|b|}},$$

$$F^{k,13} = -F^{k,01}, \quad F^{k,23} = -F^{k,02}, \quad k = 1,2,3.$$

Here, $\tau = ct$, where t is the time.

Consider now other classes of wave solutions to the Yang–Mills equations.

3.3. Expanding Waves in Yang–Mills Fields

Let us seek wave solutions to the Yang–Mills equations (3.1.1) and (3.1.2) in the form

$$A^{k,0} = u^k(y_0, y_1, y_2, y_3), \ A^{k,n} = (x^n/r)A^{k,0},$$
$$y_0 = x^0 - r, \ y_n = x^n, \tag{3.3.1}$$
$$n = 1, \ 2, \ 3, \ r = \sqrt{(x^1)^2 + (x^2)^2 + (x^3)^2},$$

where u^k are some differentiable functions of the wave phase $y_0 = x^0 - r$ and the spatial coordinates $y_n = x^n$.

Substituting expressions (3.3.1) into formula (3.1.2) for the field strengths $F^{k,\mu\nu}$, we readily find

$$F^{k,0n} = \partial u^k/\partial y_n, \ F^{k,in} = (1/r)(y_i \partial u^k/\partial y_n - y_n \partial u^k/\partial y_i),$$
$$y_n = x^n, \ i, n = 1, \ 2, \ 3. \tag{3.3.2}$$

Let us now substitute expressions (3.3.1) and (3.3.2) for $A^{k,\nu}$ and $F^{k,\mu\nu}$ into the Yang–Mills equation (3.1.1).

Then, when the index $\nu = 0$, from Eq. (3.1.1), we obtain

$$\sum_{i=1}^{3} \left(\frac{\partial^2 u^k}{\partial y_i^2} - \frac{y_i}{r}\frac{\partial^2 u^k}{\partial y_0\,\partial y_i} - g\frac{y_i}{r}\varepsilon_{klm}u^l\frac{\partial u^m}{\partial y_i} \right) = 0, \tag{3.3.3}$$

where $y_i = x^i$ and $y_0 = x^0 - r$. Further, we will denote x^i by y_i when $i = 1, 2, 3$.

When the index $\nu = n = 1, 2, 3$, from Eq. (3.1.1), we obtain after reductions

$$\frac{y_n}{r}\sum_{i=1}^{3}\left(y_i\frac{\partial^2 u^k}{\partial y_0 \partial y_i} - r\frac{\partial^2 u^k}{\partial y_i^2} + \frac{y_i}{r}\frac{\partial u^k}{\partial y_i} + g\varepsilon_{klm}y_i u^l\frac{\partial u^m}{\partial y_i} \right)$$

$$+ \frac{\partial}{\partial y_n}\left(\sum_{i=1}^{3} y_i\frac{\partial u^k}{\partial y_i} \right) = 0. \tag{3.3.4}$$

It should be noted that Eqs. (3.3.3) and (3.3.4) can be represented in the form

$$\partial_\mu F^{k,\mu 0} = -g \sum_{i=1}^{3} \frac{y_i}{r} \varepsilon_{klm} u^l \frac{\partial u^m}{\partial y_i},$$

$$\partial_\mu F^{k,\mu n} = -g \frac{y_n}{r} \sum_{i=1}^{3} \frac{y_i}{r} \varepsilon_{klm} u^l \frac{\partial u^m}{\partial y_i}. \tag{3.3.5}$$

It readily follows from Eqs. (3.3.5) that the field strengths $F^{k,\mu\nu}$ satisfy the additional condition (1.4.16) considered in Chapter 1, where we should put $J^{k,\nu} = 0$.

Let us denote

$$p^k = \sum_{i=1}^{3} y_i \frac{\partial u^k}{\partial y_i}, \quad q^k = \sum_{i=1}^{3} \frac{\partial^2 u^k}{\partial y_i^2}. \tag{3.3.6}$$

Then, from Eqs. (3.3.3) and (3.3.4), we find

$$q^k = \frac{1}{r} \left(\frac{\partial p^k}{\partial y_0} + g \varepsilon_{klm} u^l p^m \right), \quad r = \sqrt{y_1^2 + y_2^2 + y_3^2}, \tag{3.3.7}$$

$$y_n \left(\frac{1}{r} \frac{\partial p^k}{\partial y_0} - q^k + \frac{p^k}{r^2} + \frac{1}{r} g \varepsilon_{klm} u^l p^m \right) + \frac{\partial p^k}{\partial y_n} = 0, \quad n = 1, 2, 3. \tag{3.3.8}$$

As follows from (3.3.2) and (3.3.6), in the case $p^k = 0$, the considered expanding waves are transverse and when $p^k \neq 0$, these waves also have longitudinal components.

Let us substitute expression (3.3.7) for q^k into Eq. (3.3.8). Then, we readily obtain

$$y_n p^k / r^2 + \partial p^k / \partial y_n = 0, \quad n = 1, 2, 3. \tag{3.3.9}$$

As can be easily verified, these equations have the following solution:

$$p^k = s^k(y_0)/r, \tag{3.3.10}$$

where s^k are arbitrary differentiable functions of the argument y_0.

From (3.3.6), (3.3.7) and (3.3.10), we obtain

$$\sum_{i=1}^{3} y_i \frac{\partial u^k}{\partial y_i} = \frac{s^k(y_0)}{r}, \tag{3.3.11}$$

$$\sum_{i=1}^{3} \frac{\partial^2 u^k}{\partial y_i^2} = \frac{1}{r^2}\left[\dot{s}^k(y_0) + g\varepsilon_{klm}u^l s^m(y_0)\right], \quad \dot{s}^k(y_0) \equiv \frac{ds^k}{dy_0}. \tag{3.3.12}$$

It can be readily verified that Eq. (3.3.11) has the following solutions:

$$u^k(y_0, y_1, y_2, y_3) = -s^k(y_0)/r + b^k(y_0, \xi_1, \xi_2, \xi_3),$$

$$\xi_i = y_i/r, \quad r = \sqrt{y_1^2 + y_2^2 + y_3^2}, \tag{3.3.13}$$

where b^k are arbitrary differentiable functions.

Indeed, from (3.3.13), we find

$$\frac{\partial u^k}{\partial y_i} = \frac{s^k(y_0)y_i}{r^3} + \frac{1}{r}\frac{\partial b^k}{\partial \xi_i} - \frac{y_i}{r^3}\sum_{n=1}^{3} y_n \frac{\partial b^k}{\partial \xi_n}, \quad i = 1, 2, 3. \tag{3.3.14}$$

From (3.3.14), we obtain the identity $\sum_{i=1}^{3} y_i \partial u^k/\partial y_i \equiv s^k(y_0)/r$. Therefore, formula (3.3.13) gives solutions to Eq. (3.3.11).

Consider Eq. (3.3.12) and use formulas (3.3.13). For the functions $b^k(y_0, \xi_1, \xi_2, \xi_3)$, we have

$$\frac{\partial b^k}{\partial y_i} = \frac{1}{r}\sum_{j=1}^{3} \frac{\partial b^k}{\partial \xi_j}(\delta_{ij} - \xi_i\xi_j),$$

$$i = 1, 2, 3, \quad \xi_i = y_i/r, \quad \delta_{ii} = 1, \quad \delta_{ij} = 0 \text{ when } j \neq i,$$

$$\frac{\partial^2 b^k}{\partial y_i^2} = \frac{1}{r^2}\sum_{j,n=1}^{3} \frac{\partial^2 b^k}{\partial \xi_j \partial \xi_n}(\delta_{ij} - \xi_i\xi_j)(\delta_{in} - \xi_i\xi_n)$$

$$- \frac{1}{r^2}\sum_{j=1}^{3} \frac{\partial b^k}{\partial \xi_j}\left[\xi_j\left(1 - 3\xi_i^2\right) + 2\xi_i\delta_{ij}\right]. \tag{3.3.15}$$

Let us substitute expression (3.3.13) for the functions u^k into Eq. (3.3.12) and take into account that the function $1/r$ is harmonic and the constants ε_{klm} are antisymmetric. Then, using formulas (3.3.15) and the evident equality $\xi_1^2 + \xi_2^2 + \xi_3^2 = 1$, we obtain

$$\sum_{i=1}^{3} \left[(1 - \xi_i^2) \frac{\partial^2 b^k}{\partial \xi_i^2} - 2\xi_i \frac{\partial b^k}{\partial \xi_i} \right] - \sum_{\substack{i,j=1 \\ i \neq j}}^{3} \xi_i \xi_j \frac{\partial^2 b^k}{\partial \xi_i \partial \xi_j}$$

$$= \dot{s}^k(y_0) + g\varepsilon_{klm} b^l s^m(y_0). \tag{3.3.16}$$

The arguments $\xi_i = y_i/r$ of the functions b^k are not independent because of the identity $\xi_1^2 + \xi_2^2 + \xi_3^2 = 1$. Therefore, instead of ξ_1, ξ_2, ξ_3, we can choose two independent arguments related to them.

As will be seen later on, it is convenient to choose the following two arguments θ and σ:

$$b^k(y_0, \xi_1, \xi_2, \xi_3) = h^k(y_0, \theta, \sigma),$$

$$\theta = \frac{1}{2} \ln \left(\frac{1 + \xi_1}{1 - \xi_1} \right), \quad \sigma = \arctan \left(\frac{\xi_2}{\xi_3} \right), \tag{3.3.17}$$

where h^k are some differentiable functions.

Then, we have

$$\frac{\partial b^k}{\partial \xi_1} = \beta \frac{\partial h^k}{\partial \theta}, \quad \frac{\partial b^k}{\partial \xi_2} = \gamma \xi_3 \frac{\partial h^k}{\partial \sigma}, \quad \frac{\partial b^k}{\partial \xi_3} = -\gamma \xi_2 \frac{\partial h^k}{\partial \sigma},$$

$$\tag{3.3.18}$$

$$\beta = \frac{1}{1 - \xi_1^2}, \quad \gamma = \frac{1}{\xi_2^2 + \xi_3^2},$$

$$\frac{\partial^2 b^k}{\partial \xi_1^2} = \beta^2 \left(\frac{\partial^2 h^k}{\partial \theta^2} + 2\xi_1 \frac{\partial h^k}{\partial \theta} \right),$$

$$\frac{\partial^2 b^k}{\partial \xi_2^2} = \gamma^2 \xi_3 \left(\xi_3 \frac{\partial^2 h^k}{\partial \sigma^2} - 2\xi_2 \frac{\partial h^k}{\partial \sigma} \right), \tag{3.3.19}$$

$$\frac{\partial^2 b^k}{\partial \xi_3^2} = \gamma^2 \xi_2 \left(\xi_2 \frac{\partial^2 h^k}{\partial \sigma^2} + 2\xi_3 \frac{\partial h^k}{\partial \sigma} \right),$$

$$\frac{\partial^2 b^k}{\partial \xi_1 \partial \xi_2} = \beta \gamma \xi_3 \frac{\partial^2 h^k}{\partial \theta \partial \sigma},$$

$$\frac{\partial^2 b^k}{\partial \xi_1 \partial \xi_3} = -\beta \gamma \xi_2 \frac{\partial^2 h^k}{\partial \theta \partial \sigma}, \tag{3.3.20}$$

$$\frac{\partial^2 b^k}{\partial \xi_2 \partial \xi_3} = -\gamma^2 \left(\xi_2 \xi_3 \frac{\partial^2 h^k}{\partial \sigma^2} + (\xi_3^2 - \xi_2^2) \frac{\partial h^k}{\partial \sigma} \right)$$

and, as can be readily verified, the left-hand side of Eq. (3.3.16) acquires the form

$$\sum_{i=1}^{3} \left[(1 - \xi_i^2) \frac{\partial^2 b^k}{\partial \xi_i^2} - 2\xi_i \frac{\partial b^k}{\partial \xi_i} \right] - \sum_{\substack{i,j=1 \\ i \neq j}}^{3} \xi_i \xi_j \frac{\partial^2 b^k}{\partial \xi_i \partial \xi_j}$$

$$= \frac{1}{1 - \xi_1^2} \frac{\partial^2 h^k}{\partial \theta^2} + \frac{1}{\xi_2^2 + \xi_3^2} \frac{\partial^2 h^k}{\partial \sigma^2}. \tag{3.3.21}$$

Since the variables $\xi_i = y_i / r$ satisfy the equality $\xi_2^2 + \xi_3^2 = 1 - \xi_1^2$, from (3.3.16), (3.3.17) and (3.3.21), we come to the following equation:

$$\frac{\partial^2 h^k}{\partial \theta^2} + \frac{\partial^2 h^k}{\partial \sigma^2} = (1 - \xi_1^2) \left[\dot{s}^k(y_0) + g\varepsilon_{klm} h^l s^m(y_0) \right],$$

$$\xi_1 = \tanh\theta. \tag{3.3.22}$$

Let us put

$$h^k = v^k(y_0, \theta, \sigma) + \chi(y_0) s^k(y_0) \ln(\cosh\theta) + d^k(y_0), \tag{3.3.23}$$

where $v^k(y_0, \theta, \sigma)$, $\chi(y_0)$ and $d^k(y_0)$ are some functions.

Then, substituting (3.3.23) into (3.3.22) and taking into account that ε_{klm} are antisymmetric, we obtain

$$\frac{\partial^2 v^k}{\partial \theta^2} + \frac{\partial^2 v^k}{\partial \sigma^2} = (1 - \tanh^2\theta) [\dot{s}^k(y_0) - \chi(y_0) s^k(y_0) + g\varepsilon_{klm}$$

$$\times (v^l + d^l(y_0)) s^m(y_0)]. \tag{3.3.24}$$

Let us require that the four functions $\chi(y_0)$ and $d^k(y_0)$ should satisfy the following system of three algebraic equations, which are linear with respect to them:

$$\dot{s}^k(y_0) - \chi(y_0)s^k(y_0) + g\varepsilon_{klm}d^l(y_0)s^m(y_0) = 0. \qquad (3.3.25)$$

Then, from (3.3.24), we obtain

$$\frac{\partial^2 v^k}{\partial\theta^2} + \frac{\partial^2 v^k}{\partial\sigma^2} = g\left(1 - \tanh^2\theta\right)\varepsilon_{klm}v^l s^m(y_0), \quad v^k = v^k(y_0,\theta,\sigma).$$
$$(3.3.26)$$

After multiplying (3.3.25) by $s^k(y_0)$, summing it over the index k and taking into account the antisymmetry of ε_{klm}, we come to the following simple formula for the function $\chi(y_0)$:

$$\chi(y_0) = \dot{s}/s, \ (s)^2 = \sum_{k=1}^{3}(s^k)^2, \ s = s(y_0), \ \dot{s} \equiv ds/dy_0. \qquad (3.3.27)$$

From this formula, we find that in the case $s = \text{const}$, the function $\chi = 0$.

Let us seek solutions to Eq. (3.3.26) as the real part of the following complex sum:

$$v^k(y_0,\theta,\sigma) = \text{Re} \sum_{n=0}^{M} V_n^k(y_0,\theta) \exp\left(-n(\theta + i\sigma)\right), \qquad (3.3.28)$$

where $V_n^k(y_0,\theta)$ are some complex functions and n, M are nonnegative integers. Then, substituting (3.3.28) into Eq. (3.3.26), we come to the following equation:

$$\frac{\partial^2 V_n^k}{\partial\theta^2} - 2n\frac{\partial V_n^k}{\partial\theta} = g\left(1 - \tanh^2\theta\right)\varepsilon_{klm}V_n^l s^m(y_0). \qquad (3.3.29)$$

Let us choose the variable $\varphi = \tanh\theta \equiv \xi_1$ instead of θ and put

$$V_n^k = V_n^k(y_0,\varphi), \quad \varphi = \tanh\theta. \qquad (3.3.30)$$

Then, Eq. (3.3.29) acquires the following form since $d\varphi/d\theta = 1 - \tanh^2\theta = 1 - \varphi^2$:

$$(1 - \varphi^2)\frac{\partial^2 V_n^k}{\partial\varphi^2} - 2(n + \varphi)\frac{\partial V_n^k}{\partial\varphi} = g\varepsilon_{klm}V_n^l s^m(y_0), \quad -1 \le \varphi \le 1.$$
$$(3.3.31)$$

Putting

$$\eta = (1 - \varphi)/2, \ 0 \le \eta \le 1, \quad -1 \le \varphi \le 1, \tag{3.3.32}$$

from (3.3.31), we find

$$\eta(\eta - 1)\frac{\partial^2 V_n^k}{\partial \eta^2} - (n + 1 - 2\eta)\frac{\partial V_n^k}{\partial \eta} + g\varepsilon_{klm}V_n^l s^m(y_0) = 0. \tag{3.3.33}$$

We seek solutions $V_n^k(y_0, \eta)$ to Eq. (3.3.33) in the form

$$V_n^k = \sum_{j=0}^{\infty} \lambda_{j,n}^k(y_0)\eta^j, \tag{3.3.34}$$

where $\lambda_{j,n}^k$ are some complex functions of the argument y_0.

Then, substituting (3.3.34) into Eq. (3.3.33), we obtain the following recurrence relation for $\lambda_{j,n}^k$:

$$\lambda_{j+1,n}^k = \frac{j(j+1)\lambda_{j,n}^k + g\varepsilon_{klm}\lambda_{j,n}^l s^m}{(j+1)(j+1+n)}, \ j = 0, 1, 2, \ldots, \tag{3.3.35}$$

where the complex numbers $\lambda_{0,n}^k = \lambda_{0,n}^k(y_0)$ can be arbitrary.

From (3.3.35), we readily find that the sequence $\left|\lambda_{j,n}^k\right|$ is bounded for any y_0.

Indeed, let us put

$$L(y_0) = \max_{1 \le k,l \le 3} |g\varepsilon_{klm}s^m(y_0)| \tag{3.3.36}$$

and consider (3.3.35) when $j > L(y_0) - 1$ and for arbitrary y_0 and n. Then, we have

$$\max_{1\le k\le 3}\left|\lambda_{j+1,n}^k\right| \le \frac{j(j+1) + L(y_0)}{(j+1)(j+1+n)}\max_{1\le k\le 3}\left|\lambda_{j,n}^k\right| \le \max_{1\le k\le 3}\left|\lambda_{j,n}^k\right|,$$
$$j > L(y_0) - 1. \tag{3.3.37}$$

Formula (3.3.37) just shows that the sequence $\left|\lambda_{j,n}^k\right|$ is bounded for any y_0 and n.

From (3.3.37), we also obtain that the values $\max_{1\le k\le 3}\left|\lambda_{j,n}^k\right|$, $0 \le j < \infty$, are bounded by their maximum values when $0 \le j \le L(y_0)$.

Since the sequence $\left|\lambda_{j,n}^k\right|$ is bounded for any y_0 and n, the considered power series (3.3.34) is absolutely convergent when $0 \le \eta < 1$, which corresponds to the values $\theta > -\infty$.

After finding $\lambda_{j,n}^k$ and then using (3.3.28), (3.3.30), (3.3.32) and (3.3.34), we can determine the functions $v^k(y_0, \theta, \sigma)$. After determining the functions $v^k(y_0, \theta, \sigma)$ and then using formulas (3.3.13), (3.3.17), (3.3.23), (3.3.25) and (3.3.27), we can find the functions $u^k(y_0, y_1, y_2, y_3)$, describing nonabelian expanding waves, by means of formulas (3.3.1) and (3.3.2).

Let us represent the obtained wave solutions in a more convenient form by returning to the variables ξ_1, ξ_2, ξ_3, instead of θ, σ, using formulas (3.3.17). Then, from (3.3.28) and (3.3.34), we come to the following form of solutions to Eq. (3.3.26):

$$v^k = \operatorname{Re} \sum_{n=0}^{\mathrm{M}} \sum_{j=0}^{\infty} \lambda_{j,n}^k (x^0 - r) \frac{(r - x^1)^j (x^3 - ix^2)^n}{(2r)^j (r + x^1)^n}, \qquad (3.3.38)$$

where $\lambda_{j.n}^k$ satisfy the recurrence relation (3.3.35).

It should be noted that when $s^k = 0$, from the recurrence relation (3.3.35), we obtain the equalities $\lambda_{j,n}^k = 0$ when $j \ge 1$. Therefore, in this case, the sum over j in (3.3.38) has only one summand with $j = 0$.

The results obtained in this section can be formulated as the following theorem:

Theorem 3.3.1. *For the Yang–Mills equations (3.1.1)–(3.1.2), there exists a class of wave solutions in the region outside the axis x^1, which is determined by the formulas*

$$A^{k,0} = -s^k(x^0 - r)\left\{1/r + \chi(x^0 - r)\ln\left(\sqrt{(x^2)^2 + (x^3)^2}/r\right)\right\}$$

$$+ d^k(x^0 - r) + v^k(x^0 - r, x^1, x^2, x^3), \quad A^{k,l} = A^{k,0}x^l/r,$$

$$k, l = 1, 2, 3, \quad r = \sqrt{(x^1)^2 + (x^2)^2 + (x^3)^2}, \qquad (3.3.39)$$

where the functions $s^k(x^0 - r)$ are arbitrary and the four functions $\chi(x^0 - r)$ and $d^k(x^0 - r)$ are determined from the system of three linear algebraic equations (3.3.25). The functions $v^k(x^0 - r, x^1, x^2, x^3)$ have the form of series (3.3.38) which is absolutely convergent outside *the negative semi-axis x^1, in which the functions $\lambda_{j,n}^k(x^0 - r)$ are determined by the recurrence relation (3.3.35), where $\lambda_{0,n}^k(x^0 - r)$ are arbitrary complex functions.*

As stated above, the found wave solutions to the Yang–Mills equations can have longitudinal components when the functions p^k of form (3.3.10) are nonzero. This peculiarity of the obtained wave solutions to the Yang–Mills equations could be used to determine nonabelian components in cosmic radiations.

It should be noted that when $s^k = \text{const}$, we have $\chi = 0$ and then the obtained solutions (3.3.39) to the Yang–Mills equations have no singularities outside the negative semi-axis x^1. These solutions can be applied to describe Yang–Mills fields generated by streak lightning.

3.4. Transverse Progressive Waves in Yang–Mills Fields

Let us study nonabelian wave solutions describing progressive waves propagating in the direction z and seek the field potentials $A^{k,\nu}$ corresponding to them in the form

$$A^{k,\nu} = u^{k,\nu}(\theta, x, y), \ \theta = x^0 - z, \tag{3.4.1}$$

where $x^0 = ct$, $x^1 = x$, $x^2 = y$, $x^3 = z$ and t is the time.

Then, the field strengths $F^{k,\mu\nu}$ acquire the form

$$F^{k,01} = u_\theta^{k,1} + u_x^{k,0} + g\varepsilon_{klm}u^{l,0}u^{m,1},$$

$$F^{k,02} = u_\theta^{k,2} + u_y^{k,0} + g\varepsilon_{klm}u^{l,0}u^{m,2},$$

$$F^{k,03} = (u^{k,3} - u^{k,0})_\theta + g\varepsilon_{klm}u^{l,0}u^{m,3},$$

$$F^{k,12} = u_y^{k,1} - u_x^{k,2} + g\varepsilon_{klm}u^{l,1}u^{m,2},$$

$$F^{k,13} = -u_x^{k,3} - u_\theta^{k,1} + g\varepsilon_{klm}u^{l,1}u^{m,3},$$

$$F^{k,23} = -u_y^{k,3} - u_\theta^{k,2} + g\varepsilon_{klm}u^{l,2}u^{m,3}, \tag{3.4.2}$$

where

$$u_\theta^{k,1} \equiv \partial u^{k,1}/\partial\theta, \ u_x^{k,0} \equiv \partial u^{k,0}/\partial x, \ u_y^{k,0} \equiv \partial u^{k,0}/\partial y.$$

Since the considered field potentials $A^{k,\nu}$ are of form (3.4.1) and, as follows from (3.4.2), the field strengths $F^{k,\mu\nu} = F^{k,\mu\nu}(\theta, x, y)$, the

Yang–Mills equations (3.1.1) can be represented as

$$F_\theta^{k,03} - F_x^{k,01} - F_y^{k,02} + g\varepsilon_{klm}(u^{l,1}F^{m,01} + u^{l,2}F^{m,02} + u^{l,3}F^{m,03}) = 0,$$

$$(F^{k,01} + F^{k,13})_\theta - F_y^{k,12} + g\varepsilon_{klm}(u^{l,0}F^{m,01} + u^{l,2}F^{m,12} + u^{l,3}F^{m,13}) = 0,$$

$$(F^{k,02} + F^{k,23})_\theta + F_x^{k,12} + g\varepsilon_{klm}(u^{l,0}F^{m,02} - u^{l,1}F^{m,12} + u^{l,3}F^{m,23}) = 0,$$

$$F_\theta^{k,03} + F_x^{k,13} + F_y^{k,23} + g\varepsilon_{klmc}(u^{l,0}F^{m,03} - u^{l,1}F^{m,13} - u^{l,2}F^{m,23}) = 0.$$

$$(3.4.3)$$

Let us study the case of transverse waves in which

$$F^{k,03} = 0, \quad F^{k,12} = 0. \tag{3.4.4}$$

For this purpose, we put

$$u^{k,3} = u^{k,0}. \tag{3.4.5}$$

Then, from (3.4.1) and (3.4.2), we find

$$F^{k,03} \equiv 0 \tag{3.4.6}$$

and also

$$F^{k,13} \equiv -F^{k,01}, \quad F^{k,23} \equiv -F^{k,02}. \tag{3.4.7}$$

Let us now substitute formulas (3.4.4), (3.4.5) and (3.4.7) into Eqs. (3.4.3). Then, we find that the second and third equations in (3.4.3) are identically satisfied and the first and fourth equations in (3.4.3) are coinciding and give

$$F_x^{k,01} + F_y^{k,02} - g\varepsilon_{klm}(u^{l,1}F^{m,01} + u^{l,2}F^{m,02}) = 0. \tag{3.4.8}$$

From the expression for $F^{k,12}$ in (3.4.2) and the second equality in (3.4.4), we get

$$u_y^{k,1} - u_x^{k,2} + g\varepsilon_{klm}u^{l,1}u^{m,2} = 0. \tag{3.4.9}$$

Thus, we have proved the following proposition:

Proposition 3.4.1. *In the considered case of transverse progressive waves, for which the field potentials $A^{k,\nu}$ satisfy equalities (3.4.1) and (3.4.5) and the field strengths $F^{k,03}$ and $F^{k,12}$ are zero, the Yang–Mills equations (3.4.1) reduce to Eq. (3.4.8), where the field strengths $F^{k,01}$ and $F^{k,02}$ are expressed in (3.4.2) by the field potentials $u^{k,0}$ and $u^{k,1}, u^{k,2}$ satisfying Eq. (3.4.9).*

It is evident that Eqs. (3.4.8) and (3.4.9) have the following particular solution:

$$u^{k,0} = L^k(\theta) + M^k(\theta)x + N^k(\theta)y, \quad u^{k,1} = u^{k,2} = 0, \qquad (3.4.10)$$

where L^k, M^k, N^k are arbitrary functions of θ.

Indeed, then, from (3.4.2) and (3.4.10), we get

$$F^{k,01} = M^k(\theta), \quad F^{k,02} = N^k(\theta) \qquad (3.4.11)$$

and Eqs. (3.4.8) and (3.4.9) are identically satisfied.

The particular solutions (3.4.5) and (3.4.10) to the Yang–Mills equations, which describe nonabelian plane waves, were proposed by S. R. Coleman [21].

Further, we will study a wide class of wave solutions to the Yang–Mills equations which generalize Coleman's solution.

Let us seek solutions to Eqs. (3.4.8) and (3.4.9) in the form

$$u^{k,0} = L^k(\theta, \rho) + M^k(\theta, \rho)x + N^k(\theta, \rho)y, \quad \rho = \sqrt{x^2 + y^2},$$
$$(3.4.12)$$
$$u^{k,1} = V^k(\theta, \rho)x, \quad u^{k,2} = V^k(\theta, \rho)y, \qquad (3.4.13)$$

where L^k, M^k, N^k and V^a are some differentiable functions of θ and ρ.

Using (3.4.13) and the antisymmetry of ε_{klm}, we find that Eq. (3.4.9) is identically satisfied.

From (3.4.2), (3.4.12) and (3.4.13), we obtain the following expressions for the field strengths $F^{k,01}$ and $F^{k,02}$:

$$F^{k,01} = M^k + (H^k + P^k x + Q^k y)x,$$
$$F^{k,02} = N^k + (H^k + P^k x + Q^k y)y, \qquad (3.4.14)$$

where H^k, P^k, Q^k are functions of θ and ρ, which are as follows:

$$H^k = (1/\rho)L^k_\rho + g\varepsilon_{klm}L^l V^m + V^k_\theta, \qquad (3.4.15)$$
$$P^k = (1/\rho)M^k_\rho + g\varepsilon_{klm}M^l V^m,$$
$$Q^k = (1/\rho)N^k_\rho + g\varepsilon_{klm}N^l V^m. \qquad (3.4.16)$$

Substituting formulas (3.4.13) and (3.4.14) into Eq. (3.4.8), we derive

$$\rho H_\rho^k + 2H^k - \rho^2 g \varepsilon_{klm} V^l H^m$$
$$+ [(1/\rho)M_\rho^k + \rho P_\rho^k + 3P^k - g\varepsilon_{klm}V^l(M^m + \rho^2 P^m)]x$$
$$+ [(1/\rho)N_\rho^k + \rho Q_\rho^k + 3Q^k - g\varepsilon_{klm}V^l(N^m + \rho^2 Q^m)]y = 0. \tag{3.4.17}$$

Let us introduce the complex functions

$$S^k = M^k + iN^k, \ W^k = P^k + iQ^k. \tag{3.4.18}$$

Then, from (3.4.16), we find

$$W^k = (1/\rho)S_\rho^k + g\varepsilon_{klm}S^l V^m. \tag{3.4.19}$$

From Eqs. (3.4.17) and (3.4.18), we obtain

$$\rho H_\rho^k + 2H^k - \rho^2 g\varepsilon_{klm}V^l H^m = 0, \tag{3.4.20}$$
$$(1/\rho)S_\rho^k + \rho W_\rho^k + 3W^k - g\varepsilon_{klm}V^l(S^m + \rho^2 W^m) = 0. \tag{3.4.21}$$

Substituting expression (3.4.19) into Eq. (3.4.21) and using the antisymmetry of ε_{klm}, we derive

$$S_{\rho\rho}^k + (3/\rho)S_\rho^k - g\varepsilon_{klm}\{\rho V_\rho^l S^m + V^l[2\rho S_\rho^m + 4S^m$$
$$- \rho^2 g\varepsilon_{mab}V^a S^b]\} = 0. \tag{3.4.22}$$

Let us turn to Eq. (3.4.20). Multiplying it by H^k, summing it over k and using the identity $\varepsilon_{klm}H^k H^m \equiv 0$, we find

$$\rho H_\rho + 2H = 0, \ (H)^2 = \sum_{k=1}^n (H^k)^2. \tag{3.4.23}$$

This gives

$$H = (1/\rho^2)K(\theta), \tag{3.4.24}$$

where $K(\theta)$ is some function.

In order to have no singularity at $\rho = 0$, we should put $K(\theta) = 0$ and hence $H = 0$. Therefore, from (3.4.23), we get $H^k = 0$. Then, using (3.4.15), we find

$$(1/\rho)L_\rho^k + g\varepsilon_{klm}L^l V^m + V_\theta^k = 0. \tag{3.4.25}$$

Further, we will study solutions to Eqs. (3.4.22) and (3.4.25) in the regions $0 \le \rho \le \rho_0$ and $\rho \ge \rho_0$, where ρ_0 is an arbitrary nonzero radius.

First, let us study Eqs. (3.4.22) and (3.4.25) in the cylindrical region $0 \le \rho \le \rho_0$.

It should be noted that Eqs. (3.4.22) and (3.4.25) contain the arbitrary differentiable functions $V^k(\theta, \rho)$. Consider the case in which these functions are as follows:

$$V^k = \sum_{n=0}^{p} V_n^k(\theta)\rho^{2n}, \ 0 \le \rho \le \rho_0, \tag{3.4.26}$$

where p is some nonnegative integer and $V_n^k(\theta)$ are some differentiable functions.

We will seek the functions L^k and S^k in the considered region $0 \le \rho \le \rho_0$ in the form of the following power series:

$$L^k = \sum_{n=0}^{\infty} L_n^k(\theta)\rho^{2n}, \ S^k = \sum_{n=0}^{\infty} S_n^k(\theta)\rho^{2n}, \tag{3.4.27}$$

where $S_n^k(\theta) = M_n^k(\theta) + iN_n^k(\theta)$ and L_n^k, M_n^k and N_n^k are some differentiable functions of θ.

Then, substituting the expressions for V^k and L^k in (3.4.26) and (3.4.27) into Eq. (3.4.25) and taking the terms with ρ^{2n}, we obtain

$$2(n+1)L_{n+1}^k + g\varepsilon_{klm} \sum_{j=0}^{p} L_{n-j}^l V_j^m + dV_n^k/d\theta = 0, \ n \ge 0, \tag{3.4.28}$$

where $L_u^k = 0$ when $u < 0$ and $V_n^k = 0$ when $n > p$. In (3.4.28), $L_0^k(\theta)$ can be arbitrary differentiable functions.

Substituting now the expressions for V^k and S^k in (3.4.26) and (3.4.27) into Eq. (3.4.22) and taking the terms with ρ^{2n}, we find

$$4(n+1)(n+2)S_{n+1}^k - 2g\varepsilon_{klm}\sum_{j=0}^{p}(2n-j+2)V_j^l S_{n-j}^m$$

$$+ g^2\varepsilon_{klm}\varepsilon_{mab}\sum_{j=n-1-p}^{n-1} V_{n-1-j}^l \sum_{d=0}^{p}V_d^a S_{j-d}^b = 0, \ n \geq 0, \quad (3.4.29)$$

where $S_u^k = 0$ when $u < 0$. In (3.4.29), the functions $S_0^k(\theta) = M_0^k(\theta) + iN_0^k(\theta)$ can be arbitrary differentiable functions.

The obtained equalities (3.4.28) and (3.4.29) are recurrence relations for the functions $L_n^k(\theta)$ and $S_n^k(\theta)$.

Let us prove the following lemma:

Lemma 3.4.1. *There exist functions $C_1(\theta) \geq 0$, $C_2(\theta) \geq 0$ and $D_1(\theta) \geq 1$, $D_2(\theta) \geq 1$ which give the estimates*

$$\left| L_n^k(\theta) \right| \leq C_1(\theta)D_1^n(\theta)/(s_n)!, \quad \left| S_n^k(\theta) \right| \leq C_2(\theta)D_2^n(\theta)/(s_n)!,$$
(3.4.30)

where s_n are nonnegative integers satisfying the inequalities

$$s_n(p+1) \leq n < (s_n+1)(p+1). \quad (3.4.31)$$

Proof. Let us put

$$C_1(\theta) = \max_{\substack{1 \leq k \leq 3, \\ 0 \leq n \leq 2p+1}} \left| L_n^k(\theta) \right|, \quad C_2(\theta) = \max_{\substack{1 \leq k \leq 3, \\ 0 \leq n \leq 2p+1}} \left| S_n^k(\theta) \right|,$$
(3.4.32)

$$D_1(\theta) = \max(1, R_1), \quad R_1 = (9/2)f^{\max}V^{\max}(\theta), \quad (3.4.33)$$

$$D_2(\theta) = \max(1, R_2), \quad R_2 = 9f^{\max}V^{\max}(\theta)\left[\frac{1}{p+1} + \frac{9f^{\max}V^{\max}(\theta)}{4}\right],$$
(3.4.34)

where

$$f^{\max} = \max_{1 \leq k,l,m \leq 3} |g\varepsilon_{klm}|, \quad V^{\max}(\theta) = \max_{\substack{1 \leq k \leq 3 \\ 0 \leq n \leq p}} \left| V_n^k(\theta) \right|. \quad (3.4.35)$$

We will prove estimates (3.4.30) by complete induction for the chosen expressions (3.4.32)–(3.4.35) for the functions $C_1(\theta)$, $C_2(\theta)$, $D_1(\theta)$, $D_2(\theta)$.

It is evident that estimates (3.4.30) are true when $0 \leq n \leq 2p+1$ since $D_1 \geq 1$ and $D_2 \geq 1$. Let us assume that they are true when $0 \leq n \leq r$, where $r \geq 2p+1$, and prove that they are true when $n = r+1$. First, consider formula (3.4.31) for $n = r+1$. It gives

$$r+1 \geq s_{r+1}(p+1). \tag{3.4.36}$$

Therefore, we get that for $j = 0, 1, \ldots, p$,

$$r - j \geq (s_{r+1} - 1)(p+1) + p - j \geq (s_{r+1} - 1)(p+1). \tag{3.4.37}$$

From (3.4.31) and (3.4.37), we find

$$s_{r-j} \geq s_{r+1} - 1, \quad j = 0, 1, \ldots, p. \tag{3.4.38}$$

Consider the recurrence relation (3.4.28) for $n = r$. From it and (3.4.35), we obtain

$$2(r+1)\left|L_{r+1}^k(\theta)\right| \leq 9(p+1)f^{\max}V^{\max}(\theta) \max_{\substack{1 \leq k \leq 3 \\ 0 \leq j \leq p}}\left|L_{r-j}^k(\theta)\right|, \tag{3.4.39}$$

where it has been used that $r \geq 2p + 1$ and hence $V_r^k(\theta) = 0$.

Since we assume that inequalities (3.4.30) are true when $0 \leq n \leq r$ and since $D_1(\theta) \geq 1$, from inequality (3.4.39), using (3.4.38), we derive

$$\left|L_{r+1}^k(\theta)\right| \leq \frac{9(p+1)}{2(r+1)}f^{\max}V^{\max}(\theta)C_1(\theta)\frac{D_1^r(\theta)}{(s_{r+1}-1)!}. \tag{3.4.40}$$

Using (3.4.33) and (3.4.36), from inequality (3.4.40), we obtain

$$\left|L_{r+1}^k(\theta)\right| \leq C_1(\theta)\frac{D_1^{r+1}(\theta)}{(s_{r+1})!}. \tag{3.4.41}$$

Consider the recurrence relation (3.4.29) for $n = r$. From it and (3.4.35), we find

$$4(r + 1)(r + 2)\left|S_{r+1}^k(\theta)\right| \leq 36(r + 1) f^{\max} V^{\max}(\theta) \max_{\substack{1 \leq k \leq 3 \\ 0 \leq j \leq p}} \left|S_{r-j}^k(\theta)\right|$$

$$+ 81 \left((p + 1) f^{\max} V^{\max}(\theta)\right)^2 \max_{\substack{1 \leq k \leq 3 \\ 0 \leq u \leq 2p}} \left|S_{r-1-u}^k(\theta)\right|. \tag{3.4.42}$$

From (3.4.36), we have for $u = 0, 1, \ldots, 2p$:

$$r - 1 - u \geq (s_{r+1} - 2)(p + 1) + (2p - u) \geq (s_{r+1} - 2)(p + 1). \tag{3.4.43}$$

Formulas (3.4.31) and (3.4.43) give

$$s_{r-1-u} \geq s_{r+1} - 2, \quad u = 0, 1, \ldots, 2p. \tag{3.4.44}$$

Since we assume that inequalities (3.4.30) are true when $0 \leq n \leq r$ and since $D_2(\theta) \geq 1$, from inequality (3.4.42), using (3.4.38) and (3.4.44), we derive

$$4(r + 1)(r + 2)\left|S_{r+1}^k(\theta)\right| \leq 36(r + 1) f^{\max} V^{\max}(\theta) C_2(\theta) \frac{D_2^r(\theta)}{(s_{r+1} - 1)!}$$

$$+ 81 \left((p + 1) f^{\max} V^{\max}(\theta)\right)^2 C_2(\theta) \frac{D_2^r(\theta)}{(s_{r+1} - 2)!}. \tag{3.4.45}$$

Taking into account (3.4.36), from inequality (3.4.45), we obtain

$$\left|S_{r+1}^k(\theta)\right| \leq 9 f^{\max} V^{\max}(\theta) \left[\frac{1}{p + 1} + \frac{9 f^{\max} V^{\max}(\theta)}{4}\right] C_2(\theta) \frac{D_2^r(\theta)}{(s_{r+1})!}. \tag{3.4.46}$$

Using (3.4.34), from inequality (3.4.46), we find

$$\left|S_{r+1}^k(\theta)\right| \leq C_2(\theta) \frac{D_2^{r+1}(\theta)}{(s_{r+1})!}. \tag{3.4.47}$$

As follows from inequalities (3.4.41) and (3.4.47), Lemma 3.4.1 has been proved by induction.

Let us now prove the following theorem:

Theorem 3.4.1. *The considered series* (3.4.27), *where the functions* $L_n^k(\theta)$ *and* $S_n^k(\theta)$ *satisfy the recurrence relations* (3.4.28) *and* (3.4.29), *are absolutely convergent in the region* $0 \le \rho \le \rho_0$, *where* ρ_0 *is an arbitrary nonzero radius.*

Proof. From (3.4.27) and (3.4.30), we obtain that in the region $0 \le \rho \le \rho_0$,

$$
\begin{aligned}
\left| L^k \right| &\le \sum_{n=0}^{\infty} \left| L_n^k(\theta) \right| \rho_0^{2n} \le C_1(\theta) \sum_{n=0}^{\infty} \frac{\left(\rho_0^2 D_1(\theta) \right)^n}{(s_n)!}, \\
\left| S^k \right| &\le \sum_{n=0}^{\infty} \left| S_n^k(\theta) \right| \rho_0^{2n} \le C_2(\theta) \sum_{n=0}^{\infty} \frac{\left(\rho_0^2 D_2(\theta) \right)^n}{(s_n)!}.
\end{aligned} \tag{3.4.48}
$$

As follows from (3.4.31), for $m = 0, 1, 2, \ldots$,

$$
s_n = m \text{ when } m(p+1) \le n \le m(p+1) + p. \tag{3.4.49}
$$

Therefore, we have

$$
\sum_{n=0}^{\infty} \frac{\left(\rho_0^2 D_l(\theta) \right)^n}{(s_n)!} = \sum_{j=0}^{p} \left(\rho_0^2 D_l(\theta) \right)^j \sum_{m=0}^{\infty} \frac{\left(\rho_0^2 D_l(\theta) \right)^{m(p+1)}}{m!}
$$

$$
= \frac{U_l - 1}{\rho_0^2 D_l(\theta) - 1} \exp(U_l), \quad U_l = \left(\rho_0^2 D_l(\theta) \right)^{p+1}, \quad l = 1, 2. \tag{3.4.50}
$$

From (3.4.48) and (3.4.50), we find that the considered series (3.4.27) are absolutely convergent in the region $0 \le \rho \le \rho_0$, where ρ_0 is an arbitrary nonzero radius. Thus, Theorem 3.4.1 has been proved.

Let us study Eqs. (3.4.22) and (3.4.25) in the region $\rho \ge \rho_0$, where ρ_0 is an arbitrary nonzero radius, and represent the functions V^k and S^k in the form

$$
V^k = \hat{V}^k(\theta, \rho)/\rho^4, \quad S^k = \hat{S}^k(\theta, \rho)/\rho^2, \quad \rho \ge \rho_0, \tag{3.4.51}
$$

where $\hat{V}^k$ and $\hat{S}^k$ are some differentiable functions of θ and ρ.

Substituting expressions (3.4.51) into Eqs. (3.4.22) and (3.4.25), we obtain

$$\hat{S}_{\rho\rho}^{k} - (1/\rho)\hat{S}_{\rho}^{k} - (1/\rho)^{4}g\varepsilon_{klm}\{\rho\hat{V}_{\rho}^{l}\hat{S}^{m}$$

$$+ \hat{V}^{l}[2\rho\hat{S}_{\rho}^{m} - 4\hat{S}^{m} - (1/\rho)^{2}g\varepsilon_{mab}\hat{V}^{a}\hat{S}^{b}]\} = 0, \quad (3.4.52)$$

$$(1/\rho)L_{\rho}^{k} + (1/\rho)^{4}(g\varepsilon_{klm}L^{l}\hat{V}^{m} + \hat{V}_{\theta}^{k}) = 0. \quad (3.4.53)$$

Equations (3.4.52) and (3.4.53) contain the arbitrary differentiable functions $\hat{V}^{k}(\theta, \rho)$. Consider the case in which these functions have the following form:

$$\hat{V}^{k} = \sum_{n=0}^{\hat{p}} \hat{V}_{n}^{k}(\theta)/\rho^{2n}, \quad \rho \geq \rho_{0}, \quad (3.4.54)$$

where $\hat{p}$ is some nonnegative integer and $\hat{V}_{n}^{k}$ are some differentiable functions of θ.

Let us seek the functions L^{k} and $\hat{S}^{k}$ in the considered region $\rho \geq \rho_{0}$ in the form of the following series:

$$L^{k} = \sum_{n=0}^{\infty} \hat{L}_{n}^{k}(\theta)/\rho^{2n}, \quad \hat{S}^{k} = \sum_{n=0}^{\infty} \hat{S}_{n}^{k}(\theta)/\rho^{2n}, \quad (3.4.55)$$

where $\hat{S}_{n}^{k}(\theta) = \hat{M}_{n}^{k}(\theta) + i\hat{N}_{n}^{k}(\theta)$ and $\hat{L}_{n}^{k}$, $\hat{M}_{n}^{k}$ and $\hat{N}_{n}^{k}$ are some differentiable functions of θ.

Then, substituting the expressions for $\hat{V}^{k}$ and L^{k} in (3.4.54) and (3.4.55) into Eq. (3.4.53) and taking the terms with $\rho^{-2(n+2)}$, we obtain

$$2(n+1)\hat{L}_{n+1}^{k} - g\varepsilon_{klm} \sum_{j=0}^{\hat{p}} \hat{L}_{n-j}^{l}\hat{V}_{j}^{m} - d\hat{V}_{n}^{k}/d\theta = 0, \quad n \geq 0, \quad (3.4.56)$$

where $\hat{L}_{u}^{l} = 0$ when $u < 0$ and $\hat{V}_{n}^{k} = 0$ when $n > \hat{p}$. In (3.4.56), $\hat{L}_{0}^{k}(\theta)$ can be arbitrary differentiable functions.

Substituting now the expressions for $\hat{V}^k$ and $\hat{S}^k$ in (3.4.54) and (3.4.55) into Eq. (3.4.52) and taking the terms with $\rho^{-2(n+2)}$, we find

$$4(n+1)(n+2)\hat{S}^k_{n+1} + 2g\varepsilon_{klm}\sum_{j=0}^{\hat{p}}(2n-j+2)\hat{V}^l_j\hat{S}^m_{n-j}$$

$$+ g^2\varepsilon_{klm}\varepsilon_{mab}\sum_{j=n-1-\hat{p}}^{n-1}\hat{V}^l_{n-1-j}\sum_{d=0}^{\hat{p}}\hat{V}^a_d\hat{S}^b_{j-d} = 0, \ n \geq 0,$$

$$(3.4.57)$$

where $\hat{S}^k_u = 0$ when $u < 0$. In (3.4.57), the functions $\hat{S}^k_0(\theta) = \hat{M}^k_0(\theta) + i\hat{N}^k_0(\theta)$ can be arbitrary differentiable functions.

Let us put $V^k_n = -\hat{V}^k_n$, $L^k_n = \hat{L}^k_n$, $S^k_n = \hat{S}^k_n$, $p = \hat{p}$ in (3.4.28) and (3.4.29). Then, we obtain the recurrence relations (3.4.56) and (3.4.57). Therefore, we can apply Lemma 3.4.1 to them. From this lemma, we find that there exist functions $\hat{C}_1(\theta) \geq 0$, $\hat{C}_2(\theta) \geq 0$ and $\hat{D}_1(\theta) \geq 1$, $\hat{D}_2(\theta) \geq 1$ which give the estimates

$$\left|\hat{L}^k_n(\theta)\right| \leq \hat{C}_1(\theta)\hat{D}^n_1(\theta)/(\hat{s}_n)!, \ \left|\hat{S}^k_n(\theta)\right| \leq \hat{C}_2(\theta)\hat{D}^n_2(\theta)/(\hat{s}_n)!,$$

$$(3.4.58)$$

where $\hat{s}_n$ are nonnegative integers satisfying the inequalities

$$\hat{s}_n(\hat{p}+1) \leq n < (\hat{s}_n+1)(\hat{p}+1). \quad (3.4.59)$$

Here, the functions $\hat{C}_l(\theta)$ and $\hat{D}_l(\theta)$, $l = 1,2$, are determined by formulas (3.4.32)–(3.4.35), where we should put

$$C_l = \hat{C}_l, \ D_l = \hat{D}_l, V^k_n = -\hat{V}^k_n, \ L^k_n = \hat{L}^k_n, \ S^k_n = \hat{S}^k_n, \ p = \hat{p}.$$

Let us now prove the following theorem:

Theorem 3.4.2. *The considered series* (3.4.55), *where the functions* $\hat{L}^k_n(\theta)$ *and* $\hat{S}^k_n(\theta)$ *satisfy the recurrence relations* (3.4.56) *and* (3.4.57), *are absolutely convergent in the region* $\rho \geq \rho_0$, *where* ρ_0 *is an arbitrary nonzero radius.*

Proof. From (3.4.55) and (3.4.58), we obtain that in the region $\rho \geq \rho_0$,

$$\left| L^k \right| \leq \sum_{n=0}^{\infty} \left| \hat{L}_n^k(\theta) \right| \rho_0^{-2n} \leq \hat{C}_1(\theta) \sum_{n=0}^{\infty} \frac{\left(\rho_0^{-2} \hat{D}_1(\theta) \right)^n}{(\hat{s}_n)!},$$

$$\left| \hat{S}^k \right| \leq \sum_{n=0}^{\infty} \left| \hat{S}_n^k(\theta) \right| \rho_0^{-2n} \leq \hat{C}_2(\theta) \sum_{n=0}^{\infty} \frac{\left(\rho_0^{-2} \hat{D}_2(\theta) \right)^n}{(\hat{s}_n)!}.$$

$$(3.4.60)$$

As follows from (3.4.59), for $m = 0, 1, 2, \ldots$,

$$\hat{s}_n = m \text{ when } m(\hat{p} + 1) \leq n \leq m(\hat{p} + 1) + \hat{p}. \tag{3.4.61}$$

Therefore, we have

$$\sum_{n=0}^{\infty} \frac{\left(\rho_0^{-2} \hat{D}_l(\theta) \right)^n}{(\hat{s}_n)!} = \sum_{j=0}^{\hat{p}} \left(\rho_0^{-2} \hat{D}_l(\theta) \right)^j \sum_{m=0}^{\infty} \frac{\left(\rho_0^{-2} \hat{D}_l(\theta) \right)^{m(\hat{p}+1)}}{m!}$$

$$= \frac{\hat{U}_l - 1}{\rho_0^{-2} \hat{D}_l(\theta) - 1} \exp(\hat{U}_l), \ \hat{U}_l = \left(\rho_0^{-2} \hat{D}_l(\theta) \right)^{\hat{p}+1},$$

$$l = 1, 2. \tag{3.4.62}$$

From (3.4.60) and (3.4.62), we find that the considered series (3.4.55) are absolutely convergent in the region $\rho \geq \rho_0$, where ρ_0 is an arbitrary nonzero radius. Thus, Theorem 3.4.2 has been proved.

The obtained wave solutions (3.4.12), (3.4.13), (3.4.26) and (3.4.27) to the Yang–Mills equations contain the arbitrary differentiable functions $L_0^k(\theta)$, $M_0^k(\theta)$, $N_0^k(\theta)$ and $V_j^k(\theta)$, $0 \leq j \leq p$, in the region $0 \leq \rho \leq \rho_0$ and the arbitrary differentiable functions $\hat{L}_0^k(\theta)$, $\hat{M}_0^k(\theta)$, $\hat{N}_0^k(\theta)$ and $\hat{V}_j^k(\theta)$, $0 \leq j \leq \hat{p}$, in the region $\rho \geq \rho_0$, where p and $\hat{p}$ are arbitrary nonnegative integers. These arbitrary functions and integers can be chosen so as to have the considered field potentials and strengths continuous at $\rho = \rho_0$.

From formulas (3.4.4) and (3.4.7), we derive that the field strengths $F^{k,\mu\nu} = F^{k,\mu\nu}(\theta, x, y)$ satisfy the following equalities:

$$\partial_\mu F^{k,\mu 0} = \partial_\mu F^{k,\mu 3}, \;\; \partial_\mu F^{k,\mu 1} = \partial_\mu F^{k,\mu 2} = 0. \qquad (3.4.63)$$

It follows from these equalities that the obtained wave solutions satisfy the additional condition (1.4.16) proposed in Chapter 1, where we should put $J^{k,\nu} = 0$.

Chapter 4

Nonlinear Nuclear Interactions

In this chapter, we discuss the nonlinear properties of nuclear interactions in atomic nuclei and, in particular, the phenomenon of saturation of nuclear forces. To describe them, a nonlinear generalization of the Yukawa equation for the nuclear potential is proposed. In the generalized equation, the influence of the nuclear potential on the mass of nucleons is taken into account, which is determined on the basis of several fundamental principles. Then, expressions for the energy–momentum tensor and the Lagrangian of the nuclear field are found in the classical approximation. They are used to obtain dynamic equations describing the motion of atomic nuclei in this approximation under the action of nuclear and electromagnetic fields.

With the help of the proposed nonlinear equation of the nuclear field and the resulting dynamic equations, the dimensionless constant of the strong interaction of two protons is calculated for a sufficiently high collision energy. The found value of this constant turns out to be approximately equal to 15, in full agreement with the known experimental data. The proposed theory of nuclear forces is then used to determine the binding energies and radii of medium and heavy nuclei and to explain a number of their amazing properties. It is shown as a result of computer calculations that the found theoretical formulas for their binding energies and radii are in good agreement with experimental data. The obtained dynamic equations are used to describe the movement of nucleons near heavy atomic nuclei. These results are important for finding the conditions for the formation of quasi-nuclei, in which nucleons or antinucleons revolve around atomic nuclei in closed orbits.

The application of the proposed theory of nuclear forces to the study of the effect of their saturation in neutron stars is also considered. The joint action of nuclear and gravitational forces in cooled massive neutron stars with a temperature close to absolute zero is examined in detail. It is shown that in such stars, due to the effect of saturation of nuclear forces at high nuclear potentials, gravitational forces can be balanced by nuclear forces. In this chapter, the results obtained in our papers [33–36] are used.

4.1.   Mysteries of Nuclear Interactions

As is well known, nuclear forces, which bind protons and neutrons together in atomic nuclei, are a kind of strong interaction. Their properties are very peculiar and still not clear. The known simple theories describe only a small part of them. As for complicated theories, it is impossible in many cases even to derive quantitative estimates from them for nuclear forces. However, there exists a poorly explored way which, as will be seen later on, allows one to describe a number of the known properties of atomic nuclei.

Let us begin with the first reasonable theory of nuclear forces proposed by Hideki Yukawa in 1934. The years of his studies in the Kyoto University coincided with the time of the creation of quantum mechanics. At that time in Japan, there were no good experts in this field, and Yukawa acquired knowledge of quantum mechanics on his own. Despite difficulties to grow as a physicist, Yukawa, when he was 27 years old, proposed a new concept of nuclear forces. His theory was highly appreciated and in 1949, he became the first Japanese physicist who was awarded a Nobel Prize.

The Yukawa linear equation for the nuclear field potential looked very attractive. It generalized the equation for the electric field potential in the case of massive field carriers and satisfied the principles of quantum physics and the theory of relativity. The most triumphant event for Yukawa's theory was the discovery in experiments in 1947 of carriers of nuclear forces which were named pions.

It is worth noting that Yukawa's theory explained such important properties of nuclear forces as their short-range character, their independence of electric charges of nucleons and the zero spin of pions.

However, soon after the triumph of the theory, it was clear that Yukawa's theory had not only successes but also failures. It was found that in many cases, the Yukawa equation had serious contradictions with experimental data. For example, in contrast to the Yukawa equation, the dimensionless parameter of strong interaction is not a constant: It substantially depends on the energy of interacting nucleons. In the case of low-energy interactions of nucleons in atomic nuclei, the parameter of strong interaction is equal to 0.08. At the same time, in the case of high-energy interactions of nucleons, this parameter is approximately equal to 15 [37–39].

Besides, nuclear forces between nucleons are not always attractive. When the energy of interaction between nucleons is sufficiently high, nuclear forces between them can become repulsive [37–39]. This property, which is called the effect of saturation of nuclear forces, is not described by the Yukawa equation.

When these contradictions with experimental data were found, physicists realized that Yukawa's theory was not right and a new one should be found. After that, a number of other concepts were proposed [40–42]. However, they have not led to any quantitative theory of nuclear forces which can describe the properties of atomic nuclei.

Since up to now there is no good quantitative description of nuclear forces, the following question arises: Why not return to Yukawa's ideas which have a number of merits, such as a true description of pions, and make only some corrections of them?

This way is realized in this chapter. We do not change Yukawa's linear equation when the nuclear potential is small enough. But we find its nonlinear generalization in the case of sufficiently large values of this potential.

In order to come to this generalization, we turn to the Einstein formula for the relativistic mass of a particle. This formula expresses the well-known influence of the velocity of a particle on its mass. As for the mass at rest of a particle, it is regarded as an invariant. This property of the mass at rest of a particle is absolutely true when it moves under electromagnetic forces. However, when there are also nuclear forces acting on it, the invariance of its mass at rest is not evident. Moreover, there exists the well-known phenomenon of the mass defect of atomic nuclei: The mass of a nucleus is slightly less than the total mass of its protons and neutrons. This phenomenon has a qualitative explanation in relativistic physics. However, the theory of relativity gives no quantitative explanation for the mass defect of nuclei. It should be stressed that the behavior of this characteristic of nuclei is quite peculiar. In particular, the mass defect per nucleon is not constant. It peculiarly depends on the number of protons and neutrons in a nucleus and has an approximate maximum in the case of the nucleus of the iron.

Since the mass at rest of a particle is considered constant in classical physics, the following challenging question arises: Why the ratios of the masses at rest of nuclei to the numbers of their nucleons are so different, whereas the rest masses of the proton and neutron are practically coinciding?

To answer this question, one can suppose that the mass at rest of a nucleon somehow depends on the nuclear potential. In this chapter, we develop this point of view and come to a nonlinear generalization of the Yukawa theory. Our approach to the problem under consideration is based on the following principles:

(1) **Principle 1:** The Yukawa theory is correct when the nuclear potential is sufficiently small since it agrees in this case with experimental data.
(2) **Principle 2:** The Yukawa equation gives a correct description of the nuclear field outside its source since it leads to the correct values of the characteristics of the carriers of this field — pions — and is consistent with relativistic quantum mechanics.
(3) **Principle 3:** The mass at rest of a nucleon is not a constant. It depends on the nuclear potential. This principle will help us to describe the mass defect of atomic nuclei.

It is shown in this chapter that there is a theory of nuclear forces which is consistent with general physical concepts and the above three principles. It gives the sought nonlinear generalization of the Yukawa equation for the nuclear potential and dynamic equations for nucleons in the classical approximation.

The found nonlinear equation for the nuclear potential coincides with the Yukawa equation in the regions outside nuclear field sources. At the same time, it differs significantly from the Yukawa equation in the regions inside nuclear field sources when the nuclear potential is large enough.

4.2. Yukawa Theory of Nuclear Interactions and Its Nonlinear Generalization

Let us turn to the Yukawa equation describing the potential φ of a nuclear field. It can be represented in the form

$$\partial^\mu \partial_\mu \varphi + (m_\pi c/\hbar)^2 \varphi = -4\pi (G/m_p)^2 \rho_0(\mathbf{r}, t), \qquad (4.2.1)$$

$$G^2/(\hbar c) = 0.080, \qquad (4.2.2)$$

where $m_p \varphi$ is the potential energy of a proton, m_p is its rest mass, m_π is the rest mass of a neutral pion, G is the constant of nuclear interaction, $\partial_\mu = \partial/\partial x^\mu$, $x^\mu (\mu = 0, 1, 2, 3)$ are rectangular coordinates of the Minkowski space–time, ρ_0 is the rest mass density of the source of the nuclear field, $t = x^0/c$ is the time and $\mathbf{r} = (x^1, x^2, x^3)$ is the radius vector.

Hideki Yukawa
(1907–1981)

The Yukawa equation allows one to describe the short-range character of nuclear forces and properties of their carriers — pions. However, because of its linearity, it does not describe a number of nonlinear properties of nuclear forces when their potentials are sufficiently large and, in particular, their property of saturation. That is why the Yukawa equation should be replaced by some nonlinear generalization to describe nuclear fields in the case of sufficiently large values of their potentials.

To find a nonlinear generalization of the Yukawa theory, let us consider the energy–momentum tensor for a dust-like nuclear matter and the nuclear field generated by it in the classical approximation.

For the energy–momentum tensor $T_m^{\mu\nu}$ of the considered matter, we have the well-known expression [13]

$$T_m^{\mu\nu} = c^2 \rho \, dx^\mu/ds \, dx^\nu/ds, \qquad (4.2.3)$$

where ds is the space–time interval, dx^μ/ds is the 4-vector of velocity of a particle and ρ is the density of the rest mass of the matter.

Taking into account Principle 3, let us represent this density in the form

$$\rho = \rho_0 f(\varphi), \quad f(0) = 1, \tag{4.2.4}$$

where ρ_0 is the rest mass density when $\varphi = 0$ and $f(\varphi)$ is some function which should be determined.

Consider the energy–momentum tensor $T_{\text{nuc}}^{\mu\nu}$ of a nuclear field of the Yukawa type. It has the following form [43], which is consistent with Principle 2:

$$T_{\text{nuc}}^{\mu\nu} = (1/\lambda) \left[\left(g^{\mu\alpha} g^{\nu\beta} - \frac{1}{2} g^{\mu\nu} g^{\alpha\beta} \right) \partial_\alpha \varphi \partial_\beta \varphi + \frac{1}{2} g^{\mu\nu} (m_\pi c/\hbar)^2 \varphi^2 \right], \tag{4.2.5}$$

where λ is some constant and $g^{\mu\nu}$ are components of the Minkowski metric tensor.

The full energy–momentum tensor $T^{\mu\nu}$, which is the sum of expressions (4.2.3) and (4.2.5), should satisfy the following differential equation of energy and momentum conservation:

$$\partial_\mu T^{\mu\nu} = 0, T^{\mu\nu} = T_m^{\mu\nu} + T_{\text{nuc}}^{\mu\nu}. \tag{4.2.6}$$

In addition to Eq. (4.2.6), we can write down the differential equation of conservation of the rest mass of the matter with the density ρ_0 when the nuclear field is absent. This equation can be represented in the form [13]

$$\partial_\mu(\rho_0 dx^\mu/ds) = 0. \tag{4.2.7}$$

From the differential equation of energy and momentum conservation (4.2.6) and formulas (4.2.3)–(4.2.5), we obtain in an inertial frame of reference

$$\partial_\mu(T_m^{\mu\nu} + T_{\text{nuc}}^{\mu\nu})$$
$$= c^2 \partial_\mu(\rho_0 f dx^\mu/ds) dx^\nu/ds + c^2 \rho_0 f \partial_\mu(dx^\nu/ds) dx^\mu/ds$$
$$+ (1/\lambda) \left[\left(g^{\mu\alpha} g^{\nu\beta} - \frac{1}{2} g^{\mu\nu} g^{\alpha\beta} \right) \partial_\mu(\partial_\alpha \varphi \partial_\beta \varphi) \right.$$
$$\left. + g^{\mu\nu} (m_\pi c/\hbar)^2 \varphi \partial_\mu \varphi \right] = 0. \tag{4.2.8}$$

After calculations, this equation can be represented in the form

$$(1/c^2)\partial_\mu(T_m^{\mu\nu} + T_{\text{nuc}}^{\mu\nu})$$
$$= [f\partial_\mu(\rho_0 dx^\mu/ds) + \rho_0\partial_\mu f dx^\mu/ds]dx^\nu/ds + \rho_0 f d^2 x^\nu/ds^2$$
$$+ (\lambda c^2)^{-1}\partial^\nu\varphi[\partial^\mu\partial_\mu\varphi + (m_\pi c/\hbar)^2\varphi] = 0. \tag{4.2.9}$$

Using (4.2.4) and (4.2.7), from Eq. (4.2.9), we obtain

$$(1/c^2)\partial_\mu(T_m^{\mu\nu} + T_{\text{nuc}}^{\mu\nu}) = \rho[d^2 x^\nu/ds^2 + (1/f)df/dsdx^\nu/ds]$$
$$+ (\lambda c^2)^{-1}\partial^\nu\varphi[\partial^\mu\partial_\mu\varphi + (m_\pi c/\hbar)^2\varphi] = 0. \tag{4.2.10}$$

Multiplying Eq. (4.2.10) by dx_ν/ds, we find

$$(\rho/2)d(dx^\nu/dsdx_\nu/ds)/ds + (\rho/f)df/ds$$
$$+ (\lambda c^2)^{-1}[\partial^\mu\partial_\mu\varphi + (m_\pi c/\hbar)^2\varphi]d\varphi/ds = 0. \tag{4.2.11}$$

Taking into account that $dx^\nu/dsdx_\nu/ds = 1$ and, consequently, that the first summand in equality (4.2.11) is equal to zero, we come to the following equation for the nuclear potential φ:

$$\partial^\mu\partial_\mu\varphi + (m_\pi c/\hbar)^2\varphi = -\lambda c^2(\rho/f)df/d\varphi. \tag{4.2.12}$$

The substitution of Eq. (4.2.12) into (4.2.10) gives the following four dynamic equations for nuclear matter under the action of the considered nuclear field:

$$d^2 x^\nu/ds^2 + d(\ln f)/dsdx^\nu/ds - \partial^\nu(\ln f) = 0, \tag{4.2.13}$$

where x^ν are coordinates of a particle of the nuclear matter in an inertial frame of reference.

It should be noted that the four equations in (4.2.13) are not independent. Indeed, multiplying (4.2.13) by $2dx_\nu/ds$, we come to the identity

$$d(dx^\nu/dsdx_\nu/ds)/ds \equiv 0 \tag{4.2.14}$$

since the differentiated expression inside the round brackets is equal to 1. This means that only the three equations in (4.2.13) with $\nu = 1,2,3$ are independent and the equation in (4.2.13) with $\nu = 0$ is their consequence.

It should also be noted that the nuclear field equation (4.2.12) is a consequence of the differential equations of conservation of energy and momentum (4.2.6) and of conservation of rest mass (4.2.7).

As follows from this field equation, the function $f(\varphi)$ cannot be equal to a constant. Indeed, if the right-hand side of the field equation (4.2.12) had been equal to zero, then this equation would have only the trivial solution equal to zero, which cannot be true. Therefore, $f(\varphi) \neq$ const and as follows from formula (4.2.4), the rest mass density ρ should depend on the nuclear field potential φ.

Let us turn to the determination of this dependence.

4.2.1.   Determination of the dependence of nucleon's rest mass on nuclear field's potential

Consider a nucleon moving in a nuclear field, using the classical approximation. In a local inertial frame of reference comoving with the nucleon, the second law of Newtonian dynamics takes place. It can be represented as

$$\frac{d(m_0 \mathbf{v})}{dt} = m_0 \mathbf{E}_{\text{nuc}}, \quad \mathbf{E}_{\text{nuc}} = -\text{grad}\varphi, \qquad (4.2.15)$$

where m_0 is the rest mass of the nucleon, $\mathbf{v}$ is the vector of its velocity and $\mathbf{E}_{\text{nuc}}$ is the vector of the strength of the nuclear field with the potential φ. Using (4.2.4), we have the following formula for the rest mass of the nucleon:

$$m_0 = m_p f(\varphi), \quad f(0) = 1, \qquad (4.2.16)$$

where m_p is the rest mass of the proton when the nuclear field is absent and $f(\varphi)$ is an unknown function which should be determined.

Let us take this formula into account. Then, in an arbitrary inertial frame of reference, Eq. (4.2.15) of the movement of a nucleon acquires the following relativistic form:

$$\frac{d}{d\tau}\left(m_p f(\varphi)\frac{dx^\mu}{d\tau}\right) = m_p f(\varphi)\frac{\partial\varphi}{\partial x_\mu}, \quad d\tau = \sqrt{1 - v^2/c^2}\,dt, \quad (4.2.17)$$

where x^μ are rectangular space–time coordinates in the inertial frame of reference, $\mu = 0, 1, 2, 3$, $x^0 = ct$, τ is the proper time of the nucleon and $v = |\mathbf{v}|$ is its velocity.

From these equations, we obtain

$$f(\varphi)\frac{d^2x^\mu}{d\tau^2} + f'(\varphi)\frac{d\varphi}{d\tau}\frac{dx^\mu}{d\tau} = f(\varphi)\frac{\partial\varphi}{\partial x_\mu}. \qquad (4.2.18)$$

Let us multiply Eq. (4.2.18) by $dx_\mu/d\tau$, sum it over μ and take into account the following identities:

$$\frac{dx_\mu}{d\tau}\frac{dx^\mu}{d\tau} = c^2, \quad \frac{dx_\mu}{d\tau}\frac{d^2x^\mu}{d\tau^2} = \frac{1}{2}\frac{d}{d\tau}\left(\frac{dx_\mu}{d\tau}\frac{dx^\mu}{d\tau}\right) = 0, \quad \frac{\partial\varphi}{\partial x_\mu}\frac{dx_\mu}{d\tau} = \frac{d\varphi}{d\tau}. \qquad (4.2.19)$$

Then, we obtain

$$c^2 f'(\varphi) = f(\varphi). \qquad (4.2.20)$$

Since $f(0) = 1$, as indicated in (4.2.16), from Eq. (4.2.20), we find

$$f(\varphi) = \exp(\varphi/c^2). \qquad (4.2.21)$$

Thus, formula (4.2.16) for the rest mass m_0 of a nucleon in a nuclear field with potential φ acquires the form

$$m_0 = m_p \exp(\varphi/c^2). \qquad (4.2.22)$$

Using (4.2.4) and substituting formula (4.2.21) into Eq. (4.2.12), we obtain the following nonlinear generalization of the Yukawa equation for the nuclear potential φ:

$$\partial^\mu\partial_\mu\varphi + (m_\pi c/\hbar)^2\varphi = -\lambda\rho_0(\mathbf{r}, t)\exp(\varphi/c^2). \qquad (4.2.23)$$

According to Principle 1, the Yukawa equation (4.2.1) should be true for small values of the potential φ. This gives

$$\lambda = 4\pi(G/m_p)^2. \qquad (4.2.24)$$

Thus, we come to the following nonlinear generalization of the Yukawa equation:

$$\partial^\mu\partial_\mu\varphi + (m_\pi c/\hbar)^2\varphi = -4\pi(G/m_p)^2\rho_0(\mathbf{r}, t)\exp(\varphi/c^2). \qquad (4.2.25)$$

Let us now substitute formula (4.2.21) into Eqs. (4.2.13) and (4.2.18). Then, we obtain the same dynamic equations:

$$c^2 d^2x^\mu/ds^2 + d\varphi/ds dx^\mu/ds - \partial^\mu\varphi = 0, \qquad (4.2.26)$$

where $ds = cd\tau$ is the space–time interval.

Substituting formula (4.2.21) into Eq. (4.2.17), we come to the following equations for the movement of a relativistic nucleon in a nuclear field with potential φ:

$$\frac{d}{d\tau}\left(m_p e^{\varphi/c^2}\frac{dx^\mu}{d\tau}\right) = m_p e^{\varphi/c^2}\frac{\partial\varphi}{\partial x_\mu}, \qquad (4.2.27)$$

$$d\tau = \sqrt{1 - v^2/c^2}\,dt. \qquad (4.2.28)$$

From (4.2.27), we come to the following formulas for the momentum $\mathbf{p}$, kinetic energy E and mass m of a relativistic nucleon moving at velocity $\mathbf{v}$ in a nuclear field with potential φ:

$$\mathbf{p} = m\mathbf{v}, \quad E = mc^2, \quad m = \frac{m_p \exp(\varphi/c^2)}{\sqrt{1 - v^2/c^2}}. \qquad (4.2.29)$$

Expressions (4.2.29) are generalized formulas of relativistic mechanics for a nucleon which give the dependence of its mass not only on its velocity v but also on the nuclear field potential φ.

4.3. Dynamic Equations in Nuclear and Electromagnetic Fields

Let us turn now to the case in which a dust-like nuclear matter moves under the action of nuclear and electromagnetic fields. For the energy–momentum tensor $T_e^{\mu\nu}$ of an electromagnetic field, we have [13]

$$T_e^{\mu\nu} = (1/4\pi)\left[-F^{\mu\alpha}F^\nu{}_\alpha + \frac{1}{4}g^{\mu\nu}F^{\alpha\beta}F_{\alpha\beta}\right], F_{\mu\nu} = \partial_\mu A_\nu - \partial_\nu A_\mu, \qquad (4.3.1)$$

where $F_{\mu\nu}$ are its strengths. When its potentials A_μ satisfy the Lorentz gauge, the Maxwell equations are as follows:

$$\partial^\mu\partial_\mu A^\nu = 4\pi j^\nu, j^\nu = \theta_0 dx^\nu/ds, \partial_\mu A^\mu = 0. \qquad (4.3.2)$$

Here, cj^ν is the 4-vector of current densities and θ_0 is the charge density in a comoving inertial frame of reference.

Using formulas (4.2.3)–(4.2.5), (4.2.21), (4.2.24) and (4.3.1), we obtain the following expression $T^{\mu\nu}$ for the energy–momentum tensor of a charged dust-like nuclear matter and the nuclear and electromagnetic fields generated by it:

$$T^{\mu\nu} = c^2 \rho_0 \exp(\varphi/c^2) \frac{dx^\mu}{ds} \frac{dx^\nu}{ds}$$

$$+ \frac{m_p^2}{4\pi G^2} \left[\left(g^{\mu\alpha} g^{\nu\beta} - \frac{1}{2} g^{\mu\nu} g^{\alpha\beta} \right) \frac{\partial\varphi}{\partial x^\alpha} \frac{\partial\varphi}{\partial x^\beta} + \frac{1}{2} g^{\mu\nu} (m_\pi c/\hbar)^2 \varphi^2 \right]$$

$$+ \frac{1}{4\pi} \left(-F^{\mu\alpha} F^\nu_{\ \alpha} + \frac{1}{4} g^{\mu\nu} F^{\alpha\beta} F_{\alpha\beta} \right). \tag{4.3.3}$$

For the energy–momentum tensor of an electromagnetic field, we have the well-known formula [13]

$$\partial_\mu T_e^{\mu\nu} = -\theta_0 F^\nu_{\ \mu} dx^\mu/ds. \tag{4.3.4}$$

Therefore, using formulas (4.2.10), (4.2.21) and (4.2.24), we find

$$\partial_\mu (T_m^{\mu\nu} + T_{\text{nuc}}^{\mu\nu} + T_e^{\mu\nu})$$
$$= \rho_0 \exp(\varphi/c^2) (c^2 d^2 x^\nu / ds^2$$
$$+ d\varphi/ds dx^\nu/ds) + (m_p^2/4\pi G^2) [\partial^\mu \partial_\mu \varphi$$
$$+ (m_\pi c/\hbar)^2 \varphi] \partial^\nu \varphi - \theta_0 F^\nu_{\ \mu} dx^\mu/ds = 0. \tag{4.3.5}$$

Since the tensor $F^{\mu\nu}$ is antisymmetric, we have

$$dx_\nu/ds F^\nu_{\ \mu} dx^\mu/ds = F^{\nu\mu} dx_\nu/ds dx_\mu/ds = 0. \tag{4.3.6}$$

That is why multiplying Eq. (4.3.5) by dx_ν/ds, and using (4.2.19), we come to the same equation for the nuclear field, as Eq. (4.2.25) obtained in the case $A_\mu = 0$.

Therefore, Eq. (4.2.25) for the nuclear field is also true when there are electromagnetic fields.

From (4.2.25) and (4.3.5), we come to the dynamic equations

$$\rho_0 \exp(\varphi/c^2) \left(c^2 \frac{d^2 x^\mu}{ds^2} + \frac{d\varphi}{ds} \frac{dx^\mu}{ds} - \frac{\partial\varphi}{\partial x_\mu} \right) - \theta_0 F^\mu_{\ \nu} \frac{dx^\nu}{ds} = 0. \tag{4.3.7}$$

It should be noted that owing to (4.2.19) and (4.3.6), after multiplying the left-hand side of Eq. (4.3.7) by dx_μ/ds, we obtain zero.

Therefore, the first equation in (4.3.7) ($\mu = 0$) is a consequence of the other three equations ($\mu = 1, 2, 3$).

Consider now the Lagrangian of a dust-like nuclear matter and electromagnetic and nuclear fields generated by it. Let us show that in this case, the action S is determined by the following expression:

$$S = -\sum_k \left[m_{0k}c \int \exp(\varphi/c^2)ds + (q_k/c) \int A_\mu dx^\mu \right]$$
$$+ (8\pi c)^{-1} \int \left\{ -\frac{1}{2}F^{\mu\nu}F_{\mu\nu} + (m_p/G)^2[\partial^\mu\varphi\partial_\mu\varphi \right.$$
$$\left. - (m_\pi c/\hbar)^2\varphi^2] \right\} d^4x, \qquad (4.3.8)$$

where the summation is taken over all particles, m_{0k} is the mass at rest of the kth particle when $\varphi = 0$, q_k is its electric charge, the integrals for the particles are taken between fixed space–time points, the integrals for the fields are taken over the entire four-dimensional space–time between two fixed moments of time and $d^4x = dx^0dx^1dx^2dx^3$.

Let us take into account that in an inertial frame of reference, $dV_0ds = d^4x$, where dV_0 is a small part of a three-dimensional volume in an inertial frame comoving with it. Then, expression (4.3.8) for the action S can be represented by Lagrangian L in the form

$$S = (1/c) \int L d^4x,$$
$$L = -c^2\rho_0 \exp(\varphi/c^2) - \theta_0 A_\mu dx^\mu/ds - (16\pi)^{-1}F^{\mu\nu}F_{\mu\nu}$$
$$+ [m_p^2/(8\pi G^2)][\partial^\mu\varphi\partial_\mu\varphi - (m_\pi c/\hbar)^2\varphi^2], \qquad (4.3.9)$$

where ρ_0 and θ_0 are the rest mass density when $\varphi = 0$ and the charge density in a comoving local inertial frame of the considered matter, respectively.

We will take the variation of formula (4.3.8) for the action S with respect to the particle trajectories $x^\mu(s)$ and then take the variation of formula (4.3.9), which is equivalent to formula (4.3.8), with respect to the field potentials A_μ and φ.

Let us denote

$$m_{0k}^{(\varphi)} = m_{0k} \exp(\varphi/c^2), \qquad (4.3.10)$$

which, as follows from (4.2.22), presents the rest mass of the kth particle when the nuclear potential is equal to φ.

Then, the variation of expression (4.3.8) with respect to the particle trajectories $x^\mu(s)$ can be represented as

$$\delta S = -\sum_k \int \left[m_{0k}^{(\phi)} c \frac{dx_\mu d(\delta x^\mu)}{ds} + \frac{dm_{0k}^{(\varphi)}}{d\varphi} \frac{\partial \varphi}{\partial x^\mu} c\, ds\, \delta x^\mu + \frac{q_k}{c} A_\mu d(\delta x^\mu) \right.$$

$$\left. + \frac{q_k}{c} \frac{\partial A_\nu}{\partial x^\mu} \delta x^\mu dx^\nu \right]$$

$$= \sum_k \int \left[dm_{0k}^{(\varphi)} c \frac{dx_\mu}{ds} + c m_{0k}^{(\varphi)} d\left(\frac{dx_\mu}{ds}\right) - \frac{dm_{0k}^{(\varphi)}}{d\varphi} \frac{\partial \varphi}{\partial x^\mu} c\, ds \right.$$

$$\left. + \frac{q_k}{c} \left(\frac{\partial A_\mu}{\partial x^\nu} - \frac{\partial A_\nu}{\partial x^\mu}\right) dx^\nu \right] \delta x^\mu. \tag{4.3.11}$$

The variation of expression (4.3.9) with respect to the field potentials φ and A_μ gives the following:

$$\delta S = -(1/c) \int \{ [\rho_0 \exp(\varphi/c^2) + [m_p^2/(4\pi G^2)](\partial^\mu \partial_\mu \varphi$$

$$+ (m_\pi c/\hbar)^2 \varphi)] \delta\varphi + [\theta_0 dx^\mu/ds + (4\pi)^{-1} \partial F^{\mu\nu}/\partial x^\nu] \delta A_\mu \} d^4x. \tag{4.3.12}$$

From formulas (4.3.10), (4.3.11), (4.3.12) and the principle of least action [13]

$$\delta S = 0, \tag{4.3.13}$$

we come to the equations

$$m_{0k}^{(\phi)} \left(c^2 \frac{d^2 x_\mu}{ds^2} + \frac{d\varphi}{ds} \frac{dx_\mu}{ds} - \frac{\partial \varphi}{\partial x^\mu} \right) - q_k F_{\mu\nu} \frac{dx^\nu}{ds} = 0,$$

$$\frac{\partial F^{\nu\mu}}{\partial x^\nu} = 4\pi \theta_0 \frac{dx^\mu}{ds}, \tag{4.3.14}$$

$$\partial^\mu \partial_\mu \varphi + (m_\pi c/\hbar)^2 \varphi = -4\pi (G/m_p)^2 \rho_0 \exp(\varphi/c^2),$$

which coincide with the dynamic equations (4.3.7) and the equations for electromagnetic and nuclear field potentials (4.3.2) and (4.2.25).

Therefore, expression (4.3.8) presents the action for a dust-like matter interacting with the electromagnetic and nuclear fields generated by it.

4.4. High-Energy Interaction of Protons

Let us turn to the question of the value of the dimensionless constant of nuclear interaction. When the nuclear field potential is small, $|\varphi/c^2| << 1$, this constant is determined by formula (4.2.2): $G^2/(\hbar c) = 0.080$.

Consider this dimensionless constant when the nuclear field potential is large. For this purpose, let us turn to the case in which a nucleus interacts with a high-energy proton. Let us choose an arbitrary instant of time x^0 and an inertial frame of reference in which at this instant, the nucleus velocity is zero: $dx^k/dx^0 = 0, k = 1, 2, 3$. Then, for the nucleus at the chosen instant, from the dynamic equations (4.3.7) and the field equations (4.2.25) and (4.3.2), we obtain

$$e^{\varphi/c^2}\partial\varphi/\partial x^\mu + \gamma(\partial A_0/\partial x^\mu - \partial A_\mu/\partial x^0) = W_\mu, \qquad (4.4.1)$$

where

$$\begin{aligned}
&W_\mu = e^{\varphi/c^2}(d^2x_\mu/dx_0^2 + \partial\varphi/\partial x_0 dx_\mu/dx_0), \\
&dx_k/dx_0 = 0, k = 1, 2, 3, \quad \gamma = \theta_0/\rho_0, \\
&\partial^\mu\partial_\mu\varphi + (m_\pi c/\hbar)^2\varphi = -4\pi(G/m_p)^2\rho_0 e^{\varphi/c^2}, \\
&\partial^\mu\partial_\mu A_0 = 4\pi\theta_0, \quad \partial^\mu A_\mu = 0.
\end{aligned} \qquad (4.4.2)$$

From (4.4.1), we have

$$\partial^\mu W_\mu = e^{\varphi/c^2}\partial^\mu\partial_\mu\varphi + \gamma(\partial^\mu\partial_\mu A_0 - \partial_0\partial^\mu A_\mu) + U, \qquad (4.4.3)$$

where

$$U = (1/c^2)e^{\varphi/c^2}\partial^\mu\varphi\partial_\mu\varphi + \partial^\mu\gamma(\partial_\mu A_0 - \partial_0 A_\mu). \qquad (4.4.4)$$

From (4.4.2) and (4.4.3), we find

$$U - \partial^\mu W_\mu = e^{\varphi/c^2}[4\pi(G/m_p)^2\rho_0 e^{\varphi/c^2} + (m_\pi c/\hbar)^2\varphi] - 4\pi\theta_0\gamma. \quad (4.4.5)$$

Therefore,

$$4\pi\theta_0\gamma[(G/(m_p\gamma))^2 e^{2\varphi/c^2} - 1] = U - \partial^\mu W_\mu - (m_\pi c/\hbar)^2\varphi e^{\varphi/c^2}. \qquad (4.4.6)$$

It follows from this formula that inside the considered nucleus, the extreme value of the nuclear field potential $\varphi = \varphi_{\text{extr}}$ satisfies the relation

$$Ge^{\varphi/c^2}/(m_p\gamma) = 1, \tag{4.4.7}$$

since in this case, we have an infinite value of the charge density: $\theta_0 = \infty$.

Let us assume that a proton is the considered nucleus. Then, $\gamma = \theta_0/\rho_0 = e_p/m_p$, where e_p and m_p are the charge and rest mass of the proton, respectively, and from formula (4.4.7), we obtain the extreme value φ_{extr} of the nuclear field potential inside the proton:

$$\varphi_{\text{extr}} = -c^2 \ln(G/e_p). \tag{4.4.8}$$

On the other hand, outside the proton, we have [37, 38]

$$m_p\varphi = -g_s^2 \exp(-\nu_\pi r)/r, \quad \nu_\pi = m_\pi c/\hbar, \tag{4.4.9}$$

where $g_s^2/(\hbar c)$ is a dimensionless characteristic of strong interaction.

From (4.4.8) and (4.4.9), we obtain that the extreme value of the constant g_s of strong interaction is determined by the equality

$$g_s^2 \exp(-\nu_\pi r_p)/r_p = -m_p\varphi_{\text{extr}} = m_p c^2 \ln(G/e_p), \tag{4.4.10}$$

where r_p is the radius of the proton.

From (4.4.10), we find

$$g_s^2/(\hbar c) = (m_p/m_\pi)\nu r_p \exp(\nu_\pi r_p) \ln(G/e_p), \tag{4.4.11}$$

where $\nu_\pi = m_\pi c/\hbar$.

As follows from experimental data, the radius of the proton $r_p \approx$ 1.17 fm, and from (4.4.11) and (4.2.2), we obtain

$$g_s^2/(\hbar c) \approx 14.9. \tag{4.4.12}$$

It should be noted that the determined constant of strong interaction of protons at high collision energy is in good agreement with experimental data [39].

The correspondence between the theoretical and experimental values of this constant is an important confirmation of the derived exponential dependence (4.2.4) and (4.2.21) of the rest mass density of nuclear matter on the nuclear potential. Indeed, this dependence has a very significant impact on the correct form of formula (4.4.11) for the strong interaction constant of protons at high collision energy.

4.5. Properties of the Nuclear Field Potential

Let us now study Eq. (4.2.25) for the nuclear potential φ.

First, consider the stationary case and prove the following theorem:

Theorem 4.5.1. *In the stationary case, $\partial\varphi/\partial x^0 = 0$, the solutions to Eq. (4.2.25) for the nuclear potential φ that tend to zero at infinity are not positive: $\varphi \leq 0$.*

Proof. Let us assume that at some point M of the three-dimensional space, we have $\varphi(M) > 0$. Then, since $\varphi(\infty) = 0$, some point M_0 should exist in which the function φ has a positive maximum. Therefore, at the point M_0, the following inequalities take place:

$$\varphi(M_0) > 0, \partial\varphi(M_0)/\partial x^k = 0, \partial^2\varphi(M_0)/\partial x^{k^2} \leq 0, \ k = 1, 2, 3. \tag{4.5.1}$$

In the stationary case, from (4.2.25) and (4.5.1), we obtain the following inequality at the considered point M_0:

$$0 > \sum_{k=1}^{3} \frac{\partial^2\varphi}{\partial x^{k^2}} - \frac{m_\pi^2 c^2}{\hbar^2}\varphi = \lambda\rho_0 \exp\left(\frac{\varphi}{c^2}\right), \quad \lambda = 4\pi(G/m_p)^2. \tag{4.5.2}$$

From formula (4.5.2), we have $\rho_0(M_0) < 0$, which cannot take place since the matter rest mass is nonnegative.

The obtained contradiction shows that the stationary solution φ to Eq. (4.2.25) tending to zero at infinity satisfies the inequality

$$\varphi \leq 0, \tag{4.5.3}$$

which proves the theorem.

It follows from (4.3.10) and (4.5.3) that in the stationary case, the rest mass $m_0^{(\varphi)}$ of a nuclear particle when $\varphi \neq 0$ is less than its rest mass m_0 when $\varphi = 0$. This property allows one to explain the well-known fact of the existence of mass defect in atomic nuclei.

Consider Eq. (4.2.25) in the stationary spherically symmetric case. In this case, Eq. (4.2.25) acquires the following form in the spherical coordinates:

$$\varphi'' + 2\varphi'/r - \nu_\pi^2\varphi = \lambda\rho, \qquad (4.5.4)$$

where

$$\rho = \rho_0 \exp(\varphi/c^2), \ \varphi = \varphi(r), \ \rho_0 = \rho_0(r), \qquad (4.5.5)$$

$$\nu_\pi = m_\pi c/\hbar, \ r = \sqrt{(x^1)^2 + (x^2)^2 + (x^3)^2}. \qquad (4.5.6)$$

Expressing the function $\varphi(r)$ from the differential equation (4.5.4) by the function $\rho(r)$ and taking into account the condition $\varphi(\infty) = 0$ and that the value $\varphi(0)$ should be bounded, we come to the integral equation of the form

$$\varphi(r) = \frac{\lambda}{2\nu_\pi r} \int_0^\infty z\rho(z)[\exp(-\nu_\pi(r+z)) - \exp(-\nu_\pi|r-z|)]dz.$$
$$(4.5.7)$$

Differentiating (4.5.7) and then integrating it by parts, we obtain

$$\varphi'(r) = -\frac{\lambda}{2\nu_\pi^3 r^2} \int_0^\infty \rho'(z)M(r,z)dz, \qquad (4.5.8)$$

where

$$M(r,z) = K(r,-z) - K(r,z),$$
$$(4.5.9)$$
$$K(r,z) = \exp(-\nu_\pi|r-z|)(1 - \nu_\pi^2 rz + \nu_\pi|r-z|).$$

Let us note an important property of the function $M(r,z)$. As can be readily proved, the function $M(r,z)$ is nonnegative when $r, z \geq 0$:

$$M(r,z) \geq 0. \qquad (4.5.10)$$

Substituting expression (4.5.5) for ρ into Eq. (4.5.8), we obtain the following formula:

$$\varphi'(r) = -\frac{\lambda}{2\nu_\pi^3 r^2} \int_0^\infty \exp\left(\frac{\varphi(z)}{c^2}\right) M(r,z) \left[\rho_0'(z) + \rho_0(z)\frac{\varphi'(z)}{c^2}\right] dz.$$
$$(4.5.11)$$

It should be stressed that in atomic nuclei, the rest mass density $\rho_0(r)$ decreases with distance from their center: $\rho_0'(r) \leq 0$. Therefore, if the second summand in the square brackets in (4.5.11) had been absent, then this formula would give the inequality $\varphi'(r) \geq 0$ and hence nuclear forces would manifest themselves as only attractive ones. However, it is well known that nuclear forces can also be repulsive [37]. This property of nuclear forces can be explained by the existence of the second summand in the square brackets in formula (4.5.11), which appears because of the exponential dependence (4.5.5) of the rest mass density of nuclear matter on the nuclear potential φ.

Let us find a condition that provides the inequality $\varphi'(r) < 0$ in some region of a nucleus. For this purpose, let us introduce the following functions:

$$I_1(r) = -\frac{\lambda}{2\nu_\pi^3 r^2} \int_0^\infty \exp\left(\frac{\varphi(z)}{c^2}\right) M(r,z)\rho_0'(z)dz, \qquad (4.5.12)$$

$$I_2(r) = \frac{\lambda}{2\nu_\pi^3 r^2 c^2} \int_0^\infty \exp\left(\frac{\varphi(z)}{c^2}\right) M(r,z)\rho_0(z)|\varphi'(z)|dz. \quad (4.5.13)$$

From (4.5.10), we get that the function $I_2(r) > 0$. Let us assume that at some point $r = \overline{r}$, we have the inequality

$$I_2(\overline{r}) > I_1(\overline{r}). \qquad (4.5.14)$$

Then, the following theorem takes place:

Theorem 4.5.2. *If there exists a point $\overline{r}$, at which the inequality (4.5.14) is fulfilled, then there exists some region of values of the variable r where $\varphi'(r) < 0$ and hence the nuclear forces in it are repulsive.*

Proof. Let us assume that $\varphi'(r) \geq 0$ for any r. Then, from (4.5.11)–(4.5.14), we obtain at the point $\overline{r}$:

$$\varphi'(\overline{r}) = I_1(\overline{r}) - I_2(\overline{r}) < 0. \qquad (4.5.15)$$

Therefore, we come to a contradiction, which proves the theorem.

From this theorem and formulas (4.5.12)–(4.5.14), we find that the inequality $\varphi'(r) < 0$ can be fulfilled if, in particular, inside a nuclear matter, the inequalities $\rho_0' \leq 0$ and $|\varphi'|/c^2 > |\rho_0'|/\rho_0$ take place.

Thus, the exponential dependence of the rest mass density of nuclear matter on the nuclear potential allows one to interpret the well-known experiments [37], in which nuclear forces manifest themselves as not only attractive forces but also as repulsive forces.

Let us apply the proposed theory of nuclear forces to calculate the characteristics of atomic nuclei.

4.6. Investigation of Properties of Atomic Nuclei

Consider a classical particle, which interacts with nuclear and electromagnetic fields. As shown in Sections 4.2 and 4.3, its movement in these fields and the fields themselves can be described by the equations

$$\rho_0 \exp(\varphi/c^2)(c^2 d^2 x^\mu/ds^2 + d\varphi/ds\, dx^\mu/ds - \partial\varphi/\partial x_\mu)$$

$$- \theta_0 F^\mu_{\ \nu} dx^\nu/ds = 0, \quad (x^1, x^2, x^3) \in \Omega(x^0), \tag{4.6.1}$$

$$F_{\mu\nu} = \partial_\mu A_\nu - \partial_\nu A_\mu, \partial^\mu \partial_\mu A^\nu = 4\pi\theta_0 dx^\nu/ds, \partial_\mu A^\mu = 0, \tag{4.6.2}$$

$$\partial^\mu \partial_\mu \varphi + (m_\pi c/\hbar)^2 \varphi = -4\pi(G/m_p)^2 \rho_0 \exp(\varphi/c^2), \tag{4.6.3}$$

where φ is the scalar potential of nuclear forces, $G^2/(\hbar c) = 0.080$ is the dimensionless constant of nuclear interaction, A_μ are electromagnetic potentials, $\rho_0 = \rho_0(x^0, \mathbf{r})$ is the rest mass density of the particle when $\varphi = 0$, $\theta_0 = \theta_0(x^0, \mathbf{r})$ is the density of its charge, $\mathbf{r} = (x^1, x^2, x^3)$, dx^μ/ds is the 4-vector of its velocity, $\Omega(x^0)$ is the time-dependent small spatial volume occupied by the moving particle and m_π, m_p are the rest masses of the neutral pion and proton, respectively.

Let us write down a condition, which should be fulfilled for the mass of an atomic nucleus to be considered a classical particle. In accordance with quantum mechanics, this condition has the form

$$(2mE_b)^{1/2}r \gg \hbar, \tag{4.6.4}$$

where E_b, m, r are the binding energy, rest mass and radius of the nucleus, respectively.

Since $r \sim N^{1/3}$ fm, $E_b/N \sim 10^{-2} m_p c^2$ [37], where $N = m/m_p$, from (4.6.4), we obtain

$$(m/m_p)^{4/3} \gg 1. \tag{4.6.5}$$

Therefore, a nucleus can be considered a classical particle when condition (4.6.5) is fulfilled. In accordance with this condition, we will further investigate Eqs. (4.6.1)–(4.6.3) for atomic nuclei with $m/m_p \geq 20$.

Consider a nonrelativistic atomic nucleus that moves under the action of external sources of an electromagnetic field with potentials A_μ^{ext} and apply Eqs. (4.6.1)–(4.6.3) for its description. Then, in the considered nonrelativistic case, we have the equations

$$\rho_0 \exp(\varphi/c^2)(w^k + \partial\varphi/\partial x^k) + \theta_0(\partial A_0/\partial x^k - \partial A_k/\partial x^0) = 0,$$

$$k = 1, 2, 3, \quad (x^1, x^2, x^3) \in \Omega(x^0), \quad A_\mu = A_\mu^{\text{int}} + A_\mu^{\text{ext}}, \tag{4.6.6}$$

$$\Delta A_0^{\text{int}} = -4\pi\theta_0, \quad \Delta A_k^{\text{int}} = 4\pi\theta_0 v^k/c, \tag{4.6.7}$$

where $v^k = v^k(x^0)$ and $w^k = w^k(x^0)$ are components of the velocity and acceleration of the nucleus, respectively, $\theta_0 = \theta_0(x^0, x^1, x^2, x^3)$ is the density of its charge, A_μ^{int} are potentials of the electromagnetic field generated by this nucleus, Δ is the Laplace operator and $\Omega(x^0)$ is a small region occupied by the nucleus at the moment x^0.

Equations (4.6.7) give the following expressions for A_μ^{int}:

$$A_k^{\text{int}} = -A_0^{\text{int}} v^k/c, \quad v^k = v^k(x^0), \quad k = 1, 2, 3,$$

$$A_0^{\text{int}} = \int_\Omega (\theta_0/R) dy^1 dy^2 dy^3, \quad \theta_0 = \theta_0(x^0, y^1, y^2, y^3), \tag{4.6.8}$$

$$R = \sqrt{(x^1 - y^1)^2 + (x^2 - y^2)^2 + (x^3 - y^3)^2}.$$

From Eqs. (4.6.6)–(4.6.8), we come to the following equation in the considered case $|v^k/c| \ll 1$:

$$[c^2 \exp(\varphi/c^2) + \gamma A_0^{\text{int}}] w^k/c^2 + \partial[c^2 \exp(\varphi/c^2)]/\partial x^k$$

$$+ \gamma \partial A_0^{\text{int}}/\partial x^k = \gamma E_{\text{ext}}^k, \gamma = \theta_0/\rho_0, k = 1, 2, 3, \tag{4.6.9}$$

where E_{ext}^k are the strengths of the electric field generated by the sources which are external for the nucleus.

Let us assume that the external field sources are sufficiently far from the nucleus. Then, inside it, the external electromagnetic field can be regarded as homogeneous and for the nuclear field potential φ, we can put $\varphi = \varphi^{\text{int}}$, where φ^{int} is the potential of the nuclear field generated by the nucleus itself.

Since the accelerations w^k should be the same for different points of the considered nucleus and the strengths E_{ext}^k can be arbitrary but also the same inside it, from Eq. (4.6.9), we come to the following two equations:

$$\gamma = \theta_0/\rho_0 = \text{const},$$
$$c^2 \exp(\varphi^{\text{int}}/c^2) + \gamma A_0^{\text{int}} = \text{const}, \quad (x^1, x^2, x^3) \in \Omega(x^0). \tag{4.6.10}$$

Equations (4.6.9) and (4.6.10) result in the classical law of Newtonian mechanics:

$$m w^k = q E_{\text{ext}}^k, k = 1, 2, 3, \tag{4.6.11}$$

where q is the charge of the nucleus and m is its inertial mass, which is determined by the expression

$$m = m_0 \exp(\varphi^{\text{int}}/c^2) + q A_0^{\text{int}}/c^2 = \text{const},$$
$$m_0 = q\rho_0/\theta_0 = \text{const}, (x^1, x^2, x^3) \in \Omega(x^0). \tag{4.6.12}$$

It should be noted that the obtained two equations (4.6.12) are necessary for the determination of the distribution of the mass and charge inside the nucleus.

As follows from (4.6.12), the value m_0 is the nucleus mass at rest when $\varphi = 0$.

Consider now a spherical atomic nucleus that is at rest relative to an inertial frame of reference in the case $E_{\text{ext}}^k = 0$. Then, we have a stationary and spherically symmetric problem.

Let r and r_* be the distance from the nucleus center and its radius, respectively. For the region $r \leq r_*$ occupied by the nucleus, we can use Eqs. (4.6.3), (4.6.7) and (4.6.12). From them, we derive the equations

$$f'' + 2f'/r - (m_\pi c/\hbar)^2 f = 4\pi\theta_0(m_0/q)[G/(m_p c)]^2 e^f,$$
$$f = \phi^{\text{int}}(r)/c^2, \tag{4.6.13}$$

$$\beta'' + 2\beta'/r = -4\pi\theta_0/c^2, \beta = A_0^{\text{int}}(r)/c^2, \tag{4.6.14}$$

$$m_0 e^f f' + q\beta' = 0, 0 \leq r \leq r_*. \tag{4.6.15}$$

From Eqs. (4.6.14) and (4.6.15), we find

$$\beta' = -(m_0/q)e^f f', \beta'' = -(m_0/q)e^f[f'' + (f')^2], \tag{4.6.16}$$
$$4\pi\theta_0/c^2 = (m_0/q)e^f[f'' + 2f'/r + (f')^2], f = f(r). \tag{4.6.17}$$

Equations (4.6.13) and (4.6.17) give the following equation for the nuclear field potential φ^{int}:

$$(\mu^2 e^{2f} - 1)(f'' + 2f'/r) + \mu^2 e^{2f}(f')^2 + \nu^2 f = 0, 0 \leq r \leq r_*,$$
(4.6.18)

where

$$\mu = Gm_0/(m_p q), \quad \nu_\pi = m_\pi c/\hbar.$$
(4.6.19)

Examine now Eq. (4.6.3) for the nuclear field potential ϕ^{int} in the region outside the considered nucleus: $r > r_*$. Then, this equation acquires the form

$$f'' + 2f'/r - \nu_\pi^2 f = 4\pi[G/(m_p c)]^2 \varepsilon_0 e^f, f = \varphi^{\text{int}}/c^2, r > r_*,$$
(4.6.20)

where $\varepsilon_0 = \varepsilon_0(r)$ is the rest mass density of virtual pions, created in the physical vacuum near the nucleus surface $r = r_*$, owing to the influence of this surface on it.

Taking into account that the lifetime of a virtual pion is $\sim \hbar/(m_\pi c^2)$ [37] and its average velocity $|\bar{v}| \ll c$, since the considered surface $r = r_*$ is immovable, we find that the mass of the virtual pions is mainly concentrated in a very narrow region $r_* < r < r_* + \Delta_*$, where $\Delta_* \ll \hbar/(m_\pi c)$.

Therefore, the mass density of the virtual pions $\varepsilon_0(r)$ can be approximately represented by the even delta function $\delta(r)$ as

$$\varepsilon_0(r) = 2\sigma\delta(r - r_*), r > r_*, \int_0^\infty \delta(r)dr = 1/2,$$
(4.6.21)

where σ is the constant equal to the mass of the virtual pions per unit square of the nucleus surface $r = r_*$.

Let us denote

$$x = \nu_\pi r, \quad x_* = \nu_\pi r_*, \quad \nu_\pi = m_\pi c/\hbar.$$
(4.6.22)

Then, from (4.6.18) and (4.6.20), we obtain

$$(\mu^2 e^{2f} - 1)(f'' + 2f'/x) + \mu^2 e^{2f}(f')^2 + f = 0, \quad 0 \leq x \leq x_*,$$
(4.6.23)

$$f'' + 2f'/x - f = \lambda_0 e^f, \quad x > x_*,$$
(4.6.24)

$$\lambda_0 = \lambda_0(x) = 4\pi\varepsilon_0(x)[G/(m_p c\nu)]^2, \quad f = f(x), \quad f(\infty) = 0.$$
(4.6.25)

Expressing the function $f(r)$ in the left-hand side of Eq. (4.6.24) by its right-hand side and using the condition $f(\infty) = 0$, we come to the integral equation

$$f(x) = \frac{1}{2x} \left[e^x \int_\infty^x \lambda_0 e^{f-x} x dx - e^{-x} \left(\int_{x_*}^x \lambda_0 e^{f+x} x dx + D \right) \right],$$
(4.6.26)

$$D = \text{const}, x \geq x_*.$$

Equation (4.6.26) gives the following formulas for $f(x_*)$ and $f'(x_*)$:

$$f(x_*) = -\frac{1}{2x_*} \left[\int_{x_*}^\infty \lambda_0 e^{f-x} x dx e^{x_*} + De^{-x_*} \right],$$

$$f'(x_*) = \frac{1}{2x_*^2} \left[\int_{x_*}^\infty \lambda_0 e^{f-x} x dx (1 - x_*) e^{x_*} + D(1 + x_*) e^{-x_*} \right].$$
(4.6.27)

From these formulas, we find

$$f(x_*)(1 + x_*) + x_* f'(x_*) = -e^{x_*} \int_{x_*}^\infty \lambda_0 e^{f-x} x dx.$$ (4.6.28)

Relation (4.6.28) presents the condition at the point $x = x_*$, which should be satisfied for the solutions to Eqs. (4.6.23) and (4.6.24) tending to zero at infinity.

From (4.6.21), (4.6.25) and (4.6.28), we get the equality

$$f(x_*)(1 + x_*) + x_* f'(x_*) = -sx_* e^{f(x_*)} \sigma,$$ (4.6.29)

where

$$s = 4\pi [G/(m_p c)]^2 / \nu_\pi, \quad \sigma = \text{const}.$$ (4.6.30)

Consider now the electromagnetic potential A_0^{int}. For it, we have the following classical formula in the region $r \geq r_*$:

$$\beta(r) = A_0^{\text{int}}(r)/c^2 = q/(c^2 r), \quad r \geq r_*,$$ (4.6.31)

where q is the charge of the nucleus.

From Eqs. (4.6.14), (4.6.22) and (4.6.31), we obtain

$$\beta(x) = \beta(0) + 4\pi \int_0^x t(t/x - 1)\theta_0(t)dt/(c\nu_\pi)^2, \quad 0 \le x \le x_*,$$

(4.6.32)

$$\beta(x) = q\nu_\pi/(c^2 x), \quad x \ge x_*.$$
(4.6.33)

Formulas (4.6.32) and (4.6.33) give

$$\beta'(x_*) = -4\pi \int_0^{x_*} t^2\theta_0(t)dt/(c\nu_\pi x_*)^2 = -q\nu_\pi/(cx_*)^2.$$
(4.6.34)

Using formulas (4.6.17), (4.6.22) and (4.6.34), we find

$$\int_0^{x_*} x^2 e^f[f'' + 2f'/x + (f')^2]dx = q^2\nu_\pi/(m_0 c^2), f = f(x).$$
(4.6.35)

Consider Eq. (4.6.23). From it, we have

$$y'' + 2y'/x = -H(x), 0 \le x \le x_*,$$
(4.6.36)

where

$$y = e^{f(x)}, \quad H(x) = [y^2 \ln(y) + (y')^2]/[y(\mu^2 y^2 - 1)].$$
(4.6.37)

From Eq. (4.6.36), we come to the integral equation

$$y(x) = y(0) + \int_0^x t(t/x - 1)H(t)dt.$$
(4.6.38)

This equation can be represented in the form

$$y(x) = u/x + w, \quad u = \int_0^x t^2 H(t)dt, \quad w = y(0) - \int_0^x tH(t)dt.$$
(4.6.39)

From (4.6.39), we readily find

$$y'(x) = -u/x^2.$$
(4.6.40)

Consider equality (4.6.35). It can be represented as

$$\int_0^{x_*} x^2(y'' + 2y'/x)dx = q^2\nu_\pi/(m_0c^2), \quad y = e^{f(x)}. \tag{4.6.41}$$

From (4.6.36) and (4.6.41), we obtain

$$\int_0^{x_*} x^2 H(x)dx = -q^2\nu_\pi/(m_0c^2) \tag{4.6.42}$$

and, taking into account (4.6.39), we have

$$u(x_*) = -q^2\nu_\pi/(m_0c^2). \tag{4.6.43}$$

Formulas (4.6.37), (4.6.40) and (4.6.43) give the following condition:

$$x_*^2 y'(x_*) = x_*^2 e^{f(x_*)} f'(x_*) = q^2\nu_\pi/(m_0c^2). \tag{4.6.44}$$

From (4.6.29) and (4.6.44), we obtain one more condition at the point x_*:

$$f(x_*)(1 + x_*) + q^2\nu_\pi e^{-f(x_*)}/(m_0c^2 x_*) = -bx_* e^{f(x_*)}, b = s\sigma. \tag{4.6.45}$$

Using (4.6.12), (4.6.31), (4.6.22) and also the expression for f in (4.6.13), we come to the following formula for the nucleus inertial mass m:

$$m = m_0 e^{f(x_*)} + q\beta(x_*) = m_0 e^{f(x_*)} + q^2\nu_\pi/(c^2 x_*). \tag{4.6.46}$$

Let A and Z denote the number of nucleons and the number of protons in an atomic nucleus, respectively. Then, since m_0 is the rest mass of a nucleus when $\varphi = 0$, we have

$$m_0 = Zm_0^p + (A - Z)m_0^n, q = Ze_p, \tag{4.6.47}$$

where m_0^p and m_0^n are the rest masses of the proton and neutron when $\varphi = 0$, respectively, and e_p is the proton charge.

From (4.6.46) and (4.6.47), we obtain the following formula for the binding energy E_b of a nucleus:

$$E_b = c^2[Zm_p + (A - Z)m_n - m]$$
$$= c^2\{(A - Z)(m_n - m_p) + Zm_p[1 - (1 + \delta_p)e^{f(x_*)}]$$
$$+ (A - Z)m_p[1 - (1 + \delta_n)e^{f(x_*)}]\} - Z^2e_p^2\nu_\pi/x_*, \quad (4.6.48)$$

where m_p and m_n are the experimental values of the proton and neutron rest masses, respectively, and

$$m_0^p = m_p(1 + \delta_p), m_0^n = m_p(1 + \delta_n), \quad (4.6.49)$$

where δ_p and δ_n are some small constants as compared with the number 1.

4.7.   Calculation of the Binding Energies and Radii of Medium and Heavy Atomic Nuclei

Let us turn to the calculation of the binding energies and radii of medium and heavy atomic nuclei. For this purpose, consider Eq. (4.6.23) inside an atomic nucleus with conditions (4.6.44) and (4.6.45) at its boundary to determine x_* and $f(x_*)$. We will seek its solution $f(x)$ in the form of the power series

$$f(x) = f_0\left(1 + \sum_{n=1}^{\infty} d_n x^{2n}\right), 0 \leq x \leq x_*, \quad (4.7.1)$$

where f_0 and d_n are some coefficients.

Let us also represent the function $e^{2f(x)}$ inside the nucleus as the power series

$$e^{2f(x)} = e^{2f_0}\sum_{k=0}^{\infty} 2^k f_0^k \left(\sum_{n=1}^{\infty} d_n x^{2n}\right)^k \bigg/ k! = e^{2f_0}\sum_{n=0}^{\infty} p_n x^{2n}, \quad (4.7.2)$$

where

$$p_0 = 1, \quad p_n = \sum_{i=1}^{n} 2^i f_0^i q_{n,i}/i!, \quad n \geq 1, \quad \sum_{n=i}^{\infty} q_{n,i} x^{2n} = \left(\sum_{n=1}^{\infty} d_n x^{2n} \right)^i,$$

$$q_{n,i} = \sum_{j=i-1}^{n-1} d_{n-j} q_{j,i-1}, \quad 1 \leq i \leq n, q_{0,0} = 1, \quad q_{j,0} = 0, \quad j > 0.$$

$$(4.7.3)$$

Substituting (4.7.1) and (4.7.2) into Eq. (4.6.23), we obtain

$$2 \left(D^2 \sum_{n=0}^{\infty} p_n x^{2n} - 1 \right) \sum_{n=0}^{\infty} (n+1)(2n+3) d_{n+1} x^{2n} + \sum_{n=0}^{\infty} d_n x^{2n}$$

$$+ 4D^2 f_0 \sum_{n=0}^{\infty} p_n x^{2n} \sum_{n=0}^{\infty} \sum_{k=0}^{n} (k+1)(n-k+1) d_{k+1} d_{n-k+1} x^{2n+2}$$

$$= \sum_{n=0}^{\infty} H_n x^{2n} = 0, \tag{4.7.4}$$

where

$$p_0 = 1, \quad d_0 = 1, \quad D = \mu e^{f_0},$$
$$H_n = H_n(D, f_0, d_1, d_2, \ldots, d_{n+1}), \quad 0 \leq x \leq x_*. \tag{4.7.5}$$

From (4.7.4), we get the equations $H_n = 0, n \geq 0$, which give the recurrence relation

$$d_{n+1} = d_{n+1}(D, f_0, d_0, d_1, \ldots, d_n)$$
$$= -[(2n+2)(2n+3)(D^2 - 1)]^{-1}$$
$$\times \left\{ d_n + 2D^2 \sum_{i=0}^{n-1} [(i+1)d_{i+1} \left(p_{n-i}(2i+3) + 2f_0(n-i)d_{n-i} \right) \right.$$
$$\left. + 2f_0 p_{n-i} \sum_{k=0}^{i-1} (k+1)(i-k)d_{k+1}d_{i-k}] \right\}, \tag{4.7.6}$$

where $n \geq 0, \quad d_0 = 1, \quad \sum_0^{-1} = 0.$

From (4.7.6), we find

$$d_1 = -1/[6(D^2 - 1)], d_2 = 0.3(1 + 16D^2 f_0 d_1)d_1^2,$$
$$d_3 = [d_2 + 4D^2 f_0 d_1 (5 f_0 d_1^2 + 17 d_2)]d_1/7 \cdots ,$$

$$(4.7.7)$$

where, owing to (4.6.19), (4.6.47) and (4.6.49),

$$D^2 = \mu^2 e^{2f_0} = a(1 + \delta)^2 e^{2f_0} A^2/Z^2,$$
$$a = G^2/e_p^2, \quad \delta = \delta_p Z/A + \delta_n(1 - Z/A).$$

$$(4.7.8)$$

Conditions (4.6.44) and (4.6.45) give the following equations:

$$RZ^2/A = 2(1 + \delta)x_*^3 f_0 e^{f(x_*)}(d_1 + 2d_2 x_*^2 + 3d_3 x_*^4 + \cdots),$$
$$R = (e_p^2/\hbar c)m_\pi/m_p = 0.00104966,$$
$$f_0 = f(x_*)/(1 + d_1 x_*^2 + d_2 x_*^4 + d_3 x_*^6 + \cdots),$$

$$(4.7.9)$$

$$f(x_*)(1 + x_*) + Pe^{-f(x_*)} = -bx_* e^{f(x_*)}, \quad P = RZ^2/[Ax_*(1 + \delta)].$$

$$(4.7.10)$$

From Eq. (4.7.10), we can calculate $f(x_*)$ and then from formulas (4.7.6)–(4.7.9), we can determine $d_1, d_2, d_3, \ldots, f_0$ and x_* as functions of the numbers A, Z and the four dimensionless parameters a, b, δ_p, δ_n. After that from Eq. (4.6.48), we can determine the nucleus binding energy E_b as a function of these parameters.

To find the values of the parameters a, b, δ_p, δ_n, we have carried out a large amount of computer calculations of the values $E_b(A, Z, a, b, \delta_p, \delta_n)$ by means of formula (4.6.48) and numerical solutions of Eqs. (4.7.6)–(4.7.10) for different medium and heavy nuclei with $A \geq 20$ and compared them with the well-known experimental values $E_b^{\exp}(A, Z)$ [37] of nucleus binding energies.

It should be noted that the optimal value of the parameter a is very close to the number 11. As follows from formula (4.7.8) for the constant G, this value gives the well-known value $G^2/(\hbar c) = 0.080$, indicated in (4.2.2), and presenting the dimensionless constant of low-energy nuclear interaction of nucleons inside atomic nuclei [37, 38].

As a result of minimization of the mean deviation of the computed binding energies E_b from their experimental values E_b^{exp}, the following values of the parameters a, b, δ_p, δ_n are obtained:

$$a = 11, \quad b = 0.04607, \quad \delta_p = 0.01523, \quad \delta_n = 0.03153. \quad (4.7.11)$$

The results of computer calculations are presented in Table 4.1. It gives the calculated values of the radii of nuclei and the calculated and experimental values of their specific binding energies.

It is seen from Table 4.1 that the calculated values of the binding energies of nuclei E_b are close to their experimental values E_b^{exp}.

The values of the calculated radii r_* of nuclei are also close to their experimental values. As follows from Table 4.1, the radii of nuclei can

Table 4.1. Calculated values of the radii (r_*, fm) and calculated and experimental values of the specific binding energies of nuclei $(E_b/A$ and E_b^{exp}/A, MeV).

Z	A	r_*	E_b/A	E_b^{exp}/A	Z	A	r_*	E_b/A	E_b^{exp}/A
10	20	3.963	8.156	8.032	58	140	7.225	8.367	8.378
12	25	4.245	8.294	8.224	60	145	7.303	8.324	8.312
14	30	4.491	8.409	8.521	62	150	7.380	8.280	8.263
16	35	4.709	8.501	8.538	64	155	7.455	8.235	8.213
18	40	4.907	8.574	8.596	66	160	7.529	8.189	8.186
20	45	5.089	8.630	8.630	68	166	7.617	8.155	8.141
22	50	5.257	8.672	8.756	70	170	7.671	8.094	8.106
24	55	5.414	8.701	8.728	72	176	7.755	8.060	8.060
26	58	5.497	8.742	8.792	74	180	7.807	7.997	8.024
28	65	5.701	8.732	8.737	76	185	7.873	7.947	7.982
30	70	5.833	8.735	8.730	78	190	7.938	7.896	7.947
32	75	5.958	8.733	8.696	79	195	8.007	7.907	7.921
34	80	6.078	8.724	8.711	80	200	8.075	7.917	7.906
36	85	6.193	8.711	8.700	82	205	8.136	7.867	7.874
38	90	6.303	8.694	8.700	84	210	8.197	7.816	7.834
40	95	6.409	8.672	8.645	86	215	8.256	7.764	7.764
42	100	6.512	8.648	8.604	88	220	8.315	7.712	7.712
44	105	6.611	8.620	8.566	90	225	8.372	7.659	7.660
46	110	6.706	8.590	8.553	92	230	8.429	7.606	7.621
48	115	6.799	8.558	8.509	94	235	8.485	7.553	7.579
50	120	6.889	8.523	8.505	96	240	8.540	7.500	7.543
52	125	6.976	8.486	8.458	98	245	8.594	7.446	7.500
54	130	7.061	8.448	8.438	100	250	8.647	7.391	7.462
56	135	7.144	8.408	8.398	102	255	8.700	7.337	7.430

be approximately represented as

$$r_* \approx 1.4 \cdot A^{1/3}\,\text{fm}, \quad A \geq 20. \tag{4.7.12}$$

Formula (4.7.12) just accords with the known experimental data on the capture by nuclei of neutrons with kinetic energies from 20 MeV to 90 MeV. These experiments show that the radius r_* of a nucleus is approximately proportional to $A^{1/3}$ and the coefficient of this proportionality is very close to 1.4 fm, which is just described by formula (4.7.12).

Thus, as has been shown, the proposed theory of nuclear forces has a number of serious experimental confirmations:

(1) It gives correct values of the well-known dimensionless constant of nuclear interaction of nucleons in the low-energy case when the constant is equal to 0.08 and in the high-energy case when the constant is equal to 15.
(2) It results in correct values of the binding energies and radii of medium and heavy atomic nuclei.

4.8.   Movements of Relativistic Nucleons near Heavy Atomic Nuclei

Consider a relativistic nucleon with charge q moving in the plane $x^3 = 0$ under the action of the nuclear and electric fields generated by a heavy atomic nucleus at rest. Then, from Eq. (4.2.25), we have the following Yukawa solution [37] in the region outside the nucleus where $\rho_0 = 0$:

$$\varphi = -\frac{A_0}{r}\exp(-\nu_\pi r), \quad \nu_\pi = \frac{m_\pi c}{\hbar}, \tag{4.8.1}$$

where A_0 is some constant and r is the distance from the nucleus center.

Let us use the polar coordinates r and θ in the considered plane $x^3 = 0$, which satisfy the relations

$$x^1 = r\cos\theta, \quad x^2 = r\sin\theta, \quad x^3 = 0, \tag{4.8.2}$$

where $x^1 = x^2 = 0$ at the center of the nucleus.

In the plane $x^3 = 0$, for the components F_{0k} of the electric field generated by the considered nucleus at rest, we have

$$F_{01} = \frac{Ze_p}{r^2}\cos\theta, \quad F_{02} = \frac{Ze_p}{r^2}\sin\theta, \quad F_{03} = 0, \tag{4.8.3}$$

where e_p is the proton charge and Z is the number of protons in the considered nucleus. Here, the magnetic field components are zero:

$$F_{ik} = 0 \quad \text{when } i, k = 1, 2, 3. \tag{4.8.4}$$

For the relativistic nucleon, let us put

$$\dot{r} \equiv dr/d\tau, \dot{\theta} \equiv d\theta/d\tau, \ddot{r} \equiv d^2r/d\tau^2, \ddot{\theta} \equiv d^2\theta/d\tau^2, \tag{4.8.5}$$

where τ is its proper time, which is related with the time t by formula (4.2.28).

Then, from (4.8.2), we obtain

$$dx^1/d\tau = \dot{r}\cos\theta - r\dot{\theta}\sin\theta, \quad dx^2/d\tau = \dot{r}\sin\theta + r\dot{\theta}\cos\theta, \quad x^3 = 0, \tag{4.8.6}$$

$$d^2x^1/d\tau^2 = (\ddot{r} - r\dot{\theta}^2)\cos\theta - (r\ddot{\theta} + 2\dot{r}\dot{\theta})\sin\theta,$$
$$d^2x^2/d\tau^2 = (\ddot{r} - r\dot{\theta}^2)\sin\theta + (r\ddot{\theta} + 2\dot{r}\dot{\theta})\cos\theta. \tag{4.8.7}$$

Using formulas (4.8.6) and 4.2.28), we have for the velocity v of the nucleon:

$$v^2 = (dx^1/dt)^2 + (dx^2/dt)^2 = (\dot{r}^2 + r^2\dot{\theta}^2)(1 - v^2/c^2). \tag{4.8.8}$$

Formulas (4.8.8) and (4.2.28) give

$$v^2/c^2 = (\dot{r}^2 + r^2\dot{\theta}^2)/[c^2 + \dot{r}^2 + r^2\dot{\theta}^2],$$
$$dt/d\tau = (1 - v^2/c^2)^{-1/2} = \sqrt{1 + (\dot{r}^2 + r^2\dot{\theta}^2)/c^2}. \tag{4.8.9}$$

Consider the dynamic equations (4.3.7). Taking into account (4.8.1)–(4.8.4), we find that the left-hand side, as well as the right-hand side, of the fourth equation ($\mu = 3$) in (4.3.7) is zero.

Consider now Eq. (4.3.7) when $\mu = 1, 2$. Using formulas (4.8.1)–(4.8.9), we obtain that the equations in (4.3.7) with $\mu = 1$ and $\mu = 2$ give

$$P \cos \theta - Q \sin \theta = 0, \quad P \sin \theta + Q \cos \theta = 0, \qquad (4.8.10)$$

where

$$P = m_p \exp(\varphi/c^2) \left[\ddot{r} - r\dot{\theta}^2 + \frac{A_0}{c^2 r^2} \exp(-\nu_\pi r)(1 + \nu_\pi r)(\dot{r}^2 + c^2) \right]$$

$$- \frac{Ze_p q}{r^2} \sqrt{1 + (\dot{r}^2 + r^2\dot{\theta}^2)/c^2} = 0, \qquad (4.8.11)$$

$$Q = m_p \exp(\varphi/c^2) \left[r\ddot{\theta} + 2\dot{r}\dot{\theta} + \frac{A_0}{c^2 r}\dot{r}\dot{\theta} \exp(-\nu_\pi r)(1 + \nu_\pi r) \right] = 0. \qquad (4.8.12)$$

As noted in Section 4.3, the first equation ($\mu = 0$) in (4.3.7) is a consequence of the other three equations ($\mu = 1, 2, 3$). Therefore, the first equation in (4.3.7) follows from Eqs. (4.8.11) and (4.8.12).

Dividing Eq. (4.8.12) by $r\dot{\theta}$, we find that it has the first integral

$$\int \frac{d\dot{\theta}}{\dot{\theta}} = \ln |\dot{\theta}| = - \int \left[\frac{2}{r} + \frac{A_0}{c^2 r^2}(1 + \nu_\pi r) \exp(-\nu_\pi r) \right] dr. \qquad (4.8.13)$$

This gives

$$\dot{\theta} = \frac{D}{r^2} \exp(\beta/c^2), \quad D = \text{const}, \qquad (4.8.14)$$

$$\beta = A_0 \int_r^{+\infty} \left(\frac{1}{r^2} + \frac{\nu_\pi}{r} \right) \exp(-\nu_\pi r) dr. \qquad (4.8.15)$$

Integrating by parts, we find

$$\int_r^{+\infty} \frac{\nu_\pi}{r} \exp(-\nu_\pi r) dr = - \int_r^{+\infty} \frac{1}{r} d \exp(-\nu_\pi r)$$

$$= \frac{1}{r} \exp(-\nu_\pi r)$$

$$- \int_r^{+\infty} \frac{1}{r^2} \exp(-\nu_\pi r) dr. \qquad (4.8.16)$$

Using (4.8.16), from (4.8.15), we derive

$$\beta = \frac{A_0}{r}\exp(-\nu_\pi r). \tag{4.8.17}$$

From (4.8.1) and (4.8.17), we come to the equality

$$\beta = -\varphi. \tag{4.8.18}$$

Hence, formula (4.8.14) acquires the form

$$\dot\theta = \frac{D}{r^2}\exp(-\varphi/c^2), \quad D = \text{const.} \tag{4.8.19}$$

Consider the function $r = r(\theta)$ and put

$$\sigma(\theta) = \frac{1}{r(\theta)}. \tag{4.8.20}$$

Then, using (4.8.1) and (4.8.19), we find

$$\varphi = -A_0\sigma(\theta)\exp(-\nu_\pi/\sigma(\theta)), \tag{4.8.21}$$

$$\dot\theta = D\sigma^2(\theta)\exp\left(-\varphi(\theta)/c^2\right), \quad \dot r = \frac{dr}{d\theta}\dot\theta = -D\sigma'(\theta)\exp\left(-\varphi(\theta)/c^2\right), \tag{4.8.22}$$

$$\ddot r = -D^2\sigma^2(\theta)\exp\left(-\varphi(\theta)/c^2\right)\frac{d}{d\theta}\left(\sigma'(\theta)\exp\left(-\varphi(\theta)/c^2\right)\right)$$

$$= -D^2\sigma^2(\theta)\exp\left(-2\varphi(\theta)/c^2\right)[\sigma''(\theta) + (A_0/c^2)\sigma'^2(\theta)$$

$$\times (1 + \nu_\pi/\sigma(\theta))\exp(-\nu_\pi/\sigma(\theta))]. \tag{4.8.23}$$

Substituting formulas (4.8.21)–(4.8.23) into Eq. (4.8.11), we obtain

$$\sigma'' + \sigma - \frac{A_0}{D^2}(1 + \nu_\pi/\sigma)\exp(-\nu_\pi/\sigma + 2\varphi/c^2)$$

$$+ \frac{Ze_pq}{m_pD^2}\exp(\varphi/c^2)\sqrt{1 + (D^2/c^2)\exp(-2\varphi/c^2)(\sigma'^2 + \sigma^2)} = 0, \tag{4.8.24}$$

where $\sigma = \sigma(\theta), \sigma' = d\sigma/d\theta, \sigma'' = d^2\sigma/d\theta^2$ and $\varphi = \varphi(\theta)$ is determined by formula (4.8.21).

Let us introduce the dimensionless function u and parameters a, b and s:

$$u = \frac{\sigma}{\nu_\pi} = \frac{r_\pi}{r}, r_\pi = \frac{1}{\nu_\pi} = \frac{\hbar}{m_\pi c}, a = \frac{A_0}{\nu_\pi D^2}, b = \frac{Ze_pq}{\nu_\pi m_pD^2}, s = \frac{\nu_\pi A_0}{c^2}, \tag{4.8.25}$$

where r_π is the Compton radius of the neutral pion and $q = e_p$ for the proton, $q = -e_p$ for the antiproton and $q = 0$ for the neutron.

Then, from (4.8.21) and (4.8.24), we derive the following equation for the function $u = u(\theta)$:

$$u'' + u + b \exp(-su \exp(-1/u))$$
$$\times \sqrt{1 + (s/a)(u'^2 + u^2) \exp(2su \exp(-1/u))}$$
$$= a(1 + 1/u) \exp(-1/u - 2su \exp(-1/u)), \qquad (4.8.26)$$

where $b > 0$ for the proton, $b = 0$ for the neutron and $b < 0$ for the antiproton.

Equation (4.8.26) describes trajectories of nucleons or antinucleons moving near heavy atomic nuclei at rest.

Consider now solutions of Eq. (4.8.26) corresponding to movements of nucleons in circular orbits around nuclei. Then, we have $u = u_0 = \text{const}$ and u_0 satisfies the equality

$$u_0 + b \exp \left(-su_0 \exp(-1/u_0)\right)$$
$$\times \sqrt{1 + (s/a)u_0^2 \exp \left(2su_0 \exp(-1/u_0)\right)}$$
$$= a(1 + 1/u_0) \exp \left(-1/u_0 - 2su_0 \exp(-1/u_0)\right). \qquad (4.8.27)$$

In this case, from (4.8.21), (4.8.22) and (4.8.25), we find that for the angular velocity of rotation $\dot\theta = \omega_0 = \text{const}$ of a nucleon:

$$\omega_0 = D\mu^2 u_0^2 \exp \left(su_0 \exp(-1/u_0)\right). \qquad (4.8.28)$$

Consider now the case in which $u_0 \ll 1$. Then, from (4.8.27), we obtain the following approximate equation for the value u_0:

$$u_0 + b = (a/u_0) \exp(-1/u_0). \qquad (4.8.29)$$

The obtained formulas (4.8.27)–(4.8.29) describe some quasi-nuclei in which nucleons or antinucleons move in circular orbits around heavy atomic nuclei.

As shown by numerical studies of the nonlinear differential equation (4.8.26), there are wide regions of its parameters for which antiprotons can move in closed orbits around heavy atomic nuclei. In these cases, specific quasi-nuclei could be created.

4.9. Massive Neutron Stars under the Action of Gravitational and Nuclear Forces

The problem of the equilibrium of neutron stars was considered in the classical work by Oppenheimer and Volkoff [44]. They concluded that there are no stationary solutions to the equations of general relativity corresponding to cold neutron stars with a mass greater than $\sim 0.7 M_S$, where M_S is the mass of the Sun. However, this study cannot be considered complete, since the nuclear forces in neutron stars were not taken into account in it. At the same time, nuclear forces at high potentials can exhibit a repulsive character [37] and, therefore, can exert a significant counteraction to gravitational forces in cooled massive neutron stars with a high density of matter.

To eliminate this gap in the study of the problem of the equilibrium of cold massive neutron stars, we turn to the theory of nuclear forces proposed in the previous sections and generalize it to the case of strong gravitational fields.

Equation (4.2.25) obtained above for the potential φ of nuclear forces in an inertial frame of reference can be rewritten in a covariant form in an arbitrary coordinate system x^μ as [13]

$$(-g)^{-1/2}\partial_\mu[(-g)^{1/2}\partial^\mu\varphi] + (m_\pi c/\hbar)^2\varphi = -4\pi(G/m_p)^2\rho(\varphi),$$
(4.9.1)

$$g = \det(g_{\mu\nu}), \quad \rho(\varphi) = \rho_0\exp(\varphi/c^2),$$
(4.9.2)

where $g_{\mu\nu}$ are components of the metric tensor, $\rho(\varphi)$ is the density of the rest mass of nuclear matter in a local inertial frame of reference, $\rho_0 = \rho(0)$, m_π and m_p are the rest masses of the neutral pion and proton, respectively, and $G^2/(\hbar c) = 0.080$.

As shown in Section 4.2, the energy–momentum tensors of an uncharged dust-like matter and its own nuclear field $T_m^{\mu\nu}$ and $T_{\rm nuc}^{\mu\nu}$, respectively, and the full energy–momentum tensor $T^{\mu\nu}$ are as follows:

$$T_m^{\mu\nu} = c^2\rho_0\exp(\varphi/c^2)dx^\mu/ds\,dx^\nu/ds,$$

$$T_{\rm nuc}^{\mu\nu} = (m_p^2/4\pi G^2)\left[\left(g^{\mu\alpha}g^{\nu\beta} - \frac{1}{2}g^{\mu\nu}g^{\alpha\beta}\right)\partial_\alpha\varphi\partial_\beta\varphi\right.$$

$$\left. + \frac{1}{2}g^{\mu\nu}(m_\pi c/\hbar)^2\varphi^2\right],$$
(4.9.3)

$$T^{\mu\nu} = T_m^{\mu\nu} + T_{\rm nuc}^{\mu\nu}, \quad \nabla_\mu T^{\mu\nu} = 0, \quad ds^2 = g_{\mu\nu}dx^\mu dx^\nu, \quad (4.9.4)$$

where ∇_μ is the covariant derivative, which coincides with the usual derivative in a local inertial frame of reference. It should be noted that for two arbitrary differentiable tensor functions A and B, the following formula takes place [13]: $\nabla_\mu(AB) = A\nabla_\mu B + B\nabla_\mu A$. This formula is a consequence of its general covariance and obvious validity in local inertial frames of reference.

The expression in (4.9.3) for the energy–momentum tensor of dust-like matter $T_m^{\mu\nu}$ corresponds to the case in which the pressures and heat flows are absent.

Let us now determine this tensor when the pressures and heat flows are nonzero. For the considered matter in a comoving inertial frame of reference, we have the following expressions for components $T_m^{\mu\nu}$ of its energy–momentum tensor:

$$T_m^{00} = c^2\rho_0 e^{\varphi/c^2} + u, \quad T_m^{0k} = q_0^k/c, \quad T_m^{kl} = p\delta_{kl},$$
$$k, l = 1, 2, 3, dx^k/ds = 0, \tag{4.9.5}$$
$$ds^2 = (dx^0)^2 - (dx^1)^2 - (dx^2)^2 - (dx^3)^2,$$

where δ_{kl} is the Kronecker symbol, p is the pressure in the matter, q_0^k is the three-dimensional vector of heat flow and u is the density of the internal energy of the matter associated with its thermal motion.

It is easy to see that the energy–momentum tensor of matter $T_m^{\mu\nu}$, which satisfies (4.9.5), has the following form in an arbitrary coordinate system x^μ:

$$T_m^{\mu\nu} = (c^2\rho_0 e^{\varphi/c^2} + u + p)dx^\mu/ds\,dx^\nu/ds - pg^{\mu\nu}$$
$$+ (q^\mu dx^\nu/ds + q^\nu dx^\mu/ds)/c, \tag{4.9.6}$$

where q^μ is the four-dimensional vector of heat flow having the following components in a comoving local inertial frame of reference:

$$q^0 = 0, \quad q^k = q_0^k, \quad k = 1, 2, 3, \quad dx^k/ds = 0,$$
$$ds^2 = (dx^0)^2 - (dx^1)^2 - (dx^2)^2 - (dx^3)^2. \tag{4.9.7}$$

Let us calculate the covariant derivative $\nabla_\mu T_m^{\mu\nu}$, taking into account the differential equation

$$\nabla_\mu(\rho_0 dx^\mu/ds) = 0, \tag{4.9.8}$$

which is a generalization of Eq. (4.2.7), expressing the conservation of matter rest mass when $\varphi = 0$, for arbitrary coordinate systems.

From (4.9.6), (4.9.8) and the identity $\nabla_\mu \psi \equiv \partial_\mu \psi$ for an arbitrary differentiable scalar function ψ, we find

$$\nabla_\mu T_m^{\mu\nu} = \{e^{\phi/c^2}\partial_\mu\varphi + \partial_\mu[(u+p)/\rho_0]\}\rho_0 dx^\mu/ds dx^\nu/ds$$
$$+ (c^2\rho_0 e^{\varphi/c^2} + u + p)\nabla_\mu(dx^\nu/ds)dx^\mu/ds - \partial p/\partial x_\nu$$
$$+ [\nabla_\mu(q^\nu/\rho_0)\rho_0 dx^\mu/ds + \nabla_\mu q^\mu dx^\nu/ds$$
$$+ q^\mu \nabla_\mu(dx^\nu/ds)]/c. \tag{4.9.9}$$

Let us use the well-known equality [13]

$$\nabla_\mu(dx^\nu/ds)dx^\mu/ds = d^2x^\nu/ds^2 + \Gamma^\nu_{\alpha\beta}dx^\alpha/ds dx^\beta/ds, \tag{4.9.10}$$

$$\Gamma^\nu_{\alpha\beta} = \frac{1}{2}g^{\nu\gamma}(\partial_\alpha g_{\beta\gamma} + \partial_\beta g_{\alpha\gamma} - \partial_\gamma g_{\alpha\beta}),$$

where $\Gamma^\nu_{\alpha\beta}$ are the Christoffel symbols. Then, from (4.9.9), we obtain

$$\nabla_\mu T_m^{\mu\nu} = (c^2\rho_0 e^{\varphi/c^2} + u + p)(d^2x^\nu/ds^2 + \Gamma^\nu_{\alpha\beta}dx^\alpha/ds dx^\beta/ds)$$
$$+ \rho_0 e^{\varphi/c^2}d\varphi/ds dx^\nu/ds + d[(u+p)/\rho_0]/ds \rho_0 dx^\nu/ds$$
$$- \partial p/\partial x_\nu + [\nabla_\mu(q^\nu/\rho_0)\rho_0 dx^\mu/ds + \nabla_\mu q^\mu dx^\nu/ds$$
$$+ q^\mu \nabla_\mu(dx^\nu/ds)]/c. \tag{4.9.11}$$

For the energy–momentum tensor of the nuclear field $T_{nuc}^{\mu\nu}$, we have

$$\nabla_\mu T_{nuc}^{\mu\nu} = -\rho_0 \exp(\varphi/c^2)\partial\varphi/\partial x_\nu. \tag{4.9.12}$$

In a local inertial frame of reference, formula (4.9.12) follows from Eqs. (4.2.10), (4.2.24) and (4.2.25), obtained above. Since this formula has a tensor form, it is true in an arbitrary coordinate system.

From (4.9.4), (4.9.11) and (4.9.12), we obtain the dynamic equations

$$\nabla_\mu T^{\mu\nu} = (c^2\rho_0 e^{\varphi/c^2} + u + p)(d^2x^\nu/ds^2 + \Gamma^\nu_{\alpha\beta}dx^\alpha/ds dx^\beta/ds)$$
$$+ \rho_0 e^{\varphi/c^2}(d\varphi/ds dx^\nu/ds - \partial\varphi/\partial x_\nu) + [d(u+p)/ds$$
$$- (1/\rho_0)(u+p)d\rho_0/ds]dx^\nu/ds - \partial p/\partial x_\nu + (1/c)$$
$$\times [\nabla_\mu(q^\nu/\rho_0)\rho_0 dx^\mu/ds + \nabla_\mu q^\mu dx^\nu/ds + q^\mu \nabla_\mu(dx^\nu/ds)]$$
$$= 0. \tag{4.9.13}$$

Let us use the equalities

$$(d^2x^\nu/ds^2 + \Gamma^\nu_{\alpha\beta}dx^\alpha/dsdx^\beta/ds)dx_\nu/ds = 0,$$

$$\nabla_\mu(dx^\nu/ds)dx_\nu/ds = 0. \tag{4.9.14}$$

In a local inertial frame of reference, their validity follows from the formulas

$$\frac{dx_\nu}{ds}\frac{d^2x^\nu}{ds^2} = \frac{1}{2}\frac{d}{ds}\left(\frac{dx^\nu}{ds}\frac{dx_\nu}{ds}\right) = 0,$$

$$\frac{dx_\nu}{ds}\frac{\partial}{\partial x^\mu}\left(\frac{dx^\nu}{ds}\right) = \frac{1}{2}\frac{\partial}{\partial x^\mu}\left(\frac{dx^\nu}{ds}\frac{dx_\nu}{ds}\right) = 0, \tag{4.9.15}$$

where $ds^2 = (dx^0)^2 - (dx^1)^2 - (dx^2)^2 - (dx^3)^2$.

Taking into account (4.9.10), we find that formulas (4.9.14) have tensor forms. Therefore, they are true in arbitrary coordinate systems.

Multiplying Eq. (4.9.13) by dx_ν/ds, summing it over the index ν and using formulas (4.9.14), we derive

$$du/ds - (1/\rho_0)(u + p)d\rho_0/ds$$

$$+[\nabla_\mu q^\mu + \rho_0\nabla_\mu(q^\nu/\rho_0)dx^\mu/dsdx_\nu/ds]/c = 0. \tag{4.9.16}$$

To understand the physical sense of equality (4.9.16), let us choose a local inertial frame comoving with the considered matter. Then, in this coordinate system from (4.9.16) and (4.9.7), we find

$$d(u + 2q^0/c)/dt - (1/\rho_0)(u + p)d\rho_0/dt + \partial q^k/\partial x^k = 0, \tag{4.9.17}$$

$$dt = ds/c, \quad dx^k/ds = 0, \quad k = 1, 2, 3.$$

Let v_0 be a small three-dimensional volume in the comoving local inertial frame of reference. Then, we have

$$\rho_0 v_0 = dm_0 = \text{const}, \tag{4.9.18}$$

where dm_0 is a constant small rest mass.

From (4.9.18), we find

$$d(\rho_0 v_0)/dt = 0, \quad d\rho_0 = -(\rho_0/v_0)dv_0 \tag{4.9.19}$$

and from (4.9.17) and (4.9.19), we obtain

$$v_0 d(u + 2q^0/c) + (u + p)dv_0 + v_0 dt \partial q^k / \partial x^k = 0, k = 1, 2, 3.$$
(4.9.20)

Using (4.9.7), we can represent formula (4.9.20) as

$$d[(u + 2q^0/c)v_0] = -pdv_0 - v_0 dt \partial q^k / \partial x^k.$$
(4.9.21)

Let us put

$$U = (u + 2q^0/c)v_0, \quad \delta A = -pdv_0,$$

$$\delta Q = -v_0 dt \partial q^k / \partial x^k = -dt \int_{v_0} \partial q^k / \partial x^k dv_0 = -dt \int_{s_0} q_n ds_0.$$
(4.9.22)

Here, as follows from (4.9.6), U is the internal energy of the small volume v_0, δA is the work done by the pressure p, δQ is the heat entering the volume v_0 through its surface s_0 for the small time dt and q_n is the projection of the vector q^k onto the outer normal of the surface s_0. It should be noted that in (4.9.22), the well-known Gauss theorem has been used.

Applying now formulas (4.9.21) and (4.9.22), we come to the first law of thermodynamics

$$dU = \delta A + \delta Q.$$
(4.9.23)

This thermodynamic law just presents the physical sense of equality (4.9.16).

As is known, the thermodynamic law (4.9.23) cannot be deduced from Newtonian mechanics. Hence, it is worth noting that, as shown above, this law is just derived within the framework of relativistic physics and is a consequence of its differential equations (4.9.4) and (4.9.8). These equations present the differential laws of conservation of energy, momentum and mass at rest in the theory of relativity.

Let us now turn to a new investigation of massive neutron stars by using the proposed nonlinear theory of nuclear forces and apply the Einstein gravitational equations to them, which have the form [13]

$$R_\beta^\alpha - \frac{1}{2}Rg_\beta^\alpha = \kappa T_\beta^\alpha,$$
(4.9.24)

where g_β^α coincides with the Kronecker symbol, R_β^α are the components of the Ricci tensor, $R = R_\alpha^\alpha$, $R_{\alpha\beta} = R_{\beta\alpha} = \partial_\gamma\Gamma_{\alpha\beta}^\gamma - \partial_\beta\Gamma_{\alpha\gamma}^\gamma + \Gamma_{\alpha\beta}^\gamma\Gamma_{\gamma\delta}^\delta - \Gamma_{\alpha\gamma}^\delta\Gamma_{\beta\delta}^\gamma$, T_β^α are the components of the energy–momentum tensor of matter and nongravitational fields, $\kappa = 8\pi f_N/c^4$ and f_N is the Newton gravitational constant.

Consider the stationary and spherically symmetric state of a neutron star. Then, the interval ds can be represented in the form [13]

$$ds^2 = e^\nu(dx^0)^2 - r^2(d\theta^2 + \sin^2\theta d\varphi^2) - e^\lambda dr^2, \qquad (4.9.25)$$

where $\nu = \nu(r)$, $\lambda = \lambda(r)$ and r, θ, ϕ are spherical coordinates.

In this case, the Einstein equations (4.9.24) reduce to the following three equations [13]:

$$\kappa T_0^0 = e^{-\lambda}(\lambda'/r - 1/r^2) + 1/r^2,$$
$$\kappa T_1^1 = -e^{-\lambda}(\nu'/r + 1/r^2) + 1/r^2, \qquad (4.9.26)$$
$$\nabla_\alpha T_1^\alpha = 0 = (T_1^1)' + (2/r)(T_1^1 - T_2^2) + (\nu'/2)(T_1^1 - T_0^0),$$

where $x^1 = r, x^2 = \theta, x^3 = \phi$.

For the other components T_β^α, we have

$$T_3^3 = T_2^2, \quad T_\beta^\alpha = 0, \quad \alpha \neq \beta. \qquad (4.9.27)$$

In the stationary case under consideration, taking into account (4.9.27) and hence that $T_k^0 = 0, k \neq 0$, from formula (4.9.6), we get

$$q^k = 0, dx^k/ds = 0, \quad k = 1, 2, 3. \qquad (4.9.28)$$

Using now formulas (4.9.3), (4.9.6), (4.9.25) and (4.9.28), we find

$$
\begin{aligned}
T_0^0 &= c^2\rho_0 e^{\varphi/c^2} + u + 2e^{\nu/2}q^0/c + (m_p^2/(8\pi G^2)) \\
&\quad \times [e^{-\lambda}\varphi'^2 + (m_\pi c\varphi/\hbar)^2], \\
T_1^1 &= -p - (m_p^2/(8\pi G^2))[e^{-\lambda}\varphi'^2 - (m_\pi c\varphi/\hbar)^2], \\
T_2^2 &= T_3^3 = -p + (m_p^2/(8\pi G^2))[e^{-\lambda}\varphi'^2 + (m_\pi c\varphi/\hbar)^2].
\end{aligned}
\qquad (4.9.29)
$$

In the considered case, Eq. (4.9.1) for the nuclear field potential φ acquires the form

$$
\begin{aligned}
\varphi'' + 2\varphi'[1/r + (\nu - \lambda)'/4] \\
= e^\lambda[(m_\pi c/\hbar)^2\varphi + 4\pi(G/m_p)^2\rho_0 e^{\varphi/c^2}].
\end{aligned}
\qquad (4.9.30)
$$

4.10. Equilibrium of Cooled Massive Neutron Stars

Consider now an arbitrary motion of a cooled massive neutron star with an absolute temperature $T \to 0$. In this case, its nuclear forces play an important role since the star must shrink significantly when its temperature becomes extremely low.

For such a state of the star, we have

$$\lim_{T \to 0} u = 0, \lim_{T \to 0} q^\alpha = 0, \alpha = 0, 1, 2, 3. \qquad (4.10.1)$$

Equalities (4.10.1) follow from the fact that the internal energy and heat flows of a matter, which are described by the density u and vector q^α and associated with thermal motion, are absent when the absolute temperature T is equal to zero.

From (4.9.16) and (4.10.1), we easily find that also

$$\lim_{T \to 0} p = 0. \qquad (4.10.2)$$

Let us return to the stationary spherically symmetric case. Then, when the absolute temperature is very close to zero, from (4.9.29), (4.10.1), (4.10.2) and (4.9.26), we obtain

$$T_0^0 = c^2 \rho_0 e^{\varphi/c^2} + (m_p^2/(8\pi G^2))[e^{-\lambda}\varphi'^2 + (m_\pi c\varphi/\hbar)^2],$$
$$T_1^1 = (m_p^2/(8\pi G^2))[(m_\pi c\varphi/\hbar)^2 - e^{-\lambda}\varphi'^2], \qquad (4.10.3)$$
$$T_2^2 = T_3^3 = (m_p^2/(8\pi G^2))[e^{-\lambda}\varphi'^2 + (m_\pi c\varphi/\hbar)^2],$$

$$\nabla_\alpha T_1^\alpha = -c^2 \rho_0 e^{\varphi/c^2}(\nu'/2 + \varphi'/c^2) = 0, \quad 0 \le r \le r_0, \qquad (4.10.4)$$

where $dx^k/ds = 0$, $k = 1, 2, 3$, $x^1 = r$, $x^2 = \theta$, $x^3 = \phi$ and r_0 is the radius of the cooled neutron star.

From (4.10.4), we find

$$\varphi/c^2 = (\nu_0 - \nu)/2, \nu_0 = \text{const}, 0 \le r \le r_0. \qquad (4.10.5)$$

Equations (4.9.2), (4.9.30) and (4.10.5) give the following formulas for the mass density $\rho(\varphi)$:

$$\rho = \rho_0 e^{\varphi/c^2} = \rho_0 e^{(\nu_0 - \nu)/2} = -(m_p^2 c^2/(8\pi G^2))\{(m_\pi c/\hbar)^2(\nu_0 - \nu)$$
$$+ e^{-\lambda}[\nu'' + \nu'(2/r + (\nu' - \lambda')/2)]\}, \quad 0 \le r \le r_0. \qquad (4.10.6)$$

From (4.9.26), (4.10.3), (4.10.5) and (4.10.6), we obtain

$$1 - e^{-\lambda}(1 - \lambda'r) = -(\kappa m_p^2 c^4 r^2/(8\pi G^2))\left[\left(\frac{1}{2}m_\pi c/\hbar\right)^2 (\nu_0 - \nu)\right.$$

$$\left. \times (4 + \nu - \nu_0) + e^{-\lambda}\left(\nu'' + 2\nu'/r - \frac{1}{2}\nu'\lambda' + \frac{1}{4}\nu'^2\right)\right],$$

(4.10.7)

$$1 - e^{-\lambda}(1 + \nu'r) = -(\kappa m_p^2 c^4 r^2/(32\pi G^2))$$

$$\times [e^{-\lambda}\nu'^2 - (m_\pi c/\hbar)^2(\nu_0 - \nu)^2], \quad 0 \le r \le r_0.$$

Consider $\lambda(r)$ and $\nu(r)$ at the point $r = 0$. As follows from (4.10.7),

$$\lambda(0) = 0. \tag{4.10.8}$$

Differentiating the two equations (4.10.7) at the point $r = 0$ and using (4.10.8), we get

$$2\lambda'(0) + (\kappa m_p^2 c^4/(4\pi G^2))\nu'(0) = 0, \quad \lambda'(0) - \nu'(0) = 0. \tag{4.10.9}$$

From equalities (4.10.8) and (4.10.9), we have

$$\lambda(0) = \lambda'(0) = \nu'(0) = 0. \tag{4.10.10}$$

Let us introduce the following variable x and functions $f(x)$ and $g(x)$, taking into account conditions (4.10.10):

$$x = (r/r_0)^2, \quad e^{-\lambda} = 1 - xf(x), \quad \nu - \nu_0 = g(x), 0 \le x \le 1$$

(4.10.11)

and put

$$\alpha = \kappa m_p^2 c^4/(8\pi G^2), \quad \beta = m_\pi c r_0/(2\hbar). \tag{4.10.12}$$

Then, Eqs. (4.10.7) can be represented in the form

$$2xf' + 3f = \alpha[4\beta^2 g - 2(1 - xf)(2xg'' + 3g' + xg'^2)$$

$$+ 2xg'(xf' + f) + (1 - xf)xg'^2 + \beta^2 g^2],$$

(4.10.13)

$$2(1 - xf)g' - f = \alpha[(1 - xf)xg'^2 - \beta^2 g^2],$$

$$f = f(x), g = g(x), 0 \le x \le 1.$$

Since

$$\beta = 0.5 r_0/r_\pi, \quad r_\pi = \hbar/(m_\pi c), \qquad (4.10.14)$$

where r_0 is the radius of the star and r_π is the Compton radius of the pion, the value β is very large:

$$\beta \gg 1. \qquad (4.10.15)$$

Taking this into account, we find that Eqs. (4.10.13) are practically coinciding with the equations

$$2xf' + 3f = \alpha\beta^2 g(4+g), \quad f - 2(1 - xf)g' = \alpha\beta^2 g^2. \qquad (4.10.16)$$

As follows from (4.10.11) and (4.10.16), to satisfy conditions (4.10.10), we should seek the functions $f(x)$ and $g(x)$ that are bounded at the point $x = 0$. For such functions, we obtain the following condition by putting $x = 0$ in the first equation of (4.10.16):

$$f(0) = \alpha\beta^2 g(0)[4 + g(0)]/3. \qquad (4.10.17)$$

Consider now Eqs. (4.9.26), (4.9.30) and (4.10.3) when $r \geq r_0$.

As shown in Section 4.6, outside the neutron star with the radius r_0, the density ρ_0 can be represented in the form

$$\rho_0 = 2\sigma\delta(\overline{r - r_0}), r > r_0, \qquad (4.10.18)$$

where $\delta(x)$ denotes the even delta function, $\sigma = \mathrm{const}$ and $\overline{r - r_0}$ is the distance between the point with the coordinates (r, θ, ϕ) and the sphere $r = r_0$, determined in the comoving local inertial frame of reference that is chosen in a small vicinity of this sphere.

Formula (4.10.18) expresses the density of virtual pions generated in the physical vacuum near the surface $r = r_0$ because of its influence on the vacuum.

The value of the constant σ, which is determined in Section 4.7, is given by formulas

$$\sigma = 0.04607/s, \quad s = 4\pi\hbar G^2/(m_p^2 m_\pi c^3), G^2/(\hbar c) = 0.080. \qquad (4.10.19)$$

When $d\theta = d\phi = 0$, from (4.9.25), we have

$$ds^2 = e^\nu (dx^0)^2 - e^\lambda dr^2 = (d\overline{x}^0)^2 - d\overline{r}^2, \qquad (4.10.20)$$

where $\overline{x}^0, \overline{r}$ are the corresponding coordinates in a comoving local inertial frame of reference.

For the comoving coordinate system $\bar{x}^0, \bar{r}$, from (4.10.20), we readily obtain

$$d\bar{x}^0 = e^{\nu/2}dx^0, d\bar{r} = e^{\lambda/2}dr. \qquad (4.10.21)$$

From (4.10.18) and (4.10.21), we find

$$\rho_0 = 2\sigma\delta[e^{\lambda/2}(r - r_0)] = 2\sigma e^{-\lambda/2}\delta(r - r_0), r > r_0. \qquad (4.10.22)$$

Formulas (4.9.30) and (4.10.22) give the following equation for the nuclear field potential φ outside the star:

$$\varphi'' + 2a\varphi'/r - b^2\varphi = w, \quad \varphi = \varphi(r), \quad r \geq r_0, \qquad (4.10.23)$$

where

$$a = 1 + (\nu - \lambda)'r/4, \quad b = e^{\lambda/2}m_\pi c/\hbar,$$

$$w = 8\pi(G/m_p)^2\sigma e^{\lambda/2+\varphi/c^2}\delta(r - r_0), \quad \int_0^\infty \delta(x)dx = 1/2. \qquad (4.10.24)$$

Equation (4.10.23) can be rewritten in the form

$$\varphi''(z) + 2a(z)\varphi'(z)/z - 4\beta^2 e^{\lambda(z)}\varphi(z) = r_0^2 w(z),$$
$$z = r/r_0, z \geq 1, \qquad (4.10.25)$$

where β is determined by formula (4.10.14), $\beta \gg 1$, and

$$a(z) = 1 + [\nu'(z) - \lambda'(z)]z/4. \qquad (4.10.26)$$

To solve Eq. (4.10.25), let us introduce two functions $p(z)$ and $q(z)$, satisfying the equations

$$p + q = 2a(z)/z, \quad p' + pq = -4\beta^2 e^{\lambda(z)}. \qquad (4.10.27)$$

Consider the function

$$y(z) = \varphi'(z) + p(z)\varphi(z). \qquad (4.10.28)$$

Then, from (4.10.27) and (4.10.28), we find

$$y' + qy = \varphi'' + (p + q)\varphi' + (p' + pq)\varphi = \varphi'' + 2a\varphi'/z - 4\beta^2 e^\lambda\varphi$$
$$(4.10.29)$$

and from (4.10.25), we obtain

$$y' + qy = r_0^2 w, \ z \geq 1. \tag{4.10.30}$$

Thus, the differential equation of second order (4.10.25) is equivalent to the differential equations of first order (4.10.27), (4.10.28) and (4.10.30).

From (4.10.27), we get

$$(p_0' + 2ap_0/z)/\beta - p_0^2 = -4e^\lambda, \quad p_0 = p/\beta. \tag{4.10.31}$$

Since the value β, determined by formula (4.10.14), is very large, we find the following approximate solution to Eq. (4.10.31), which is practically coinciding with its exact solution:

$$p_0^2 \approx 4e^\lambda, \quad p_0 = p/\beta, \quad \beta \gg 1. \tag{4.10.32}$$

As follows from (4.10.27) and (4.10.32), we can put

$$p \approx 2\beta e^{\lambda/2}, \quad q \approx 2(a/z - \beta e^{\lambda/2}) \approx -2\beta e^{\lambda/2}. \tag{4.10.33}$$

From (4.10.28), (4.10.30), (4.10.33) and (4.10.24), we readily find the following solutions $\phi(z)$ and $y(z)$, tending to zero at infinity:

$$\varphi(z) = \exp\left(-\int_1^z p\,dz\right)\left(\int_\infty^z y \exp\left(\int_1^z p\,dz\right)dz + D\right), \quad D = \text{const},$$

$$y(z) = r_0^2 \exp\left(-\int_1^z q\,dz\right)\int_\infty^z w \exp\left(\int_1^z q\,dz\right)dz, \tag{4.10.34}$$

where $\varphi(\infty) = y(\infty) = 0, z \geq 1$.

Formulas (4.10.28) and (4.10.34) give

$$\varphi'(1) + p(1)\varphi(1) = y(1) = -r_0^2\int_1^\infty w \exp\left(\int_1^z q\,dz\right)dz, \tag{4.10.35}$$

where $\varphi = \varphi(z), z = r/r_0$.

From (4.10.24), (4.10.33) and (4.10.35), we obtain

$$\varphi'(1) + 2\beta e^{\lambda(1)/2}\varphi(1) = -4\pi r_0\sigma(G/m_p)^2 e^{\lambda(1)/2+\varphi(1)/c^2},$$
$$\varphi = \varphi(z), \quad \lambda = \lambda(z). \tag{4.10.36}$$

Returning again to the variable $x = (r/r_0)^2 = z^2$, from (4.10.5), (4.10.11) and (4.10.36), we come to the following condition at the point $x = 1(r = r_0)$:

$$g'(1) + \beta g(1)(1 - f(1))^{-1/2}$$
$$= 4\pi r_0 \sigma (G/(m_p c))^2 (1 - f(1))^{-1/2} e^{-g(1)/2},$$
$$g = g(x), \quad f = f(x), \quad x = (r/r_0)^2. \tag{4.10.37}$$

Since β is very large, from (4.10.37) and (4.10.14), we obtain

$$g(1)e^{g(1)/2} \approx 4\pi r_0 \sigma G^2/(m_p^2 c^2 \beta) = 8\pi \sigma \hbar G^2/(m_p^2 m_\pi c^3), \beta \gg 1. \tag{4.10.38}$$

Formulas (4.10.19) and (4.10.38) give

$$g(1)e^{g(1)/2} = 0.09214. \tag{4.10.39}$$

This equation has the solution

$$g(1) = 0.08817. \tag{4.10.40}$$

Consider now the functions $\lambda(r)$ and $\nu(r)$ when $r \geq r_0$. From (4.9.26), we obtain

$$e^{-\lambda} = 1 - r_g/r - (\kappa/r) \int_\infty^r r^2 T_0^0 dr, \nu = \int_\infty^r [(e^\lambda - 1)/r - \kappa r e^\lambda T_1^1] dr. \tag{4.10.41}$$

The Einstein gravitational constant κ and the gravitational radius r_g, which are contained in (4.10.41), are determined by the expressions [13]

$$\kappa = 8\pi f_N/c^4, \quad r_g = 2f_N M/c^2. \tag{4.10.42}$$

where f_N is the Newtonian gravitational constant and M is the mass of the considered neutron star.

When $r > r_0$, from (4.10.24), (4.10.33) and (4.10.34), we find

$$w(z) = 0, \quad y(z) = 0, \quad \varphi(z) = D \exp\left(-2\beta \int_1^z e^{\lambda/2} dz\right),$$
$$z = r/r_0, z > 1. \tag{4.10.43}$$

From (4.10.14) and (4.10.43), we obtain

$$\varphi'(r) = -(m_\pi c/\hbar)e^{\lambda/2}\varphi(r), r > r_0, \qquad (4.10.44)$$

and from (4.10.3), (4.10.12), (4.10.22) and (4.10.44), it follows that

$$T_0^0 = 2c^2\sigma e^{\varphi/c^2-\lambda/2}\delta(r-r_0) + 8\varphi^2/(\varepsilon\kappa r_0^2 c^4), T_1^1 = 0, \quad r > r_0, \qquad (4.10.45)$$

where

$$\varepsilon = 1/(\alpha\beta^2). \qquad (4.10.46)$$

From (4.10.12), (4.10.42) and (4.10.46), we obtain

$$\varepsilon^{-1} = f_N(m_p m_\pi c r_0)^2/(2\hbar G)^2 = (r_0/r_g)^2(f_N^{3/2} m_p m_\pi M)^2/(c\hbar G)^2. \qquad (4.10.47)$$

This formula can be represented in the form

$$\varepsilon^{-1/2} = \chi(M/M_S)(r_0/r_g), \quad \chi = f_N^{3/2} m_p m_\pi M_S/(c\hbar G), \qquad (4.10.48)$$

where f_N is the Newtonian gravitational constant, M_S is the mass of the Sun and χ is a dimensionless constant.

Using formula (4.10.19) for the constant G and the well-known values of the constants $f_N, m_p, m_\pi, c, \hbar, M_S$, it is easy to calculate the constant χ. Its value is as follows:

$$\chi = 0.2747. \qquad (4.10.49)$$

Thus, from (4.10.48) and (4.10.49), we find

$$\varepsilon^{-1/2} = 0.2747(M/M_S)(r_0/r_g). \qquad (4.10.50)$$

When $r \geq r_0$, formulas (4.10.41) and (4.10.45) give

$$e^{-\lambda} = 1 - r_g/r + \kappa c^2 r_0\sigma e^{\varphi/c^2-\lambda/2}N(r-r_0)$$
$$+ 8(\varepsilon c^4 r_0^2 r)^{-1}\int_r^\infty r^2\varphi^2 dr, \quad r \geq r_0, \qquad (4.10.51)$$

where

$$N(0) = 1, \quad N(r-r_0) = 0, r > r_0. \qquad (4.10.52)$$

Since β is very large, from (4.10.43), we obtain

$$\varphi(r) \approx 0, r > r_0, \beta \gg 1. \tag{4.10.53}$$

Therefore, from (4.10.51) and (4.10.53), we find

$$e^{-\lambda} = 1 - r_g/r + \kappa c^2 r_0 \sigma e^{\varphi/c^2 - \lambda/2} N(r - r_0), r \geq r_0. \tag{4.10.54}$$

From (4.10.12), (4.10.19) and (4.10.46), we have

$$\kappa c^2 r_0 \sigma = 0.04607 m_p^2 m_\pi c^5 r_0 \kappa/(4\pi \hbar G^2) = 0.18428 \alpha \beta = 0.18428/(\beta \varepsilon), \tag{4.10.55}$$

where ε is determined by (4.10.50).

Taking again into account that β is very large, from (4.10.54) and (4.10.55), we derive

$$e^{-\lambda} \approx 1 - r_g/r, r \geq r_0, \beta \gg 1. \tag{4.10.56}$$

From (4.10.41), (4.10.45) and (4.10.56), it follows that

$$\nu = - \int_r^\infty (e^\lambda - 1)/r dr \approx \ln(1 - r_g/r), r \geq r_0. \tag{4.10.57}$$

Thus, from (4.9.25), (4.10.56) and (4.10.57), we obtain the Schwarzschild interval ds [13] outside the neutron star.

When $r = r_0$, formulas (4.10.11), (4.10.40), (4.10.56) and (4.10.57) give

$$r_g/r_0 = f(1), \quad \nu_0 = \ln(1 - r_g/r_0) - 0.08817. \tag{4.10.58}$$

Let us put

$$h(x) = \varepsilon f(x). \tag{4.10.59}$$

For the functions $h(x)$ and $g(x)$, from (4.10.16), (4.10.17), (4.10.40), (4.10.46) and (4.10.59), we obtain the following equations:

$$2xh'(x) + 3h(x) = g(x)(4 + g(x)),$$
$$2(\varepsilon - xh(x))g'(x) = h(x) - g^2(x), \quad 0 \leq x \leq 1, \tag{4.10.60}$$

$$g(0) = k, \quad h(0) = k(4+k)/3, \quad g(1) = 0.08817, \tag{4.10.61}$$

where k is some constant dependent on ε. To determine k, we should use the third equality of (4.10.61) for $g(1)$.

The obtained equations (4.10.60) and (4.10.61) can be numerically solved for different values of the parameter ε.

Thus, the considered problem is reduced to Eqs. (4.10.60) and (4.10.61). After obtaining numerical solutions to these equations, we can find the functions $\lambda(r), \nu(r), \varphi(r)$ and $\rho(r)$ inside the neutron star by means of formulas (4.10.5), (4.10.6), (4.10.11), (4.10.58) and (4.10.59).

The values r_0/r_g and M/M_S can be determined from the following formulas, which are consequences of formulas (4.10.50), (4.10.58) and (4.10.59):

$$r_0/r_g = \varepsilon/h(1), \quad M/M_S = 3.640 h(1)/\varepsilon^{3/2}, \qquad (4.10.62)$$

where r_g and M are the gravitational radius and mass of the neutron star, respectively, and M_S is the mass of the Sun.

By means of our computer calculations, the numerical integration of Eqs. (4.10.60) and (4.10.61) is carried out for different values of

Table 4.2. Constants k, radii r_0 and masses M of cooled neutron stars for different values of the parameter ε.

ε	k	M/M_S	r_0/r_g
10.000	$8.2761 \cdot 10^{-2}$	0.0135	85.3876
4.000	$7.5382 \cdot 10^{-2}$	0.0514	35.4381
2.000	$6.4792 \cdot 10^{-2}$	0.1371	18.7693
1.000	$4.8618 \cdot 10^{-2}$	0.3502	10.3954
0.500	$2.8925 \cdot 10^{-2}$	0.8396	6.1308
0.200	$8.1870 \cdot 10^{-3}$	2.4008	3.3902
0.100	$1.6565 \cdot 10^{-3}$	4.9623	2.3196
0.070	$5.4233 \cdot 10^{-4}$	7.0683	1.9464
0.050	$1.5032 \cdot 10^{-4}$	9.7469	1.6702
0.040	$5.5626 \cdot 10^{-5}$	11.9765	1.5196
0.030	$1.2715 \cdot 10^{-5}$	15.4750	1.3580
0.025	$4.3945 \cdot 10^{-6}$	18.0941	1.2723
0.020	$1.0298 \cdot 10^{-6}$	21.7481	1.1835
0.017	$3.1911 \cdot 10^{-7}$	24.7172	1.1295
0.015	$1.2048 \cdot 10^{-7}$	27.1707	1.0939
0.013	$3.6440 \cdot 10^{-8}$	30.1303	1.0596
0.012	$1.7926 \cdot 10^{-8}$	31.8448	1.0435
0.011	$8.0070 \cdot 10^{-9}$	33.7434	1.0285
0.010	$3.1713 \cdot 10^{-9}$	35.8431	1.0155

the parameter ε. For each value of ε, the parameter k is chosen so as to satisfy the condition $g(1) = 0.08817$ in (4.10.61).

The results of the computer calculations of solutions to Eqs. (4.10.60) and (4.10.61) and then the values r_0/r_g and M/M_S are given in Table 4.2.

As follows from this table, Eqs. (4.10.60) and (4.10.61) have solutions for large values of the parameter M/M_S. This implies that the nuclear forces, owing to their property of saturation and repulsive character for sufficiently large values of the nuclear field potential, can balance the gravitational forces in cooled massive neutron stars.

Chapter 5

Relativistic Quantum Equations for Nucleons and Light Atomic Nuclei

In this chapter, a generalization of the Dirac equation is sought to describe the well-known quark structure and anomalous magnetic moments of protons and neutrons. For this purpose, two matrices are inserted into the Dirac equation, which reflect the quark structure of the charge and mass of a nucleon and are consistent with the fundamental principles of quantum mechanics. As a result, we arrive at a new quantum equation, describing a 12-component wave function of nucleons. Its consequence is three differential equations of conservation for the three quark charges of these particles. The proposed equation is applied to describe the interaction of nucleons with a magnetic field. It is shown that the anomalous magnetic moments of the proton and neutron arising from this equation are close to their experimental values. With the help of matrix representations of the Clifford algebra, a generalization of the obtained quantum equation for an individual nucleon is given, which makes it possible to describe the quark structure of light nuclei. A detailed study of the mathematical properties of this equation is carried out and, as a result, expressions are determined for quark currents in light nuclei that satisfy the differential equations of conservation of quark charges. The results obtained in our papers [45–48] are used in this chapter.

5.1.   Anomalous Properties of Nucleons

Let us turn to the description of the properties of nucleons. They are quite unusual. For example, neutrons having zero electrical charge interact with magnetic fields. The neutron has a magnetic moment of -1.91 nuclear magnetons. As for the proton, its magnetic moment is equal to $+2.79$ [49]. The existence of anomalous magnetic moments of nucleons shows that they have a complex structure. According to the well-known theory of quarks, the proton consists of two u-quarks and one d-quark, while the neutron has one u-quark and two d-quarks. The electric charge of the u-quark is $+2/3e_p$ and the charge of the d-quark is $-1/3e_p$, where e_p is the charge of the proton.

Quarks can exist only inside nucleons and other hadrons. As for the question of their nature, it still remains without a clear answer. Another question to be solved is how to describe the wave function of the nucleon.

As is well known, in the nonrelativistic case, the wave function of an elementary particle is described by the Schrödinger equation. However, in the relativistic case, the state of affairs with this question is not so favorable. For the relativistic electron, there is the Dirac equation, which gives a very good quantum description of it, while for the relativistic nucleons, there are no equally satisfactory quantum equations.

Indeed, trivial generalizations of the Dirac equation, in which the mass and charge of an electron would simply be replaced by the corresponding parameters of nucleons, do not allow us to give their correct description, since they would lead to incorrect magnetic moments of the proton and neutron. As for the well-known generalization of the Dirac equation, which introduces a nonminimal interaction of a nucleon with a magnetic field [49], it has serious shortcomings, namely, there is no reflection of the quark structure of nucleons in it. Moreover, the magnitudes of the anomalous magnetic moments of nucleons cannot be derived from such a generalization of the Dirac equation and they are simply inserted into this equation as coefficients, the values of which are known from experiments.

Thus, the problem arises of finding a new relativistic equation for nucleons, which would not have the defects indicated above. Such a new generalization of the Dirac equation is proposed in this chapter.

The main idea of our approach to the description of nucleons is as follows. As is known, the relativistic electron is described in Dirac's theory by means of four wave functions. As for quarks, they have spin 1/2, like an electron, and can also be described by four wave functions. Therefore, a nucleon consisting of three quarks can naturally be characterized by 12 wave functions.

Thus, our goal is to find a quantum equation for the wave function with 12 components that could correspond to a relativistic nucleon.

5.2.   Generalization of the Dirac Equation for the Description of Nucleons

Consider the well-known Dirac equation for the relativistic electron. It has the following form [49]:

$$[\gamma^\mu(i\hbar c\partial_\mu - eA_\mu) - m_e c^2]\psi = 0, \tag{5.2.1}$$

where ψ is the column consisting of four wave functions, e and m_e are the electron charge and rest mass, respectively, A_μ are the potentials of an external electromagnetic field, $\partial_\mu \equiv \partial/\partial x^\mu$, x^μ are space–time coordinates of the Minkowski geometry and γ^μ are the four 4×4 Dirac matrices satisfying the relation

$$\gamma^\mu\gamma^\nu + \gamma^\nu\gamma^\mu = 2g^{\mu\nu}I, \tag{5.2.2}$$

where $g^{\mu\nu}$ are components of the Minkowski metric tensor, μ, $\nu = 0, 1, 2, 3$ and I is the unit 4×4 matrix.

The Dirac matrices γ^μ have the following standard representation:

$$\gamma^0 = \begin{pmatrix} 1 & 0 & 0 & 0 \\ 0 & 1 & 0 & 0 \\ 0 & 0 & -1 & 0 \\ 0 & 0 & 0 & -1 \end{pmatrix}, \gamma^1 = \begin{pmatrix} 0 & 0 & 0 & 1 \\ 0 & 0 & 1 & 0 \\ 0 & -1 & 0 & 0 \\ -1 & 0 & 0 & 0 \end{pmatrix},$$

$$\gamma^2 = \begin{pmatrix} 0 & 0 & 0 & -i \\ 0 & 0 & i & 0 \\ 0 & i & 0 & 0 \\ -i & 0 & 0 & 0 \end{pmatrix}, \gamma^3 = \begin{pmatrix} 0 & 0 & 1 & 0 \\ 0 & 0 & 0 & -1 \\ -1 & 0 & 0 & 0 \\ 0 & 1 & 0 & 0 \end{pmatrix}. \tag{5.2.3}$$

In the case of a free electron when $A_\mu = 0$, after multiplying Eq. (5.2.1) by $(i\hbar\gamma^\mu\partial_\mu + m_e c)$ and using (5.2.2), we come to the

Klein–Gordon equation [49]

$$[\partial^\mu \partial_\mu + (m_e c/\hbar)^2]\psi = 0. \tag{5.2.4}$$

Paul Dirac
(1902–1984)

As is known, the Dirac equation of form (5.2.1), which gives a very good description of electrons, is inapplicable to nucleons. This is due to the fact that it cannot describe their quark structure and anomalous magnetic moments. That is why a modification of the Dirac equation was proposed, including a more complex, nonminimal interaction of nucleons with an electromagnetic field [49]. However, as mentioned above, the anomalous magnetic moments of nucleons cannot be derived theoretically from this equation. Their experimental values are inserted into it as given coefficients [49]. Moreover, this equation does not allow one to describe the quark structure of nucleons.

For these reasons, our goal is to find a new generalization of the Dirac equation for describing the relativistic nucleon, which would not have the above disadvantages. To do this, we will proceed from the following requirements for it:

(1) The desired equation should describe the quark structure of nucleons.
(2) The known experimental values of the anomalous magnetic moments of nucleons should be derived from this equation.

We will seek such a generalization of the Dirac equation in the following form:

$$[\Gamma^\mu(i\hbar c\partial_\mu - be_p A_\mu) - a\,mc^2 \exp(\varphi/c^2)]\Psi = 0, \tag{5.2.5}$$

where Ψ is a wave function having four components Ψ_k, each of them being a column consisting of three components $\Psi_{kl}(1 \leq k \leq 4, 1 \leq l \leq 3)$. These 12 components of the wave function Ψ are introduced to describe the quark structure of nucleons.

The symbols Γ^μ in (5.2.5) denote 4×4 matrices consisting of the elements Γ^μ_{ij}, which are 3×3 matrices of the form $\gamma^\mu_{ij} \cdot 1$, where γ^μ_{ij} are the elements of the Dirac matrices γ^μ and 1 is the unit 3×3

matrix, A_μ are the potentials of an external electromagnetic field, φ is the scalar nuclear field potential, e_p is the proton charge, m is the nucleon rest mass when $\varphi = 0$ and a, b are some 3×3 matrices, which will be determined later on.

The 12×12 matrices Γ^μ consist of the elements $\Gamma_{ij}^\mu = \gamma_{ij}^\mu \cdot 1$. Therefore, they satisfy the following relations of form (5.2.2):

$$\Gamma^\mu \Gamma^\nu + \Gamma^\nu \Gamma^\mu = 2g^{\mu\nu} I, \tag{5.2.6}$$

where I is the unit 12×12 matrix.

The product $a\Psi$ in Eq. (5.2.5) means the column consisting of the elements $a\Psi_k$, which are the product of the 3×3 matrix a and the column Ψ_k consisting of three elements. In the product $\Gamma^\mu b$, each element of the matrix Γ^μ, which is a 3×3 matrix, is multiplied by the 3×3 matrix b.

The scalar nuclear field potential φ is described by the following nonlinear generalization of the Yukawa equation proposed in the previous chapter:

$$\partial^\mu \partial_\mu \varphi + (m_\pi c / \hbar)^2 \varphi = -4\pi (G/m_p)^2 \rho(\varphi),$$
$$\rho(\varphi) = \rho(0) \exp(\varphi/c^2), \tag{5.2.7}$$

where m_π, m_p are the rest masses of the neutral pion and proton, respectively, $G^2/(\hbar c) = 0.080$, $\rho(\varphi)$ is the density of the nucleon rest mass, which, as shown in the previous chapter, exponentially depends on the potential φ. This exponential dependence has been taken into account in Eq. (5.2.5).

To have a differential equation of second order, which accords with the Klein–Gordon equation, and also the equation of charge conservation that follows from Eq. (5.2.5), we impose the following requirements to the choice of the matrices a and b:

$$a^+ = a, \; b^+ = b, \; ab = ba, \; a^2 = 1, \; 1 = \begin{pmatrix} 1 & 0 & 0 \\ 0 & 1 & 0 \\ 0 & 0 & 1 \end{pmatrix}, \tag{5.2.8}$$

where the sign '+' means the Hermitian conjugate.

Then multiplying Eq. (5.2.5) by

$$[\Gamma^\mu(i\hbar c\partial_\mu - b e_p A_\mu) + a m c^2 \exp(\varphi/c^2)]$$

and taking into account relations (5.2.6) and (5.2.8), we obtain the generalization of the Klein–Gordon equation of the form

$$[i\hbar c\partial^\mu - b\,e_p A^\mu)(i\hbar c\partial_\mu - b\,e_p A_\mu) - m^2 c^4 \exp(2\varphi/c^2)]\Psi = 0. \quad (5.2.9)$$

To come to the equation of charge conservation, let us multiply Eq. (5.2.5) by the matrix $\Gamma^0 k$, where k is a 3×3 matrix satisfying the equalities

$$k^+ = k, \quad ak = ka, \quad bk = kb. \quad (5.2.10)$$

Owing to (5.2.8), the following matrices k satisfy equalities (5.2.10): $k = b^l$, $k = ab^l$, $l = 0, 1, 2, \ldots$

Multiplying Eq. (5.2.5) by $\Gamma^0 k$ and taking into account relations (5.2.8), (5.2.10) and the well-known equalities $(\gamma^0\gamma^\mu)^+ = \gamma^0\gamma^\mu$, $(\gamma^0)^+ = \gamma^0$ [49], this equation and its Hermitian conjugate acquire the form

$$\Gamma^0 k[\Gamma^\mu(i\hbar c\partial_\mu - b\,e_p A_\mu) - a\,m c^2 \exp(\varphi/c^2)]\Psi = 0,$$
$$\Psi^+\Gamma^0 k[\Gamma^\mu(i\hbar c\partial_\mu + b\,e_p A_\mu) + a\,m c^2 \exp(\varphi/c^2)] = 0,$$
$$\Psi^+\partial_\mu \equiv \partial_\mu\Psi^+. \quad (5.2.11)$$

Multiplying the first equation in (5.2.11) on the left by Ψ^+ and the second equation in it on the right by Ψ and then adding them, we obtain a series of equations of conservation of the form

$$\partial_\mu(\bar{\Psi}\Gamma^\mu k\Psi) = 0, \quad (5.2.12)$$

where

$$\bar{\Psi} = \Psi^+\Gamma^0 \quad (5.2.13)$$

and k is any 3×3 matrix satisfying equalities (5.2.10).

As will be shown later on, the choice $k = b$ for the matrix k in Eq. (5.2.12) gives the equation of conservation of nucleons' charges.

Let us note the following important property of the proposed equation (5.2.5) for the description of nucleons: it is covariant under the Lorentz transformations as well as the Dirac equation. Owing to properties (5.2.8), the proof of this covariance of Eq. (5.2.5) is absolutely identical to that for the Dirac equation.

Consider now the energy–momentum tensor T^μ_ν of nucleons described by Eq. (5.2.5). As will be shown here, it has the following form analogous to the well-known expression for the Dirac equation:

$$T^\mu_\nu = \frac{i\hbar c}{2} \left\{ \bar\Psi \Gamma^\mu \left[\left(\partial_\nu + \frac{ie_p}{c\hbar} bA_\nu \right) \Psi \right] - \left[\bar\Psi \left(\partial_\nu - \frac{ie_p}{c\hbar} bA_\nu \right) \right] \Gamma^\mu \Psi \right\},$$

$$\bar\Psi \partial_\nu \equiv \partial_\nu \bar\Psi. \tag{5.2.14}$$

To verify this, let us calculate $\partial_\mu T^\mu_\nu$.

$$\partial_\mu T^\mu_\nu = i\hbar[c\partial_\mu \bar\Psi \Gamma^\mu \partial_\nu \Psi + c\bar\Psi \Gamma^\mu \partial_\nu \partial_\mu \Psi - \partial_\nu \partial_\mu (\bar\Psi \Gamma^\mu \Psi) c/2$$

$$+ i(A_\nu \partial_\mu J^\mu + J^\mu \partial_\mu A_\nu)/(c\hbar)], \tag{5.2.15}$$

where

$$J^\mu = e_p c \bar\Psi \Gamma^\mu b \Psi. \tag{5.2.16}$$

After substitution 1 and 1, b, instead of the matrix k, in Eqs. (5.2.11) and (5.2.12), respectively, we find

$$i\hbar c \partial_\mu \bar\Psi \Gamma^\mu = -\bar\Psi[\Gamma^\mu b\, e_p A_\mu + a\, mc^2 \exp(\varphi/c^2)],$$

$$i\hbar c \Gamma^\mu \partial_\mu \Psi = [\Gamma^\mu b\, e_p A_\mu + a\, mc^2 \exp(\varphi/c^2)]\Psi, \tag{5.2.17}$$

$$\partial_\mu(\bar\Psi \Gamma^\mu \Psi) = 0, \quad \partial_\mu J^\mu = 0. \tag{5.2.18}$$

From Eqs. (5.2.15)–(5.2.18), we obtain

$$\partial_\mu T^\mu_\nu = -\bar\Psi\{[\Gamma^\mu b\, e_p A_\mu + a\, mc^2 \exp(\varphi/c^2)]\partial_\nu \Psi$$

$$- \partial_\nu[\Gamma^\mu b\, e_p A_\mu \Psi + a\, mc^2 \exp(\varphi/c^2)\Psi]\} - J^\mu \partial_\mu A_\nu/c. \tag{5.2.19}$$

It follows from (5.2.19) that

$$\partial_\mu T^\mu_\nu = F_{\nu\mu} J^\mu/c + G_\nu I^0,$$

$$F_{\nu\mu} = \partial_\nu A_\mu - \partial_\mu A_\nu, G_\nu = \exp(\varphi/c^2)\partial_\nu \varphi, \tag{5.2.20}$$

$$J^\mu = e_p c \bar\Psi \Gamma^\mu b \Psi, \quad I^0 = m\bar\Psi a \Psi. \tag{5.2.21}$$

In the case $\varphi = 0$, Eqs. (5.2.20) present the classical formula for the divergence of the energy–momentum tensor of a matter in an electromagnetic field, in which the 4-vector J^μ should be identified with the 4-vector of the nucleon current densities [13]. The term $G_\nu I^0$ in (5.2.20) corresponds to the nonlinear nuclear field considered in the previous chapter and, as follows from (4.2.25) and (4.3.5), the scalar I^0 should be identified with the nucleon rest mass density when $|\varphi/c^2| \ll 1$.

Thus, the validity of the obtained formula (5.2.20) for $\partial_\mu T_\nu^\mu$ confirms the correctness of the choice of expression (5.2) as the nucleon energy−momentum tensor T_ν^μ.

Owing to the quark structure of a nucleon, its current densities J^μ, which are determined by formula (5.2.21), should be represented in the form

$$J^\mu = \sum_{l=1}^{3} \theta_l J_{(l)}^\mu, \qquad (5.2.22)$$

where $\theta_l e_p$ are the quark charges and

for the proton: $\theta_1 = \theta_2 = 2/3, \qquad \theta_3 = -1/3,$

for the neutron: $\theta_{1\cdot} = \theta_2 = -1/3, \quad \theta_3 = 2/3.$ (5.2.23)

The 4-vectors $J_{(l)}^\mu$ should correspond to the quark current densities and satisfy the differential equations of conservation of the quark charges:

$$\partial_\mu J_{(l)}^\mu = 0. \qquad (5.2.24)$$

In addition to J^μ, let us introduce the 4-vector M^μ:

$$M^\mu = m\bar{\Psi}\Gamma^\mu a\Psi. \qquad (5.2.25)$$

Since the matrix $k = a$ satisfies relations (5.2.10), Eq. (5.2.12) with $k = a$ is valid and has the form

$$\partial_\mu M^\mu = 0. \qquad (5.2.26)$$

In the nonrelativistic case, the components Ψ_3, Ψ_4 are small and the value M^0 should coincide with the density I^0 of the proton rest mass when $|\varphi/c^2| \ll 1$, which is determined by formula (5.2.21).

Thus, Eq. (5.2.25) can be interpreted as the equation of conservation of the nucleon rest mass.

From formulas (5.2.23) and Eqs. (5.2.24) and (5.2.26), which express the conservation of the quark charges and the nucleon mass m, we come to the following conserved integrals, which give the conditions of normalization for the wave function Ψ:

$$\int J^0_{(l)}\, dv = ce_p, \quad \int M^0\, dv = m, \qquad (5.2.27)$$

where dv is an infinitely small volume in the three-dimensional space and the integrals are taken over the entire space. It should be noted that from (5.2.22), (5.2.23) and (5.2.27), we just obtain the correct values of the nucleon charges, which are equal to e_p for the proton and to zero for the neutron.

As will be shown here, the following matrices a and b satisfy conditions (5.2.8) and (5.2.22)–(5.2.27):

For the proton, they have the form

$$a = b = \begin{pmatrix} 2/3 & -1/3 & 2/3 \\ -1/3 & 2/3 & 2/3 \\ 2/3 & 2/3 & -1/3 \end{pmatrix}. \qquad (5.2.28)$$

For the neutron, they are as follows:

$$a = \begin{pmatrix} 2/3 & -1/3 & 2/3 \\ -1/3 & 2/3 & 2/3 \\ 2/3 & 2/3 & -1/3 \end{pmatrix}, \quad b = \begin{pmatrix} -1/3 & 2/3 & -1/3 \\ 2/3 & -1/3 & -1/3 \\ -1/3 & -1/3 & 2/3 \end{pmatrix}. \qquad (5.2.29)$$

It should be noted that expression (5.2.28) satisfies the following requirement: the proton rest mass density is proportional to its charge density.

The substitution of expressions (5.2.28) and (5.2.29) for the matrices a and b into relations (5.2.8) shows that these equalities are satisfied. It is easy to verify that the matrices a and b have the following properties:

$$ab = \pm b^2, \quad a^2 = 1, \quad b^3 = b, \qquad (5.2.30)$$

where the sign '+' corresponds to the proton and the sign '−' corresponds to the neutron.

Let us note that the matrix a has the eigenvalues 1, 1, −1 and $\det(a) = -1$. For the proton, the matrix $b = a$ and for the neutron, the matrix b has the eigenvalues 0, 1, −1 and $\det(b) = 0$.

Let us now turn to conditions (5.2.22)–(5.2.27).

As will be shown in the following, the 4-vectors $J^\mu_{(l)}$ have the form

$$J^\mu_{(1)} = e_p c \bar\Psi \Gamma^\mu (p + \varepsilon u)\Psi, \quad J^\mu_{(2)} = e_p c \bar\Psi \Gamma^\mu (p - \varepsilon u)\Psi, \quad J^\mu_{(3)} = e_p c \bar\Psi \, \Gamma^\mu q \Psi,$$

$$p = \frac{1}{2}\begin{pmatrix} 1 & 0 & 1 \\ 0 & 1 & 1 \\ 1 & 1 & 0 \end{pmatrix}, \quad u = \begin{pmatrix} 0 & 1 & 1 \\ 1 & 0 & 1 \\ 1 & 1 & 0 \end{pmatrix}, \quad q = \begin{pmatrix} 0 & 1 & 0 \\ 1 & 0 & 0 \\ 0 & 0 & 1 \end{pmatrix},$$

$$(5.2.31)$$

where ε is some number.

This can be established from the following considerations. It is easy to verify, taking into account (5.2.21), (5.2.23), (5.2.28)–(5.2.29) and expressions (5.2.31), that equality (5.2.22) is satisfied. Besides, the matrices $k = p$, $k = u$, $k = q$ and $k = a$ satisfy equalities (5.2.10) and, therefore, Eq. (5.2.12) is satisfied for these matrices.

That is why Eqs. (5.2.24) and (5.2.26) are also satisfied since they are equivalent to Eq. (5.2.12) when $k = p$, $k = u$, $k = q$ and $k = a$.

It is also easy to show that the 4-vectors $J^\mu_{(l)}$ and M^μ are independent of the choice of a gauge for the electromagnetic potentials A_μ. This follows from the commutativity of the matrices p, u, q, a with the matrix b in both cases (5.2.28) and (5.2.29).

Since $a = \frac{4}{3}p - \frac{1}{3}q$, the second condition of normalization in (5.2.27) follows from the first condition in it. As for the first condition, it acquires the form

$$\int \Psi^+ p \Psi dv = 1, \quad \int \Psi^+ u \Psi dv = 0, \quad \int \Psi^+ q \Psi dv = 1. \quad (5.2.32)$$

From formulas (5.2.32) and the equalities $a = \frac{4}{3}p - \frac{1}{3}q$, $1 = 2p + q - u$, we find

$$\int \Psi^+ a \Psi dv = 1, \quad \int \Psi^+ \Psi dv = 3. \quad (5.2.33)$$

5.3.   Interaction of Nucleons with Magnetic Fields

First, consider Eq. (5.2.5) in the case of a free nucleon when $A_\mu = 0$ and $\varphi = 0$. Then this equation has the following simple solution:

$$\Psi = \exp\left(-ia(Et - p_1 x - p_2 y - p_3 z)/\hbar\right) B, \qquad (5.3.1)$$

where E, p_1, p_2, p_3 are constants equal to the energy and momentum components of the nucleon in an inertial frame of reference with the coordinates $x^0 = ct$, $x^1 = x$, $x^2 = y$, $x^3 = z$ and B is a constant column consisting of 12 elements.

Indeed, substituting (5.3.1) into Eq. (5.2.5), we find

$$\left(\gamma^0 E - c(\gamma^1 p_1 - \gamma^2 p_2 - \gamma^3 p_3) - mc^2\right) B = 0, \quad A_\mu = 0, \quad \varphi = 0, \qquad (5.3.2)$$

where γ^μ are the Dirac matrices.

Equation (5.3.2) can be rewritten as

$$(i\hbar\gamma^\mu \partial_\mu - mc)\,\Psi_0 = 0, \quad \Psi_0 = B \exp\left(-i(Et - p_1 x - p_2 y - p_3 z)/\hbar\right). \qquad (5.3.3)$$

Therefore, the components of the column Ψ_0 describe arbitrary de Broglie waves [49] of a free nucleon satisfying the classical Dirac equation.

It should be stressed that owing to the equality $a^2 = 1$, where 1 is the unit 3×3 matrix, we have the formula

$$\exp(ia\lambda) = \cos(a\lambda) + i\sin(a\lambda) = 1 \cdot \cos\lambda + ia \cdot \sin\lambda. \qquad (5.3.4)$$

Therefore, formula (5.3.1) can be represented as

$$\Psi = (1 \cdot \cos\vartheta - ia \cdot \sin\vartheta)B, \quad \vartheta = (Et - p_1 x - p_2 y - p_3 z)/\hbar. \quad (5.3.5)$$

Consider now a nonrelativistic movement of a nucleon in a constant homogeneous magnetic field, in which the potentials A_μ have the form

$$A_0 = 0, \quad A_1 = \frac{1}{2}yH, \quad A_2 = -\frac{1}{2}xH, \quad A_3 = 0, \quad H = \text{const}, \qquad (5.3.6)$$

where H is the strength of the magnetic field, which is directed along the axis z, and x, y, z are rectangular spatial coordinates.

Let us represent the wave function Ψ in the form

$$\Psi = \exp(-ia\, mc^2 t/\hbar)\phi, \tag{5.3.7}$$

where t is time and ϕ has four components $\phi_k (k = 1, 2, 3, 4)$, each of them consisting of three elements. Then from (5.2.5), we obtain when neglecting the small value φ of the proper nuclear field of the nucleon:

$$(i\hbar\partial/\partial t - e_p A_0 b)\begin{pmatrix}\phi_1\\\phi_2\end{pmatrix} = c(\boldsymbol{\sigma}, \mathbf{P})\begin{pmatrix}\phi_3\\\phi_4\end{pmatrix},$$

$$[2mc^2 a + i\hbar\partial/\partial t - e_p A_0 b]\begin{pmatrix}\phi_3\\\phi_4\end{pmatrix}$$

$$= c(\boldsymbol{\sigma}, \mathbf{P})\begin{pmatrix}\phi_1\\\phi_2\end{pmatrix}, \tag{5.3.8}$$

where $\mathbf{P} = (P_1, P_2, P_3)$ has the sense of the momentum operator,

$$P_k = -i\hbar\partial_k + e_p b A_k/c, \quad \boldsymbol{\sigma} = (\sigma_1, \sigma_2, \sigma_3),$$

$$(\boldsymbol{\sigma}, \mathbf{P}) = \sigma_1 P_1 + \sigma_2 P_2 + \sigma_3 P_3 \tag{5.3.9}$$

and σ_k are the Pauli matrices of the form

$$\sigma_1 = \begin{pmatrix}0 & 1\\1 & 0\end{pmatrix}, \quad \sigma_2 = \begin{pmatrix}0 & -i\\i & 0\end{pmatrix}, \quad \sigma_3 = \begin{pmatrix}1 & 0\\0 & -1\end{pmatrix}, \tag{5.3.10}$$

in which the values $d = 0, \pm 1, \pm i$ denote the matrices $d \cdot 1$, where 1 is the unit 3×3 matrix.

In the considered nonrelativistic case, from the second equation in (5.3.8), we find, taking into account (5.3.6) and the equality $a^{-1} = a$, since $a^2 = 1$,

$$\begin{pmatrix}\phi_3\\\phi_4\end{pmatrix} = \frac{1}{2mc} a(\boldsymbol{\sigma}, \mathbf{P})\begin{pmatrix}\phi_1\\\phi_2\end{pmatrix}. \tag{5.3.11}$$

It is easy to verify that

$$(\boldsymbol{\sigma}, \mathbf{P})(\boldsymbol{\sigma}, \mathbf{P}) = P^2 - e_p \hbar b(\boldsymbol{\sigma}, \mathbf{H})/c, \tag{5.3.12}$$

where $P^2 = |\mathbf{P}|^2$ and $\mathbf{H}$ is the vector of the magnetic field strength. That is why substituting expression (5.3.11) into the first equation in (5.3.8), we obtain

$$(i\hbar\partial/\partial t - e_p A_0 b)f = (2m)^{-1}[a(P^2 - e_p\hbar b(\boldsymbol{\sigma}, \mathbf{H})/c)]f, \qquad (5.3.13)$$

where

$$f = \begin{pmatrix} \phi_1 \\ \phi_2 \end{pmatrix}. \qquad (5.3.14)$$

As follows from (5.3.6), Eq. (5.3.13) acquires the form

$$i\hbar\partial f/\partial t = (2m)^{-1}a(P^2 - e_p\hbar b\sigma_3 H/c)f. \qquad (5.3.15)$$

Let us seek the solution to Eq. (5.3.15) that describes the stationary state of a nucleon in the following form:

$$f = \exp(ie_p ab\sigma_3 Ht/(2mc))f_1, \qquad (5.3.16)$$

where f_1 is the column consisting of some functions of t, x, y, z.

Then substituting (5.3.16) into Eq. (5.3.15) and taking into account the equality $ab = ba$, we obtain

$$i\hbar\partial f_1/\partial t = (2m)^{-1}P^2 a f_1. \qquad (5.3.17)$$

Let us represent the sought solution f_1 to Eq. (5.3.17) corresponding to the stationary state of the nucleon in the form

$$f_1 = \exp(-iaE_1 t/\hbar)f_2, \quad f_2 = f_2(x, y, z), \qquad (5.3.18)$$

where E_1 is the constant corresponding to the nucleon kinetic energy and f_2 is the column consisting of functions depending only on the spatial coordinates x, y, z. Then substituting (5.3.18) into Eq. (5.3.17), we come to the equation

$$E_1 f_2 = (2m)^{-1}P^2 f_2. \qquad (5.3.19)$$

From (5.3.6) and (5.3.9), we obtain the following expression for the operator $P^2 = P_1^2 + P_2^2 + P_3^2$:

$$P^2 = -\hbar^2\Delta + i\hbar(e_p/c)bH(x\partial/\partial y - y\partial/\partial x)$$
$$+ \frac{1}{4}(e_p/c)^2 b^2(x^2 + y^2)H^2, \qquad (5.3.20)$$

where $\Delta = \partial^2/\partial x^2 + \partial^2/\partial y^2 + \partial^2/\partial z^2$ is the Laplace operator.

Let us use the cylindrical coordinates r, θ, z. Then we find

$$\Delta = \partial^2/\partial r^2 + (1/r)\partial/\partial r + (1/r)^2\partial^2/\partial\theta^2 + \partial^2/\partial z^2,$$
$$x\partial/\partial y - y\partial/\partial x = \partial/\partial\theta, \quad x = r\cos\theta, \quad y = r\sin\theta. \tag{5.3.21}$$

Substituting (5.3.20) and (5.3.21) into Eq. (5.3.19), we obtain

$$\left[\frac{\partial^2}{\partial r^2} + \frac{1}{r}\frac{\partial}{\partial r} + \frac{1}{r^2}\frac{\partial^2}{\partial\theta^2} + \frac{\partial^2}{\partial z^2} + \frac{2mE_1}{\hbar^2} \right.$$
$$\left. - \frac{ie_p bH}{\hbar c}\frac{\partial}{\partial\theta} - \left(\frac{e_p rH}{2\hbar c}\right)^2 b^2 \right] f_2 = 0. \tag{5.3.22}$$

We will seek a solution to Eq. (5.3.22) in the form

$$f_2 = b\exp(-ibl\theta)\exp(iak_3 z)R(r)C, \tag{5.3.23}$$

where $R(r)$ is some function, C is a constant column, l is the azimuth quantum number taking integer values and $\hbar k_3 = p_3 = $ const. As is seen from (5.3.1), the multiplier $\exp(iak_3 z)$ in (5.3.23) describes the free movement of the considered nucleon along the axis z with the constant projection onto it of its momentum vector equal to $\hbar k_3$.

After the substitution of (5.3.23) into Eq. (5.3.22) and the use of equalities (5.3.30), we come to the equation

$$\left(\frac{d^2}{dr^2} + \frac{1}{r}\frac{d}{dr} - \frac{l^2}{r^2} - k_3^2 + \frac{2mE_1}{\hbar^2} - 2\eta l - \eta^2 r^2 \right) R(r) = 0, \tag{5.3.24}$$

where

$$\eta = e_p H/(2\hbar c). \tag{5.3.25}$$

For definiteness, we will further assume that the value η is positive.

After the choice of the variable $\rho = \eta r^2$ instead of r, Eq. (5.3.24) acquires the form

$$\left(\rho \frac{d^2}{d\rho^2} + \frac{d}{d\rho} + \lambda - \frac{l}{2} - \frac{\rho}{4} - \frac{l^2}{4\rho} \right) R(\rho) = 0, \qquad (5.3.26)$$

where

$$\lambda = \frac{2mE_1 - (\hbar k_3)^2}{4\eta\hbar^2}, \qquad \rho = \eta r^2. \qquad (5.3.27)$$

Equation (5.3.26) is well known. Its bounded solutions that tend to zero at infinity are proportional to the Laguerre functions $I_{ns}(\rho)$:

$$R(\rho) = \text{const} \times I_{ns}(\rho) = \text{const} \times e^{-\rho/2} \rho^{\frac{n-s}{2}} Q_s^{n-s}(\rho), \qquad (5.3.28)$$

where $Q_s^k = e^\rho \rho^{-k} \frac{d^s}{d\rho^s}(e^{-\rho}\rho^{k+s})$ are the associated Laguerre polynomials [50] and the parameters λ and l should satisfy the relation

$$\lambda - l - 1/2 = s, \quad s = 0, 1, 2, \ldots \qquad (5.3.29)$$

From (5.3.25), (5.3.27) and (5.3.29), we obtain the discrete levels of the kinetic energy E_1 of the considered nucleon:

$$E_1 = \frac{e_p \hbar H}{mc}\left(n + \frac{1}{2}\right) + \frac{(\hbar k_3)^2}{2m}, \quad n = l + s, \qquad (5.3.30)$$

where $H > 0$ and n should be a nonnegative integer.

5.4. Magnetic Moments of Nucleons

Consider the full energy E of a nonrelativistic nucleon. From (5.2) and (5.3.6) ($A_0 = 0$), we obtain the following classical formula of quantum mechanics for the energy E:

$$E = i\hbar \int \Psi^+ \frac{\partial \Psi}{\partial t} dv, \qquad (5.4.1)$$

where dv is an infinitesimally small three-dimensional volume and the integral is taken over the entire space.

As follows from (5.3.7) and (5.2.33), formula (5.4.1) can be represented in the form

$$E = mc^2 + i\hbar \int \phi^+ \frac{\partial \phi}{\partial t} dv. \tag{5.4.2}$$

Taking into account (5.3.11) and (5.3.14), from (5.4.2), we obtain in the considered nonrelativistic case:

$$E = mc^2 + i\hbar \int f^+ \frac{\partial f}{\partial t} dv. \tag{5.4.3}$$

Using formulas (5.3.16) and (5.3.18), from (5.4.3), we find

$$E = mc^2 + E_1 \int f^+ a f dv - \frac{e_p \hbar H}{2mc} \int f^+ \sigma_3 a b f \ dv. \tag{5.4.4}$$

This gives the following expression for the proper magnetic moment μ of the nucleon:

$$\mu = \frac{e_p \hbar}{2mc} \int f^+ \sigma_3 a b f \ dv. \tag{5.4.5}$$

Let us turn to the conditions (5.2.32) of normalization for the wave function Ψ. In the considered nonrelativistic case, they can be represented as

$$\int f^+ p f \ dv = 1, \quad \int f^+ u f \ dv = 0, \quad \int f^+ q f \ dv = 1 \tag{5.4.6}$$

and conditions (5.2.33), following from (5.2.32), acquire the form

$$\int f^+ a f \ dv = 1, \quad \int f^+ f \ dv = 3. \tag{5.4.7}$$

Taking into account (5.4.7) and the form $\sigma_3 = \begin{pmatrix} 1 & 0 \\ 0 & -1 \end{pmatrix}$ of the Pauli matrix, from (5.4.4), we find the two stationary values, which depend on the orientation of the nucleon spin,

$$E = mc^2 + E_1 \mp (m_p/m)\mu_0 H \int f^+ a b f \ dv, \tag{5.4.8}$$

where the value μ_0 is the nuclear magneton,

$$\mu_0 = \frac{e_p \hbar}{2m_p c}. \tag{5.4.9}$$

In (5.4.8), the sign '−' corresponds to the nucleon spin directed along the axis z, and hence along the vector of the magnetic field strength, and the sign '+' corresponds to the opposite spin.

As follows from (5.2.28)–(5.2.31),

$$\text{for the proton}: \quad ab = 1,$$

$$\text{for the neutron}: \quad ab = a + \frac{2}{3}(q - p) - 1. \tag{5.4.10}$$

Thus, from (5.4.6) and (5.4.7), we obtain

$$\text{for the proton:} \quad \int f^{+}abf \; dv = 3,$$

$$\text{for the neutron:} \quad \int f^{+}abf \; dv = -2. \tag{5.4.11}$$

From (5.4.8) and (5.4.11), we determine the energies $E^{(p)}$ and $E^{(n)}$ of the proton and neutron, respectively:

$$E^{(p)} = m^{(p)}c^2 + E_1^{(p)} \mp 3(m_p/m^{(p)})\mu_0 H,$$

$$E^{(n)} = m^{(n)}c^2 + E_1^{(n)} \pm 2(m_p/m^{(n)})\mu_0 H, \tag{5.4.12}$$

where $E_1^{(p)}$, $m^{(p)}$ and $E_1^{(n)}$, $m^{(n)}$ are the kinetic energies E_1 and masses m of the proton and neutron.

From (5.4.12), we obtain the following values μ_p and μ_n of the proper magnetic moments of the proton and neutron, respectively:

$$\mu_p = 3(m_p/m^{(p)})\mu_0, \quad \mu_n = -2(m_p/m^{(n)})\mu_0. \tag{5.4.13}$$

As follows from (5.3.30) and (5.4.9), formulas (5.4.12) can be represented as

$$E^{(p)} = m^{(p)}c^2 + \frac{(\hbar k_3)^2}{2m^{(p)}} + \frac{e_p \hbar H}{2m^{(p)}c}(2n + 1 \mp 3),$$

$$E^{(n)} = m^{(n)}c^2 + \frac{(\hbar k_3)^2}{2m^{(n)}} + \frac{e_p \hbar H}{2m^{(n)}c}(2n + 1 \pm 2). \tag{5.4.14}$$

Consider now a free nucleon. From (5.2.9), it follows that its energy at rest E_0 has the form

$$E_0 = mc^2 \exp(\bar{\varphi}/c^2), \tag{5.4.15}$$

where $\bar{\varphi}$ is some mean value of the proper nuclear potential φ of the nucleon.

Taking into account that for free nucleons $|\bar{\varphi}/c^2| \ll 1$, from (5.4.15), we obtain for the proton and neutron as follows:

$$m_p = m^{(\mathrm{p})}\left(1 + \bar{\varphi}^{(\mathrm{p})}/c^2\right), \quad m_n = m^{(\mathrm{n})}\left(1 + \bar{\varphi}^{(\mathrm{n})}/c^2\right), \qquad (5.4.16)$$

where m_p, $\bar{\varphi}^{(\mathrm{p})}$ and m_n, $\bar{\varphi}^{(n)}$ are the rest masses and mean values of the proper nuclear potentials of the proton and neutron, respectively.

From (5.4.13) and (5.4.16), we find, taking into account that $m_n \approx m_p$,

$$\mu_p = 3\mu_0(1 + \bar{\varphi}^{(\mathrm{p})}/c^2), \quad \mu_n = -2\mu_0(1 + \bar{\varphi}^{(\mathrm{n})}/c^2), \qquad (5.4.17)$$

where μ_0 is the nuclear magneton determined by formula (5.4.9).

As is well known [49], the experimental values μ_p^{exp} and μ_n^{exp} of the proper magnetic moments of the proton and neutron are as follows:

$$\mu_p^{\mathrm{exp}} = 2.79\mu_0, \quad \mu_n^{\mathrm{exp}} = -1.91\mu_0. \qquad (5.4.18)$$

As for the mean value $m_p\bar{\varphi}$ of the nuclear potential energy of a free nucleon, it is negative and of several percent of the value m_pc^2 [37].

Therefore, the obtained formulas (5.4.17) for the proper magnetic moments of the proton and neutron are in accordance with their experimental values (5.4.18).

Let us note the relation between the wave equation (5.2.5) and the well-known quark model [37]. Equation (5.2.5) describes the wave function Ψ having 12 components. This function can be regarded as three four-component wave functions of the three fermions presenting the quarks. As follows from (5.2.22), the current densities J^μ of nucleons are the sums of the three summands $\theta_l J^\mu_{(l)}$. These summands satisfy the differential equations of charge conservations (5.2.24) and correspond to the current densities of the two u-quarks with the charge $+\frac{2}{3}e_p$ and the one d-quark with the charge $-\frac{1}{3}e_p$ for the proton and the two d-quarks and the one u-quark for the neutron.

5.5. Generalization of the Dirac Equation for Light Atomic Nuclei

Let us now consider a generalization of the relativistic quantum equation proposed in the previous section, which describes an individual nucleon, for a system of nucleons and light atomic nuclei.

Let there be a system of N nucleons located close to each other. Then, to describe them, it is necessary to introduce a wave function that depends on the set of coordinates of all particles of this system.

In our paper [45], we proposed a generalization of the Dirac equation in the case of a system of closely spaced electrons, which, taking into account the results of the previous section, can also be applied to the case of a system of nucleons. This will be done in the following.

To describe a system of N closely spaced nucleons, we introduce a wave function Φ that depends on the coordinates of all nucleons and has 4^N components Φ_k that are columns of three elements $\Phi_{kl}(1 \leq k \leq 4^N, 1 \leq l \leq 3)$.

We will seek a generalization of the Dirac equation in this case in the following form:

$$\left\{ \sum_{s=0}^{4N-1} [\widehat{\Gamma}^s (i\hbar c \partial_s - b_s e_p A_s)] - N^{-1/2} ac^2 \sum_{j=1}^{N} m_j \exp(\varphi_j/c^2) \right\} \Phi = 0,$$

$$(5.5.1)$$

where e_p and m_j are the proton charge and the rest mass of the jth nucleon when $\varphi_j = 0$, respectively, φ_j is the nuclear field potential for the jth nucleon, A_s are electromagnetic field potentials. The indices j and s take the values $1 \leq j \leq N, 0 \leq s \leq 4N - 1$. The coordinates x^s and potentials A_s when $4(j-1) \leq s \leq 4j - 1$ correspond to the jth nucleon and $\partial_s = \partial/\partial x_s$.

In (5.5.1), a and b_s are 3×3 matrices introduced in Section 5.2:

$$a = \begin{pmatrix} 2/3 & -1/3 & 2/3 \\ -1/3 & 2/3 & 2/3 \\ 2/3 & 2/3 & -1/3 \end{pmatrix} \qquad (5.5.2)$$

and the matrix b_s is determined as follows:

If the coordinate x^s corresponds to the proton, then

$$b_s = a, \qquad (5.5.3)$$

and if the coordinate x^s corresponds to the neutron, then

$$b_s = \begin{pmatrix} -1/3 & 2/3 & -1/3 \\ 2/3 & -1/3 & -1/3 \\ -1/3 & -1/3 & 2/3 \end{pmatrix}. \qquad (5.5.4)$$

As indicated in Section 5.2, the matrices a and b_s satisfy the equalities

$$a^2 = 1, \quad ab_s = b_s a. \tag{5.5.5}$$

The symbols $\widehat{\Gamma}^s$ in (5.5.1) are the square matrices of the dimension 4^N consisting of the elements $\widehat{\Gamma}_{ij}^s$, which are 3×3 matrices of the form $\Gamma_{ij}^s \cdot 1$, where 1 is the unit 3×3 matrix and Γ_{ij}^s are numeric elements of square matrices Γ^s with the dimension 4^N, on which we impose the following requirements:

$$\Gamma^n \Gamma^m + \Gamma^m \Gamma^n = 2g^{mn} I, \tag{5.5.6}$$

where $g^{mn} = 0$, $n \neq m$; $g^{nn} = 1$, $n = 4l$; $g^{nn} = -1$, $n \neq 4l$, l is an integer and I is the unit matrix having the dimension 4^N.

Equation (5.5.1) and formulas (5.5.2)–(5.5.6) are a generalization of the corresponding formulas of Section 5.2 and in the case $N = 1$, coincide with them.

When $a = b_s = 1$ and $\varphi_j = 0$, Eq. (5.5.1) becomes the equation for a system of electrons proposed in our paper [45].

The product $a\Phi$ in (5.5.1) means the column consisting of the 4^N elements $a\Phi_k$, which are the product of the 3×3 matrix a and the column Φ_k consisting of the three elements Φ_{kl}. In the product $\widehat{\Gamma}^s b_s$, each element of the matrix $\widehat{\Gamma}^s$, which is a 3×3 matrix, is multiplied by the 3×3 matrix b_s.

After the multiplication of Eq. (5.5.1) by the operator

$$\left\{ \sum_{s=0}^{4N-1} [\widehat{\Gamma}^s (i\hbar c \partial_s - b_s e_p A_s)] + N^{-1/2} ac^2 \sum_{j=1}^{N} m_j \exp(\varphi_j/c^2) \right\}$$

and the use of relations (5.5.5) and (5.5.6), we obtain the following generalization of the Klein–Gordon equation:

$$\left\{ g^{sn}(i\hbar c \partial_s - e_p B_s)(i\hbar c \partial_n - e_p B_n) - i\hbar c e_p \Gamma^s \Gamma^n \tilde{F}_{sn} \right.$$

$$\left. -(c^4/N) \left[\sum_{j=1}^{N} m_j \exp(\varphi_j/c^2) \right]^2 \right\} \Phi = 0, \tag{5.5.7}$$

$$B_s = b_s A_s, \quad \tilde{F}_{sn} = \partial_s B_n - \partial_n B_s.$$

In the nonrelativistic stationary case, let us represent the wave function $\Phi(x^s)$ with the 4^N components $\Phi_k(x^s)$, $0 \le s \le 4N - 1$, corresponding to the system of N nucleons, in the form

$$\Phi(x^s) = \exp\left(-\frac{ia}{\hbar c}\sum_{j=1}^{N}(m_j c^2 + E_j)x^{4(j-1)}\right)\Theta(x^r),$$

$$1 \le r \le 4N - 1, \quad r \ne 4l, \quad |E_j| \ll m_j c^2, \tag{5.5.8}$$

where l is an integer and E_j has the sense of the nonrelativistic energy of the jth nucleon.

Then when $\varphi_j = 0$, $A_s = 0$ and $m_1 = m_2 = \cdots = m_N = m$, from (5.5.7) and (5.5.8) in the considered nonrelativistic case, we obtain the well-known Schrödinger equation for a system of N free identical particles:

$$\left[E - \frac{1}{2}(\hbar^2/m)\partial^r\partial_r\right]\Theta = 0,$$

$$1 \le r \le 4N - 1, \quad r \ne 4l, \quad E = \sum_{j=1}^{N}E_j, \tag{5.5.9}$$

Let us turn to the square matrices Γ^n, satisfying relations (5.5.6), and, instead of them, introduce the matrices G^n:

$$G^n = \Gamma^n, \quad n = 4l; \quad G^n = i\Gamma^n, \quad n \ne 4l, \tag{5.5.10}$$

forming the Clifford algebra:

$$(G^n)^2 = I, \quad G^n G^m = -G^m G^n, \quad n \ne m. \tag{5.5.11}$$

Further, we will obtain the concrete form of the square matrices G^n, satisfying relations (5.5.11).

5.6. General Properties of the Matrices of the Quantum Equation for Light Atomic Nuclei

To describe the introduced above matrices G^n, let us construct the set $\bar{G}_M$ of the following $2M$ square matrices of the order 2^M by the recurrence way:

$$\bar{G}_k = \left(\begin{pmatrix} I_{k-1} & O_{k-1} \\ O_{k-1} & -I_{k-1} \end{pmatrix}, \begin{pmatrix} O_{k-1} & I_{k-1} \\ I_{k-1} & O_{k-1} \end{pmatrix}, \begin{pmatrix} O_{k-1} & i\bar{G}_{k-1} \\ -i\bar{G}_{k-1} & O_{k-1} \end{pmatrix} \right),$$

(5.6.1)

where

$$2 \leq k \leq M, \quad \bar{G}_1 = \left(\begin{pmatrix} 1 & 0 \\ 0 & -1 \end{pmatrix}, \begin{pmatrix} 0 & 1 \\ 1 & 0 \end{pmatrix} \right),$$

and I_{k-1} and O_{k-1} are the unit and zero matrices, respectively, of the order 2^{k-1}.

The last matrix of the set $\bar{G}_k$ in (5.6.1) means the set of all matrices of the form

$$\begin{pmatrix} O_{k-1} & iF_{k-1} \\ -iF_{k-1} & O_{k-1} \end{pmatrix},$$

where $F_{k-1} \in \bar{G}_{k-1}$.

The following theorem is obtained.

Theorem 5.6.1. *The matrices G^n, $0 \leq n \leq 2M-1$, of the order 2^M, which belong to the set $\bar{G}_M$, defined by the recurrence formula (5.6.1), satisfy relations (5.5.11) and any other set of matrices $\bar{H}_M$, satisfying the same relations, is determined by the formula of similarity $\bar{H}_M \sim \bar{G}_M$, that is $\bar{H}_M = T_M^{-1} \bar{G}_M T_M$, where T_M is a nonsingular matrix of the order 2^M.*

Proof. Let $G^n (0 \leq n \leq 2M - 1)$ be matrices of the set $\bar{H}_M$, satisfying relations (5.5.11). Then we have the formulas

$$(G^0)^2 = I_M, \quad G^0 = (G^1)^{-1}(-G^0)G^1, \quad (G^0 \sim -G^0), \qquad (5.6.2)$$

from which we find that the Jordan form J^0 of the matrix G^0 is as follows:

$$J^0 = \begin{pmatrix} I_{M-1} & O_{M-1} \\ O_{M-1} & -I_{M-1} \end{pmatrix}. \qquad (5.6.3)$$

Therefore, there is a transformation of similarity of the matrices G^n, which transforms the matrix G^0 into J^0. After the application of this transformation, which leaves relations (5.5.11) unchanged, owing to its property of similarity, we can put $G^0 = J^0$, keeping the previous notations for the transformed matrices G^n. Then from (5.5.11), we find

$$G^1 = \begin{pmatrix} O_{M-1} & S_{M-1} \\ S_{M-1}^{-1} & O_{M-1} \end{pmatrix}, \tag{5.6.4}$$

where S_{M-1} is some nonsingular matrix of the order 2^{M-1}.

Let us now apply the transformation of similarity to the matrices G^n:

$$G^n \to T_M^{-1} G^n T_M$$

with the transformation matrix T_M of the form

$$T_M = \begin{pmatrix} I_{M-1} & O_{M-1} \\ O_{M-1} & S_{M-1}^{-1} \end{pmatrix}. \tag{5.6.5}$$

Then the matrices G^0 and G^1 are as follows:

$$G^0 = \begin{pmatrix} I_{M-1} & O_{M-1} \\ O_{M-1} & -I_{M-1} \end{pmatrix}, \quad G^1 = \begin{pmatrix} O_{M-1} & I_{M-1} \\ I_{M-1} & O_{M-1} \end{pmatrix}. \tag{5.6.6}$$

It follows from (5.5.11) that the other matrices G^n should have the form

$$y \quad G^n = \begin{pmatrix} O_{M-1} & iG_{M-1}^n \\ -iG_{M-1}^n & O_{M-1} \end{pmatrix}, \quad n \geq 2, \tag{5.6.7}$$

where the matrices G_{M-1}^n of the order 2^{M-1} should also satisfy (5.5.11).

Thus, the initial set $\bar{H}_M$ is similar to the following set of $2M$ matrices:

$$\begin{pmatrix} I_{M-1} & O_{M-1} \\ O_{M-1} & -I_{M-1} \end{pmatrix}, \quad \begin{pmatrix} O_{M-1} & I_{M-1} \\ I_{M-1} & O_{M-1} \end{pmatrix}, \quad \begin{pmatrix} O_{M-1} & i\bar{H}_{M-1} \\ -i\bar{H}_{M-1} & O_{M-1} \end{pmatrix},$$

$$\tag{5.6.8}$$

where to have relations (5.5.11) fulfilled, the $2(M-1)$ matrices $\bar{H}_{M-1}$ of the order 2^{M-1} should also satisfy (5.5.11).

If we apply the transformation of similarity to the set of matrices (5.6.8) with the transformation matrix

$$T_M = \begin{pmatrix} T_{M-1} & O_{M-1} \\ O_{M-1} & T_{M-1} \end{pmatrix}, \qquad (5.6.9)$$

then these set of matrices transforms into a set of the same form (5.6.8), in which only $\bar{H}_{M-1}$ changes to $T_{M-1}^{-1}\bar{H}_{M-1}T_{M-1}$.

Therefore, the proof of the theorem for the $2M$ matrices $\bar{H}_M$ of the order 2^M is reduced to its proof for the $2(M-1)$ matrices $\bar{H}_{M-1}$ of the order 2^{M-1}.

Continuing this process and reducing the matrix order to 2, we get as a result that $\bar{H}_M \sim \bar{G}_M$ and the matrices of the set $\bar{G}_M$ satisfies relations (5.5.11).

Hence, the theorem has been proved.

It should be noted that as follows from relations (5.5.11), to the set of the considered matrices G^n, $0 \leq n \leq 2M-1$, one can add one more matrix $G^{2M} = G^0 G^1 \cdots G^{2M-1}$, also corresponding to these relations.

Let us also note that the value $M = 2N$ corresponds to the set of the matrices $\widehat{\Gamma}^s$ in the considered equation (5.5.1).

It is easy to prove that the Hermitian conjugate $\bar{G}_M^+$ of the set $\bar{G}_M$, defined by formula (5.6.1), has the property

$$\bar{G}_M^+ = \bar{G}_M. \qquad (5.6.10)$$

For a set of matrices with such a property, we will prove a special theorem, which as the previous theorem, is a generalization of the well-known properties of the Dirac matrices [49].

Theorem 5.6.2. *Any set $\bar{H}_M$ of $2M$ matrices of the order 2^M, satisfying relations (5.5.11) and (5.6.10): $\bar{H}_M^+ = \bar{H}_M$, is related with the set of matrices $\bar{G}_M$ of form (5.6.1) by the formula*

$$\bar{H}_M = T_M^+ \bar{G}_M T_M, \qquad T_M^+ = T_M^{-1}. \qquad (5.6.11)$$

Proof.   First, let us prove the following lemma:

Lemma 5.6.1. *Any matrix S_M commuting with all matrices of the set $\bar{G}_M$ is proportional to the unit matrix: $S_M = \lambda \cdot I_M$, where λ is some number.*

Proof.   To prove the lemma, let us represent the matrix S_M in the form

$$S_M = \begin{pmatrix} S_{M-1}^{11} & S_{M-1}^{12} \\ S_{M-1}^{21} & S_{M-1}^{22} \end{pmatrix}, \tag{5.6.12}$$

where S_{M-1}^{ij} are square matrices of the order 2^{M-1}. Then from the commutativity of the multiplication of S_M by the matrices of the set $\bar{G}_M$, defined by formula (5.6.1), we have

$$S_{M-1}^{12} = S_{M-1}^{21} = O_{M-1},$$
$$S_{M-1}^{11} = S_{M-1}^{22}, \quad S_{M-1}^{11}\bar{G}_{M-1} = \bar{G}_{M-1}S_{M-1}^{11}. \tag{5.6.13}$$

Repeating the same considerations, as for S_M, for the matrix S_{M-1}^{11} and then for the matrices $S_{M-2}^{11}, S_{M-3}^{11}, \ldots, S_1^{11}$, we come to the conclusion that the matrix S_M is diagonal, and the same numbers are on its main diagonal. This just proves the lemma.

Let us return to the proof of Theorem 5.6.2. As follows from Theorem 5.6.1, $\bar{H}_M = C_M^{-1}\bar{G}_M C_M$, where C_M is some nonsingular matrix.

Since $\bar{H}_M^+ = \bar{H}_M$ and $\bar{G}_M^+ = \bar{G}_M$, we get

$$\bar{H}_M^+ = C_M^+ \bar{G}_M (C_M^+)^{-1} = C_M^{-1}\bar{G}_M C_M = \bar{H}_M. \tag{5.6.14}$$

From here, we find

$$S_M \bar{G}_M = \bar{G}_M S_M, \quad S_M = C_M C_M^+. \tag{5.6.15}$$

Therefore, from the proved lemma, it follows that

$$C_M C_M^+ = \lambda \cdot I_M, \tag{5.6.16}$$

where λ is some number.

Taking into account that the matrices of the form $C_M C_M^+$ have only positive real numbers on their main diagonal, as can be readily verified, the number λ in formula (5.6.16) is real and positive.

Putting $T_M = C_M/\sqrt{\lambda}$ and using (5.6.14) ad (5.6.16), we come to formula (5.6.11).

Thus, the theorem has been proved.

5.7.   Differential Equations of Charge Conservation in Light Atomic Nuclei

Let us now turn to the derivation of the differential equations of charge conservation from Eq. (5.5.1). For this purpose, we need a matrix R satisfying the relations

$$R^+ = R, \quad (R\Gamma^n)^+ = R\Gamma^n, \quad 0 \le n \le 4N - 1, \qquad (5.7.1)$$

which will be determined later on. Then let us introduce the matrix $\widehat{R}$ of the order 4^N with the elements $\widehat{R}_{ij}$, which are the 3×3 matrices of the form $R_{ij} \cdot 1$, where R_{ij} are the elements of the matrix R and 1 is the unit 3×3 matrix.

Multiplying now Eq. (5.5.1) on the left by the matrix $\widehat{R}$ and after that taking its Hermitian conjugate, using relations (5.7.1), we obtain

$$k \left\{ \sum_{s=0}^{4N-1} [\widehat{R}\widehat{\Gamma}^s (i\hbar c \partial_s - b_s e_p A_s)] - N^{-1/2} a \widehat{R} c^2 \sum_{j=1}^{N} m_j \exp(\varphi_j / c^2) \right\}$$
$$\Phi = 0,$$

$$k \Phi^+ \left\{ \sum_{s=0}^{4N-1} [\widehat{R}\widehat{\Gamma}^s (i\hbar c \partial_s + b_s e_p A_s)] + N^{-1/2} a \widehat{R} c^2 \sum_{j=1}^{N} m_j \exp(\varphi_j / c^2) \right\}$$
$$= 0, \qquad (5.7.2)$$

where $\Phi^+ \partial_s \equiv \partial_s \Phi^+$, $k = 1, a, b_n$ and b_n is the matrix b_s for neutrons.

Multiplying the first equation in (5.7.2) on the left by Φ^+ and the second equation in it on the right by Φ and then adding them, we come to the differential equations of quark charges conservation, which generalize the analogous equations of Section 5.2:

$$\partial_s J^s = 0, \quad J^s = e_p \bar{\Phi} \widehat{\Gamma}^s k \Phi, \quad k = 1, a, b_n, \qquad (5.7.3)$$

where

$$\bar{\Phi} = \Phi^+ \widehat{R} \qquad (5.7.4)$$

and J^s are the densities of the corresponding quark currents.

Let us require that the current densities should be of the vector type:

$$J^n = \frac{\partial x^n}{\partial x^{k'}} J^{k'}, \tag{5.7.5}$$

where J^k and $J^{k'}$ are the current densities in two arbitrary inertial frames of reference with coordinates x^k and $x^{k'}$.

Let us now turn to the problem of finding the matrix R, satisfying relations (5.7.1) and ensuring the fulfillment of the transformation law (5.7.5) for the multicomponent wave function Φ in the transition from the coordinates x^k to the new coordinates $x^{k'}$.

We start with the imposing condition (5.6.10) on the matrices G^n, related with Γ^n by formulas (5.5.10), to have them Hermitian. As stated above, this condition is valid for the matrices of the set $\bar{G}_{2N}$, which is determined by the recurrence formula (5.6.1).

Then from (5.5.10) and (5.6.10), we get

$$(\Gamma^n)^+ = \Gamma^n \text{ when } n = 4l \text{ and } (\Gamma^n)^+ = -\Gamma^n$$

$$\text{when } n \neq 4l, l \text{ is an integer.} \tag{5.7.6}$$

Let us seek the law of the transformation of the multicomponent wave function $\Phi(x^k)$ to the function $\Phi'(x^{k'})$ in the transition from the coordinates x^k to the coordinates $x^{k'}$ in the form

$$\Phi'(x^{k'}) = \widehat{F}\Phi(x^k), \tag{5.7.7}$$

where $\widehat{F}$ is the matrix of the order 4^N with the elements $\widehat{F}_{ij}$, which are the 3×3 of the form $F_{ij} \cdot 1$, where F_{ij} are the elements of some nonsingular matrix F, and 1 is the unit 3×3 matrix.

Then from the requirement of relativistic covariance of Eq. (5.5.1), we have

$$F^{-1}\Gamma^{n'} \frac{\partial x^k}{\partial x^{n'}} F = \Gamma^k. \tag{5.7.8}$$

Consider the infinitesimal transformation $x^k \to x^{k'}$ that conserves the following space–time interval ds of the examined system consisting of N nucleons:

$$ds^2 = g_{mn}dx^m dx^n, \quad m, n = 0, 1, \ldots, 4N - 1, \tag{5.7.9}$$

where $g_{mn} = 0$, $m \neq n$; $g_{nn} = 1$, $n = 4l$; $g_{nn} = -1$, $n \neq 4l$; l is an integer.

It is easy to verify that for such a transformation, we have

$$\frac{\partial x^k}{\partial x^{n'}} = \delta_n^k + \varepsilon h_n^k, \quad h_{kn} = -h_{nk}, \quad \varepsilon \to 0, \quad F = I + \varepsilon\Omega, \quad (5.7.10)$$

where δ_n^k is the Kronecker symbol, $h_{kn} = g_{kk}h_n^k$ and I is the unit matrix. Here, h_{kn} are the elements of a skew-symmetric matrix and Ω is some matrix depending on them, which will be determined in the following.

Then from (5.7.8), we obtain

$$(I - \varepsilon\Omega)\Gamma^n(\delta_n^k + \varepsilon h_n^k)(I + \varepsilon\Omega)$$
$$= \Gamma^k + \varepsilon(\Gamma^k\Omega - \Omega\Gamma^k + \Gamma^n h_n^k) = \Gamma^k, \quad (5.7.11)$$

and from this formula, we find

$$[\Omega, \Gamma^k] \equiv \Omega\Gamma^k - \Gamma^k\Omega = \Gamma^n h_n^k. \quad (5.7.12)$$

Relation (5.7.12), regarded as an equation for the matrix Ω, has the following solution, which is easy to verify using (5.5.6):

$$\Omega = \mu \cdot I - \Gamma^n\Gamma^m h_{nm}/4, \quad (5.7.13)$$

where μ is an arbitrary number.

Indeed, from (5.7.13), (5.5.6) and the antisymmetry of h_{kn}, we get

$$[\Omega, \Gamma^k] = -\Gamma^n\Gamma^m\Gamma^k h_{nm}/4 + \Gamma^k\Gamma^n\Gamma^m h_{nm}/4$$
$$= (-\Gamma^k\Gamma^m\Gamma^k h_{km} - \Gamma^n\Gamma^k\Gamma^k h_{nk} + \Gamma^k\Gamma^k\Gamma^m h_{km} + \Gamma^k\Gamma^n\Gamma^k h_{nk})/4$$
$$= (\Gamma^m h_m^k + \Gamma^n h_n^k + \Gamma^m h_m^k + \Gamma^n h_n^k)/4 = \Gamma^n h_n^k. \quad (5.7.14)$$

If some matrix Ω_1 satisfies Eq. (5.7.12), as well as Ω, then the matrix $\Omega' = \Omega_1 - \Omega$ commutates with all the matrices Γ^k. Therefore, from the proved Lemma 5.6.1, it follows that $\Omega' = \mu' \cdot I$, where μ' is some number.

Thus, formula (5.7.13) is the general solution of Eq. (5.7.12).

Let us now turn to formulas (5.7.3)–(5.7.5) and (5.7.7). From them, using the well-known formula $(F\Phi)^+ = \Phi^+ F^+$, we obtain

$$F^+ R\Gamma^{n'} \frac{\partial x^k}{\partial x^{n'}} F = R\Gamma^k. \qquad (5.7.15)$$

From (5.7.8) and (5.7.15), we find

$$F^+ R = RF^{-1}. \qquad (5.7.16)$$

Consider the infinitesimal transformation (5.7.10). From it, (5.7.16) and (5.7.13), we obtain

$$\Omega^+ R = -R\Omega, \quad \Omega = \mu \cdot I - \Gamma^n \Gamma^m h_{nm}/4. \qquad (5.7.17)$$

Thus, we should find a matrix R satisfying relations (5.7.1) and (5.7.17). This gives the system of the following relations:

$$\mu = i\nu, \quad \text{Im}\,\nu = 0, \quad R^+ = R, \quad (R\Gamma^n)^+ = R\Gamma^n,$$
$$(\Gamma^m \Gamma^n)^+ R = -R\Gamma^m \Gamma^n, \quad m \neq n. \qquad (5.7.18)$$

As can be verified, the matrix R of the following form satisfies relations (5.7.18):

$$R = i^{N(N-1)/2} Q^0 Q^1 \cdots Q^{N-1}, \qquad (5.7.19)$$

where

$$Q^s = \Gamma^{4s} \text{ when } N = 2l + 1,$$
$$Q^s = \Gamma^{4s+1}\Gamma^{4s+2}\Gamma^{4s+3} \text{ when } N = 2l, \qquad (5.7.20)$$
$$l \text{ is an integer}, \quad 0 \leq s \leq N - 1.$$

Let us show this. First, from (5.5.6), (5.7.6) and (5.7.20), we have

$$(Q^s)^+ = Q^s, \quad Q^s Q^t = -Q^t Q^s, \quad s \neq t, \qquad (5.7.21)$$

and hence,

$$R^+ = (-i)^{N(N-1)/2} Q^{N-1} Q^{N-2} \cdots Q^0 = i^{N(N-1)/2}$$
$$Q^0 Q^1 \cdots Q^{N-1} = R. \qquad (5.7.22)$$

Let us note that here we have used the well-known formula $(A_1 A_2 \cdots A_n)^+ = A_n^+ A_{n-1}^+ \cdots A_1^+$, where $A_1, A_2, \ldots, A_n$ are matrices.

Using (5.7.22), we obtain

$$(R\Gamma^n)^+ = (\Gamma^n)^+ R = i^{N(N-1)/2}(\Gamma^n)^+ Q^0 Q^1 \cdots Q^{N-1}. \qquad (5.7.23)$$

When $n = 4k$, where k is an integer, and for any N, we have, using (5.5.6), (5.7.6), (5.7.20) and (5.7.23),

$$(R\Gamma^n)^+ = i^{N(N-1)/2}\Gamma^n Q^0 Q^1 \cdots Q^{N-1} = i^{N(N-1)/2} Q^0 Q^1 \cdots Q^{N-1}$$

$$\Gamma^n = R\Gamma^n. \qquad (5.7.24)$$

When $n \neq 4k$, relations (5.5.6), (5.7.6), (5.7.20) and (5.7.23) give

$$(R\Gamma^n)^+ = -i^{N(N-1)/2}\Gamma^n Q^0 Q^1 \cdots Q^{N-1} = i^{N(N-1)/2} Q^0 Q^1 \cdots Q^{N-1}$$

$$\Gamma^n = R\Gamma^n. \qquad (5.7.25)$$

Consider the last relation in (5.7.18).

Let $n = 4k_1$ and $m = 4k_2$, where k_1, k_2 are nonnegative integers, $m \neq n$. Then from relations (5.5.6), (5.7.6), (5.7.22) and (5.7.24), we obtain

$$(\Gamma^m\Gamma^n)^+ R = \Gamma^n\Gamma^m R = \Gamma^n R\Gamma^m = R\Gamma^n\Gamma^m = -R\Gamma^m\Gamma^n. \qquad (5.7.26)$$

Let $n = 4k_1$ and $m \neq 4k_2$. Then from (5.5.6), (5.7.6), (5.7.22), (5.7.24) and (5.7.25), we find

$$(\Gamma^m\Gamma^n)^+ R = -\Gamma^n\Gamma^m R = \Gamma^n R\Gamma^m = R\Gamma^n\Gamma^m = -R\Gamma^m\Gamma^n. \qquad (5.7.27)$$

When $n \neq 4k_1$ and $m = 4k_2$, we analogously have

$$(\Gamma^m\Gamma^n)^+ R = -\Gamma^n\Gamma^m R = -\Gamma^n R\Gamma^m = R\Gamma^n\Gamma^m = -R\Gamma^m\Gamma^n. \qquad (5.7.28)$$

When $n \neq 4k_1$ and $m \neq 4k_2$, $m \neq n$, we get

$$(\Gamma^m\Gamma^n)^+ R = \Gamma^n\Gamma^m R = -\Gamma^n R\Gamma^m = R\Gamma^n\Gamma^m = -R\Gamma^m\Gamma^n. \qquad (5.7.29)$$

Therefore, the matrix R, determined by formulas (5.7.19)–(5.7.20), satisfies relations (5.7.18). Thus, formulas (5.7.3)–(5.7.4) and (5.7.19)–(5.7.20) give expressions for the quark current densities J^n satisfying the differential equations of conservation of the quark charges.

In this case, as follows from (5.7.5), the components J^n transform as a vector for arbitrary infinitesimal coordinate transformations of form (5.7.10), which conserve the space–time interval ds, determined by formula (5.7.9). This is also true for any set of such transformations. Therefore, the components J^n transform as a vector for any transformations that conserve the space–time interval (5.7.9).

Consider the coordinate rotation in the plane x^k, x^n by a finite angle θ_n^k. Such a rotation can be carried out by performing L times the infinitesimal transformation (5.7.10) with $h_n^k = \theta_n^k$ and $\varepsilon = 1/L$, where L is a sufficiently large positive integer. Then from (5.7.7), (5.7.10), (5.7.13) and (5.7.18), we obtain the following transformation formula for the multicomponent wave function $\Phi \to \Phi'$ in the case of the coordinate rotation by the angle θ_n^k in the plane x^k, x^n:

$$\Phi' = \exp\left(id - \frac{1}{2} g_{kk} \widehat{\Gamma}^k \widehat{\Gamma}^n \theta_n^k \right) \Phi, \ \mathrm{Im}\, d = 0, \tag{5.7.30}$$

where d is a real number and k, n are fixed integers.

It can be verified that the components $A^{l_1 l_2 \cdots l_n}$ of the form

$$A^{l_1 l_2 \cdots l_n} = \bar{\Phi} \widehat{\Gamma}^{l_1} \widehat{\Gamma}^{l_2} \cdots \widehat{\Gamma}^{l_n} \Phi, \quad \bar{\Phi} = \Phi^+ \widehat{R} \tag{5.7.31}$$

are a tensor.

Indeed, from (5.7.7), (5.7.8) and (5.7.16), we obtain the following transformation tensor formula of these components for the coordinate transformation (5.7.10):

$$
\begin{aligned}
A^{l_1 l_2 \cdots l_n} &= \Phi^+ \widehat{R} \left(\widehat{F}^{-1} \Gamma^{k_1'} \frac{\partial x^{l_1}}{\partial x^{k_1'}} \widehat{F} \right) \cdots \left(\widehat{F}^{-1} \Gamma^{k_n'} \frac{\partial x^{l_n}}{\partial x^{k_n'}} \widehat{F} \right) \Phi \\
&= \Phi^+ \widehat{F}^+ \widehat{R} \Gamma^{k_1'} \cdots \widehat{\Gamma}^{k_n'} \frac{\partial x^{l_1}}{\partial x^{k_1'}} \cdots \frac{\partial x^{l_n}}{\partial x^{k_n'}} \widehat{F} \Phi \\
&= A^{k_1' k_2' \cdots k_n'} \frac{\partial x^{l_1}}{\partial x^{k_1'}} \cdots \frac{\partial x^{l_n}}{\partial x^{k_n'}},
\end{aligned}
\tag{5.7.32}
$$

where $A^{k_1 \cdots k_n}$ and $A^{k_1' \cdots k_n'}$ are the considered components for the coordinates x^k and $x^{k'}$, respectively.

Let us also note that as follows from Theorem 5.6.2, these tensors and, in particular, the quark currents J^n are independent of a concrete choice of the matrices Γ^n that satisfy the conditions (5.5.10)–(5.5.11) and (5.7.6).

Integrating the differential equation (5.7.3) over the spatial coordinates of all the nucleons but the kth nucleon, choosing the same values for their time coordinates and using the Gauss theorem, we obtain the following equation for each value of $k(1 \leq k \leq N)$:

$$\frac{\partial(I^{k0} + V^{k0})}{c\partial t} + \frac{\partial I^{km}}{\partial x_{(k)}^m} = 0, \quad m = 1, 2, 3, \quad 1 \leq k \leq N, \quad (5.7.33)$$

where

$$I^{kn} = \int J^{4(k-1)+n} \prod_{j=1, j \neq k}^{N} d^3 x_{(j)}, \quad V^{k0} = \sum_{j=1, j \neq k}^{N} I^{j0}, \quad (5.7.34)$$

$x_{(j)}^n$ are the coordinates of the jth nucleon, $x_{(j)}^0 = ct, 1 \leq j \leq N$ and the integrals are taken over all the spatial coordinates from $-\infty$ to $+\infty$.

The values I^{kn} present the quark current densities of the kth nucleon which obey the differential equations of conservation (5.7.33) of the nucleon quark charges.

Integrating this equation over the spatial coordinates of the kth nucleon from $-\infty$ to $+\infty$ and using the Gauss theorem again, we come to the integral law of conservation of the sum of the quark charges:

$$\sum_{k=1}^{N} q_k = \text{const}, \quad q_k = \int I^{k0} d^3 x_{(k)}. \quad (5.7.35)$$

Let us choose the values $1, a, b_n$ for the 3×3 matrix k in formula (5.7.3) for the quark currents J^s. Then from formula (5.7.35), using (5.7.34), we obtain the laws of conservation of the three total quark charges.

It should be noted that putting $a = 1$, $b_s = 1$, $\varphi_s = 0$ in (5.5.1) and replacing the proton charge e_p by the electron charge $e = -e_p$, we come to the following generalization of the Dirac equation proposed in [45] for the system of N electrons occupying a small spatial region:

$$[\Gamma^s(i\hbar c\partial_s - eA_s) - \sqrt{N}m_e c^2]\Phi_e = 0, \quad 0 \leq s \leq 4N - 1, \quad (5.7.36)$$

where m_e is the rest mass of the electron and Φ_e is the wave function of the system of N electrons, having 4^N components.

This equation presents a particular case of Eq. (5.5.1) and all the results obtained above are valid for it. Since the 3×3 matrices a and b_s are absent in it, Eq. (5.7.35) for a system of electrons gives only one equality presenting the law of conservation of their total charge.

Chapter 6

New Approaches in General Relativity

This chapter deals with the problems of describing noninertial elastically deformed frames of reference and propagation of gravitational waves relative to them. First, a simpler class of such frames, which are called perfect, is studied. They are frames of reference in which one can neglect their own elastic deformations. Applying the Einstein principle of equivalence and taking into account the general requirements for frames of reference, we arrive at nonlinear differential equations for the metric tensor in the considered perfect frames.

As a result of studying the obtained differential equations, we find their exact solutions for a number of particular cases. Then, we turn to the study of elastically deformed frames of reference and find the deviation of the metric tensor with respect to them from the metric tensor in the perfect frames comoving with them. Their difference is determined by the introduced strain tensor in four-dimensional space–time. It is expressed by a formula that is a relativistic generalization of the well-known relation of the classical theory of elasticity.

The resulting equations for perfect and elastically deformed frames of reference are used to study the propagation of gravitational waves. The form of gravitational-wave solutions of Einstein's equations is found relative to the frames of reference under consideration, and their features and anomalous properties are investigated.

It is shown that under certain conditions, a significant increase in the amplitude of gravitational waves becomes possible, which greatly facilitates their detection.

At the end of this chapter, we consider the problem of the determination of the energy and momentum of the gravitational field, which is very important to study the properties of gravitational waves.

In this chapter, the results obtained in our papers [51, 52] are used.

6.1. The Problem of Describing Frames of Reference in the Einstein Gravitational Theory

In this chapter, we study new aspects of the Einstein theory of general relativity. This theory is one of the greatest achievements of the physics of the 20th century which explains a number of relativistic effects in gravitation and astronomy. At the same time, certain problems of general relativity remain still unsolved. In particular, to them, one can attribute the problem of description of gravitational waves and their energy and momentum.

In local inertial frames of reference, the gravitational waves become zero. As for their description in other frames of reference, it is not quite clear how to do this. Indeed, the Einstein gravitational equations are fulfilled in any system of space–time coordinates. That is why one does not know what a coordinate system could correspond to a concrete frame of reference. As a result, the expressions for the components of the metric tensor in the frame derived from the Einstein equations contain four arbitrary functions of space–time coordinates. To determine these components, one needs four specific conditions additional to the Einstein equations that could describe the frame under consideration. However, it is not clear how to find these conditions.

Often in order to describe gravitational waves, so-called harmonic coordinate conditions are used [13, 53–55]. These conditions are beautiful enough and simplify the Einstein equations. However, they do not have any clear physical sense.

We will study the problem of describing gravitational waves relative to the class of elastically deformed frames of reference. The problem of their description is divided into two parts.

At first, we examine and describe the simpler class of frames of reference in which elastic deformations are negligibly small. We call such frames perfect.

Then, we turn to the description of elastically deformed frames and find a relation between the metric tensors in them and in perfect frames comoving with them by introducing their strain tensor in four-dimensional space–time. The found relation generalizes the well-known formula of the nonrelativistic theory of elasticity.

Let us dwell on perfect frames of reference in which elastic deformations are very small. To them, one can attribute the following two types of frames:

(1) local perfect frames with sufficiently small intrinsic stresses,
(2) extended perfect frames that are sets of local perfect frames. In them, neighboring frames should continue each other.

Using the above definition of perfect frames and the equivalence principle of general relativity, we come to four nonlinear differential equations for the metric tensor in perfect frames. These equations contain one unknown numerical parameter λ.

In order to determine this parameter, we find a simple exact solution of the four differential equations corresponding to a uniform heating of a perfect frame. Using this solution and the requirement that measurements of space–time coordinates in a frame should be independent of instruments chosen in it, we find the sought parameter λ. Its value is the following: $\lambda = 3/4$.

It is interesting to note that the obtained four equations for perfect frames of reference would coincide with the harmonic coordinate conditions if the parameter λ had been equal to $1/2$. However, since $\lambda = 3/4$, the obtained equations for perfect frames considerably differ from the harmonic coordinate conditions.

Though the proposed equations describing perfect frames are substantially nonlinear, they have a number of beautiful exact solutions. These solutions are obtained by us in the following cases:

(1) **Relativistic rotation of a perfect frame**
It looks mysterious enough that there is success in finding an exact solution in this case only for the obtained value of the parameter $\lambda = 3/4$. The found solution could be applied to study

the lifetimes of charged particles rotating very fast under the action of powerful magnetic fields.

(2) **The gravitational field outside a spherical source in an extended perfect frame inertial at infinity**
It is interesting to note that the obtained exact solution has no singularity outside any finite spherical source. Moreover, this is true only when $\lambda = 3/4$.

(3) **Gravitational waves in perfect frames**
The obtained exact wave solutions have a surprising property. Namely, they could significantly increase under certain conditions. This property could become important for detecting gravitational waves.

6.2.   Description of Perfect Frames of Reference

As stated in Section 4.9, the Einstein equations of the theory of general relativity can be represented in the form [13]

$$R_{\mu\nu} - \frac{1}{2}Rg_{\mu\nu} = \kappa T_{\mu\nu}, \qquad (6.2.1)$$

Albert Einstein
(1879–1955)

where $R_{\mu\nu}$ is the Ricci tensor, $T_{\mu\nu}$ is the energy–momentum tensor of matter and nongravitational fields and κ is the Einstein constant. Owing to their tensor character, they are valid in an arbitrary coordinate system. Therefore, their general solution depends on four arbitrary functions of space–time coordinates since it can be obtained from the solution in some coordinate system by an arbitrary transformation of its four coordinates [13].

In view of this, the problem arises of finding four differential equations, in addition to Eq. (6.2.1), describing specific frames of reference with a chosen method for measuring spatial distances and time, which arbitrarily move in gravitational fields. It should be noted that the description of astronomical frames of reference for the theoretical study of relativistic effects in the motion of celestial bodies and the study of the propagation of gravitational waves are of particular interest.

Consider this problem for the following class of frames of reference, which we will call perfect:

Definition 6.2.1. We will call a frame of reference perfect if we can neglect elastic deformations in the bodies with which it is associated.

There are two main types of perfect frames:

(1) Local perfect frames of reference
They include reference frames of small size and, accordingly, small mass, in which internal stresses can be neglected.
(2) Extended perfect frames of reference
We consider them as a large set of local perfect frames of reference, in which neighboring systems are at rest relative to each other.

Thus, such frames are an extension of their local perfect frames.

In perfect frames of reference, we introduce orthogonal spatial axes associated with them and use Cartesian coordinates x^1, x^2, x^3. At the nodes of the coordinate grid, we place identical clocks that measure time t and, accordingly, the time coordinate $x^0 = ct$. These space–time coordinates $x^\mu, \mu = 0, 1, 2, 3$, will be called rectangular.

To establish a clock synchronization procedure, let us use the Einstein principle of equivalence. Consider first a local frame of reference moving with an arbitrary acceleration. Then, by virtue of this principle, the considered noninertial frame is equivalent to a local inertial frame around which some gravitational field acts. Therefore, the synchronization procedure should be the same as in the local inertial frame. At the same time, the method of clock synchronization in the inertial frame, which we choose as a standard, should be independent of the gravitational field acting around it.

The following procedure for setting synchronized clocks has just such a property: It is necessary to set identically running clocks at one point of the frame of reference and then slowly, applying small forces, transfer them from this point to other points of the frame.

In view of the principle of equivalence, this method of synchronization can also be extended to arbitrary local noninertial frames of reference.

Let us now turn to clock synchronization in the considered extended frames of reference, which are a set of local systems that continue each other. In them, we can synchronize clocks as follows: In each local frame, synchronization must be carried out in the manner indicated above, while in the common part of two neighboring local frames, the clocks used for them must be common.

We now investigate the problem of describing the space–time metric in perfect frames of reference with rectangular coordinates x^μ chosen in them in the above way. To do this, consider two local perfect frames of reference moving with acceleration in some small spatial region, and let the second frame have zero acceleration relative to the first one. Then, in accordance with the principle of equivalence, their relative motion will be the same as for two inertial frames around which a gravitational field acts.

Therefore, as in two inertial frames, the rectangular coordinates x^μ and $\bar{x}^\mu$ in the considered local noninertial frames with zero relative acceleration should be related by a linear Lorentz transformation.

As a result, we arrive at the following requirement for four differential equations that should describe perfect frames of reference:

Requirement 1. *Four differential equations describing perfect frames of reference should be covariant under linear transformations of space–time coordinates x^μ in these frames.*

A frame of reference that satisfies Requirement 1 can be characterized by two dimensionless tensors: $g_{\mu\nu}$ and $S_{\mu\nu}/E$, where $g_{\mu\nu}$ are the components of the metric tensor in this frame, $S_{\mu\nu}$ is the stress tensor in it and E is its modulus of elasticity. Since in perfect frames of reference, their stresses are small and the expressions $S_{\mu\nu}/E$ can be neglected in them, we add the following requirement:

Requirement 2. *Perfect frames of reference can be described by differential equations for only the components $g_{\mu\nu}$ of the metric tensor.*

Another requirement that we introduce for perfect frames of reference is of a general nature and should be applied to any frame of reference:

Requirement 3. *The equations describing perfect frames of reference should be independent of the design of instruments for measuring time and distances in these frames.*

It is easy to verify that the following four differential equations satisfy Requirements 1–3:

$$g^{\alpha\beta}\frac{\partial g_{\mu\alpha}}{\partial x^\beta} = \lambda g^{\alpha\beta}\frac{\partial g_{\alpha\beta}}{\partial x^\mu}, \tag{6.2.2}$$

where $g_{\mu\nu}$ are components of the metric tensor in a perfect frame of reference, x^μ are rectangular coordinates in it and λ is some unknown constant which will be determined later on by using Requirement 3.

Indeed, for the linear coordinate transformations

$$x^\mu = a^\mu_\nu \tilde{x}^\nu + b^\mu, \quad a^\mu_\nu = \text{const}, \quad b^\mu = \text{const}, \tag{6.2.3}$$

we have the tensor transformations

$$\tilde{g}_{\mu\nu} = a^\alpha_\mu a^\beta_\nu g_{\alpha\beta}, \quad \frac{\partial \tilde{g}_{\mu\nu}}{\partial \tilde{x}^\gamma} = a^\alpha_\mu a^\beta_\nu \frac{\partial g_{\alpha\beta}}{\partial x^\sigma}\frac{\partial x^\sigma}{\partial \tilde{x}^\gamma} = a^\alpha_\mu a^\beta_\nu a^\sigma_\gamma \frac{\partial g_{\alpha\beta}}{\partial x^\sigma}, \tag{6.2.4}$$

where $\tilde{g}_{\mu\nu}$ and $g_{\mu\nu}$ are components of the metric tensor in the two coordinate systems $\tilde{x}^\mu$ and x^μ, related by the linear transformations (6.2.3). Thus, for linear coordinate transformations, both the metric tensor and its derivatives transform as tensors. Therefore, the differential equations (6.2.2) are covariant under linear coordinate transformations.

It is necessary to note that the Minkowski metric tensor, corresponding to inertial frames of reference in the absence of gravitational fields, satisfies Eq. (6.2.2) since its derivatives are zero.

Let $\tilde{x}^\mu = f_\mu(x^\nu)$, where $\tilde{x}^\mu$ are space–time coordinates in a coordinate system with certain components of its metric tensor $\tilde{g}_{\mu\nu}$ and x^μ are coordinates in a perfect frame of reference. Then, since the components $g_{\mu\nu}$ of the metric tensor in the coordinates x^μ are expressed by the first derivatives of the functions $f_\mu(x^\nu)$, Eq. (6.2.2) presents four nonlinear differential equations for $f_\mu(x^\nu)$. As can be easily seen, these equations are differential equations of the second order which are linear with respect to the second derivatives of the functions $f_\mu(x^\nu)$.

This property of Eq. (6.2.2) is important for the correctness of the differential equations for the functions $f_\mu(x^\nu)$ and boundary value problems for them since otherwise their solutions would be ambiguous. Moreover, the differential equations (6.2.2) are the only possible construction that has this property and satisfies Requirements 1–3 since only these four equations for the metric tensor

$g_{\mu\nu}(x^\gamma)$ are covariant under linear coordinate transformations and contain derivatives of $g_{\mu\nu}$ not higher than first order.

It should be noted that a construction similar to Eq. (6.2.2) but containing derivatives of the contravariant components $g^{\mu\nu}$ of the metric tensor does not represent new equations since they follow from Eq. (6.2.2).

Let us show this. For this purpose, we can use the well-known formula relating the derivatives of the covariant and contravariant components of the metric tensor [13]:

$$\frac{\partial g_{\mu\nu}}{\partial x^\gamma} = -g_{\mu\alpha}g_{\nu\beta}\frac{\partial g^{\alpha\beta}}{\partial x^\gamma}, \qquad (6.2.5)$$

and express the derivatives of $g_{\mu\alpha}$ and $g_{\alpha\beta}$ in (6.2.2) by the derivatives of the contravariant components of the metric tensor. Then, from (6.2.2) and (6.2.5), we obtain

$$-g^{\alpha\beta}g_{\mu\gamma}g_{\alpha\sigma}\frac{\partial g^{\gamma\sigma}}{\partial x^\beta} = -\lambda g^{\alpha\beta}g_{\alpha\gamma}g_{\beta\sigma}\frac{\partial g^{\gamma\sigma}}{\partial x^\mu}. \qquad (6.2.6)$$

From (6.2.6), we readily find that

$$g_{\mu\gamma}\frac{\partial g^{\gamma\sigma}}{\partial x^\sigma} = \lambda g_{\gamma\sigma}\frac{\partial g^{\gamma\sigma}}{\partial x^\mu}. \qquad (6.2.7)$$

Thus, we come to the differential equations (6.2.7) which are equivalent to Eq. (6.2.2) but contain derivatives of the contravariant components of the metric tensor.

Let us apply the following known formula [13]:

$$g^{\alpha\beta}\frac{\partial g_{\alpha\beta}}{\partial x^\mu} = -g_{\alpha\beta}\frac{\partial g^{\alpha\beta}}{\partial x^\mu} = \frac{1}{g}\frac{\partial g}{\partial x^\mu}, \quad g = \det(g_{\mu\nu}). \qquad (6.2.8)$$

From here, it follows that Eq. (6.2.2) can be represented in the form

$$g^{\alpha\beta}\frac{\partial g_{\mu\alpha}}{\partial x^\beta} = \frac{\lambda}{g}\frac{\partial g}{\partial x^\mu}. \qquad (6.2.9)$$

Let us multiply Eq. (6.2.7) by the expression $gg^{\mu\nu}$. Then, we find

$$g\frac{\partial g^{\nu\sigma}}{\partial x^\sigma} = \lambda gg^{\mu\nu}g_{\gamma\sigma}\frac{\partial g^{\gamma\sigma}}{\partial x^\mu}. \qquad (6.2.10)$$

From (6.2.8) and (6.2.10), we get

$$g\frac{\partial g^{\mu\nu}}{\partial x^\mu} = -\lambda g^{\mu\nu}\frac{\partial g}{\partial x^\mu}, \quad g = \det(g_{\mu\nu}). \qquad (6.2.11)$$

The obtained equations (6.2.11) are equivalent to the differential equations (6.2.2) and follow from them by the transition from the derivatives of the covariant components of the metric tensor to the derivatives of its contravariant components and using formula (6.2.8) for the derivative of the metric tensor determinant.

Let us now consider the case of a homogeneous expansion of a perfect frame of reference due to the influx of heat into it and an increase in temperature. As we shall see, this case admits a simple exact solution.

6.3. Homogeneous Thermal Expansion of a Perfect Frame of Reference

We assume that there is no gravitational field and choose an inertial frame of reference with coordinates $\bar{x}^\mu$ in which the four-dimensional interval ds has the form

$$ds^2 = (d\bar{x}^0)^2 - (d\bar{x}^1)^2 - (d\bar{x}^2)^2 - (d\bar{x}^3)^2. \qquad (6.3.1)$$

Consider the following transformations of the coordinates $\bar{x}^\mu$:

$$\bar{x}^0 = \beta(x^0), \ \bar{x}^1 = \alpha_1(x^0)x^1, \ \bar{x}^2 = \alpha_2(x^0)x^2, \ \bar{x}^3 = \alpha_3(x^0)x^3, \quad (6.3.2)$$

where β and α_i are some functions of the time coordinate x^0.

Let x^μ be the coordinates in a perfect frame of reference which is subject to the homogeneous expansion relative to the inertial frame with $\bar{x}^\mu$ of form (6.3.2) due to the influx of heat into it and an increase in temperature. Then, the functions $\alpha_i \geq 1$.

The considered functions $\beta(x^0)$ and $\alpha_i(x^0)$ should be such that the metric tensor $g_{\mu\nu}(x^\gamma)$ in the coordinates x^μ of the perfect frame satisfies Eq. (6.2.2), which describes it.

Let us turn to the determination of the form of these functions.

For this purpose, let us substitute expressions (6.3.2) into formula (6.3.1) for the space–time interval and find the components

$g_{\mu\nu}(x^\gamma)$ of the metric tensor:

$$ds^2 = (d\bar{x}^0)^2 - (d\bar{x}^1)^2 - (d\bar{x}^2)^2 - (d\bar{x}^3)^2 = g_{\mu\nu}dx^\mu dx^\nu, \qquad (6.3.3)$$

where

$$g_{00} = \beta'^2 - \sum_{i=1}^{3}(\alpha'_i x^i)^2, \quad g_{0k} = -\alpha'_k\alpha_k x^k, \quad g_{kk} = -\alpha_k^2,$$

$$g_{ik} = 0, \quad i \neq k, \quad i,k = 1,2,3 \qquad (6.3.4)$$

and the derivatives are taken with respect to the coordinate x^0.

Knowing the covariant components $g_{\mu\nu}(x^\gamma)$ of the metric tensor, we can also determine its contravariant components $g^{\mu\nu}(x^\gamma)$. It can be readily verified that they have the following form:

$$g^{00} = 1/\beta'^2, g^{0k} = -\alpha'_k x^k/(\alpha_k\beta'^2),$$

$$g^{ik} = (\alpha'_i\alpha'_k x^i x^k - \delta_{ik}\beta'^2)/(\alpha_i\alpha_k\beta'^2), i,k = 1,2,3, \qquad (6.3.5)$$

$$\delta_{ii} = 1, \delta_{ik} = 0, i \neq k.$$

Let us now substitute formulas (6.3.4) and (6.3.5) for $g_{\mu\nu}$ and $g^{\mu\nu}$ into Eq. (6.2.2). Then, the first equation in (6.2.2) ($\mu = 0$) gives

$$(1 - 2\lambda)\sum_{i=1}^{3}\beta'^2\alpha'_i/\alpha_i + (1-\lambda)(\beta'^2)'$$

$$+ \sum_{i=1}^{3}(x^i)^2(2\alpha'^2_i - \alpha_i\alpha''_i)\alpha'_i/\alpha_i = 0. \qquad (6.3.6)$$

From (6.3.6), we obtain the equations

$$(\beta'^2)'/\beta'^2 = [(2\lambda - 1)/(1-\lambda)]\sum_{i=1}^{3}\alpha'_i/\alpha_i, \qquad (6.3.7)$$

$$\alpha''_k/\alpha'_k = 2\alpha'_k/\alpha_k, \quad k = 1,2,3. \qquad (6.3.8)$$

The other three equations in (6.2.2) ($\mu = 1,2,3$) give

$$\alpha_k\alpha''_k = 2\alpha'^2_k, \quad k = 1,2,3. \qquad (6.3.9)$$

It is interesting to note that Eqs. (6.3.9) are not new since they coincide with Eqs. (6.3.8). As will be seen in the following, this important fact just allows us to find the exact solution to Eq. (6.2.2) that corresponds to the coordinate transformations of form (6.3.2).

Integrating Eqs. (6.3.7) and (6.3.8), to which Eq. (6.2.2) leads in the considered case, we obtain

$$2\ln(\beta'/b_0) = [(2\lambda - 1)/(1 - \lambda)]\ln(\alpha_1\alpha_2\alpha_3),$$
$$\ln(\alpha'_k/\alpha_{0k}) = \ln\alpha_k^2, \quad k = 1, 2, 3, \tag{6.3.10}$$

where b_0 and α_{0k} are some constants.

From (6.3.10), we find

$$\beta' = b_0(\alpha_1\alpha_2\alpha_3)^{(2\lambda-1)/(2-2\lambda)}, \quad (1/\alpha_k)' = -\alpha_{0k}. \tag{6.3.11}$$

Therefore, we get

$$\beta(x^0) = b_0 \int (\alpha_1\alpha_2\alpha_3)^{(2\lambda-1)/(2-2\lambda)} dx^0,$$
$$\alpha_k(x^0) = u_{0k}/(1 + v_{0k}x^0), \quad k = 1, 2, 3, \tag{6.3.12}$$

where b_0, u_{0k}, v_{0k} are some constants.

The obtained formulas (6.3.2), (6.3.4) and (6.3.12) present the sought exact solution to Eq. (6.2.2), which corresponds to a homogeneous thermal expansion of a perfect frame of reference.

Let us apply this solution to find the value of the unknown parameter λ in Eq. (6.2.2).

6.4. Determination of the Unknown Parameter in the Equations for Perfect Frames of Reference

Let $\Delta\bar{t}, \Delta\bar{V}$ and $\Delta t, \Delta V$ be a small time interval and a small three-dimensional volume at the moment $x^0 = \text{const}$, respectively, in the above-considered inertial frame with coordinates $\bar{x}^\mu$ and perfect frame with coordinates x^μ, related with $\bar{x}^\mu$ by formulas (6.3.2) and (6.3.12). Then, from these formulas, we get

$$\Delta\bar{V}/\Delta V = \alpha_1\alpha_2\alpha_3, \quad \Delta\bar{t}/\Delta t = \beta' = b_0(\alpha_1\alpha_2\alpha_3)^{(2\lambda-1)/(2-2\lambda)}. \tag{6.4.1}$$

This gives

$$\Delta\bar{t}/\Delta t = \text{const} \times (\Delta\bar{V}/\Delta V)^{(2\lambda-1)/(2-2\lambda)}. \tag{6.4.2}$$

Let us now choose a procedure for measuring time in a perfect frame of reference. In accordance with Requirement 3 formulated above, the description of perfect frames of reference and, consequently, the parameter λ in Eq. (6.2.2) should be independent of this choice.

To measure time, it is necessary to set a unit of time and, as it, we choose the period of a certain physical–chemical process in a unit volume. For example, as a unit of time in the considered perfect system, we choose the period required for the evaporation of a unit volume of water at a temperature of $100°$C.

Consider a perfect frame of reference with coordinates x^{μ}, which was initially at rest relative to the inertial frame with coordinates $\overline{x}^{\mu}$, and then began to slowly expand relative to it as a result of gradual uniform heating. Let us choose in this perfect frame of reference a unit time interval Δt and a unit spatial volume ΔV:

$$\Delta t = 1, \quad \Delta V = 1. \tag{6.4.3}$$

Since the velocities of the considered perfect frame are small relative to the inertial frame with coordinates $\overline{x}^{\mu}$, in this inertial frame, the corresponding period $\Delta \overline{t}$ of evaporation of the volume $\Delta \overline{V}$ of water at temperature $100°$C is proportional to this volume.

Therefore, for a unit time interval Δt and a unit volume ΔV in the considered perfect frame and for their values $\Delta \overline{t}$ and $\Delta \overline{V}$ in the inertial frame with coordinates $\overline{x}^{\mu}$, we obtain

$$\Delta t = 1, \quad \Delta V = 1, \quad \Delta \overline{t} = \text{const} \times \Delta \overline{V}. \tag{6.4.4}$$

It should be noted that the proportionality of $\Delta \overline{t}$ and $\Delta \overline{V}$ easily follows from the following reasoning. In an inertial frame, the heat $\Delta \overline{Q}$ required to evaporate water is proportional to its volume $\Delta \overline{V}$. On the other hand, the heat $\Delta \overline{Q}$ received in this volume is proportional to the time $\Delta \overline{t}$ of its heating. Therefore, the time $\Delta \overline{t}$ is proportional to the volume $\Delta \overline{V}$.

Let us now turn to formula (6.4.2). From it, we obtain the following result for the considered unit interval Δt and unit volume ΔV in the perfect frame:

$$\Delta t = 1, \quad \Delta V = 1, \quad \Delta \overline{t} = \text{const} \times (\Delta \overline{V})^{(2\lambda-1)/(2-2\lambda)}. \tag{6.4.5}$$

Comparing formulas (6.4.4) and (6.4.5), we obtain the following equation for the determination of the parameter λ:

$$(2\lambda - 1)/(2 - 2\lambda) = 1. \tag{6.4.6}$$

From Eq. (6.4.6), we easily find the value of this parameter:

$$\lambda = 3/4. \tag{6.4.7}$$

Thus, the obtained formula (6.4.7) gives the sought value of the parameter λ in the differential equations (6.2.2), which describe perfect frames of reference, and these equations acquire the form

$$g^{\alpha\beta}\frac{\partial g_{\mu\alpha}}{\partial x^\beta} = \frac{3}{4}g^{\alpha\beta}\frac{\partial g_{\alpha\beta}}{\partial x^\mu}. \tag{6.4.8}$$

Let us now turn to the solution to these equations for a freely rotating perfect frame of reference, including the case of relativistic velocities.

6.5. Free Relativistic Rotation of a Perfect Frame of Reference

For solving this problem, let us apply the obtained equations (6.4.8) for perfect frames of reference, which coincide with Eq. (6.2.2), taking into account that the found value $\lambda = 3/4$. We assume that there is no gravitational field.

Consider a perfect frame of reference freely rotating in the plane $\overline{x}\,\overline{y}$, perpendicular to the axis $\overline{z}$. Let the coordinates $\overline{x}^\mu(c\overline{t}, \overline{x}, \overline{y}, \overline{z})$ correspond to an inertial frame of reference and the coordinates $x^\mu(ct, x, y, z)$ to the rotating frame. The axes $\overline{z}$ and z are considered coinciding.

We will seek the relation between the coordinates $\overline{x}^\mu$ and x^μ in the following form:

$$\overline{x} = f(r)(x\cos\omega t - y\sin\omega t), \quad \overline{y} = f(r)(x\sin\omega t + y\cos\omega t),$$
$$\overline{z} = z, \quad \overline{t} = t, \quad r = \sqrt{x^2 + y^2}, \tag{6.5.1}$$

where $f(r)$ is some positive function, which should be determined, and ω is a constant angular velocity of rotation of the perfect frame.

Taking into account that the coordinates $\overline{x}^\mu(c\overline{t}, \overline{x}, \overline{y}, \overline{z})$ correspond to the inertial frame and using formulas of the coordinate transformations (6.5.1), we can find the components of the metric tensor $g_{\mu\nu}(x^\gamma)$ in the coordinates $x^\mu(ct, x, y, z)$ of the considered perfect frame:

$$ds^2 = c^2 d\overline{t}^2 - d\overline{x}^2 - d\overline{y}^2 - d\overline{z}^2 = g_{\mu\nu}dx^\mu dx^\nu, \tag{6.5.2}$$

where

$$g_{00} = 1 - f^2\omega^2 r^2/c^2, \quad g_{01} = f^2\omega y/c, \quad g_{02} = -f^2\omega x/c,$$

$$g_{11} = -px^2 - f^2, \quad g_{12} = -pxy, \quad g_{22} = -py^2 - f^2, \qquad (6.5.3)$$

$$p = f'(f' + 2f/r), \quad g_{03} = g_{13} = g_{23} = 0, \quad g_{33} = -1.$$

Using the found covariant components $g_{\mu\nu}(x^\gamma)$ of the metric tensor, we can also determine its contravariant components $g^{\mu\nu}(x^\gamma)$. Calculating them, we obtain the following expressions:

$$g^{00} = 1, \quad g^{01} = \omega y/c, \quad g^{02} = -\omega x/c,$$

$$g^{11} = [(1 - f^2\omega^2 r^2/c^2)(f^2 + y^2 p) + f^4\omega^2 x^2/c^2]/g,$$

$$g^{12} = [f^4\omega^2/c^2 - (1 - f^2\omega^2 r^2/c^2)p]xy/g, \qquad (6.5.4)$$

$$g^{22} = [(1 - f^2\omega^2 r^2/c^2)(f^2 + x^2 p) + f^4\omega^2 y^2/c^2]/g,$$

$$g = \det(g_{\mu\nu}) = -f^2(f + rf')^2, \quad p = f'(f' + 2f/r).$$

As shown above, Eq. (6.2.2) describing perfect frames of reference is equivalent to Eq. (6.2.11), which has the following form, taking into account the found value (6.4.7) of the parameter λ ($\lambda = 3/4$):

$$g\frac{\partial g^{\mu\nu}}{\partial x^\mu} = -\frac{3}{4}g^{\mu\nu}\frac{\partial g}{\partial x^\mu}, \quad g = \det(g_{\mu\nu}). \qquad (6.5.5)$$

Equations (6.5.5) are equations for perfect frames of reference, which are represented as differential relations for contravariant components of the metric tensor and its determinant.

Let us substitute expressions (6.5.4) for $g^{\mu\nu}$ and g into Eq. (6.5.5). Then, we obtain that the first ($\nu = 0$) and fourth ($\nu = 3$) equations in (6.5.5) are identically satisfied for any differentiable function $f(r), r = \sqrt{x^2 + y^2}$. As for the second ($\nu = 1$) and third ($\nu = 2$) equations in (6.5.5), they give the same equation for the function $f(r)$, which can be represented in the form

$$2(f^4\omega^2 r/c^2 - rf'^2 + f^2 f'^2\omega^2 r^3/c^2 + 2f^3 f'\omega^2 r^2/c^2)$$

$$\times (f + rf') = f(3ff' + rf'^2 + rff''), \quad r = \sqrt{x^2 + y^2}. \quad (6.5.6)$$

Let us introduce the new variable u and the function $q(u)$:

$$u = \omega^2 r^2/c^2, \quad q(u) = uf^2(u). \qquad (6.5.7)$$

Using them, we find

$$f = (q/u)^{1/2}, \quad \frac{df}{dr} = \frac{df}{du}\frac{du}{dr}$$

$$= \frac{\omega}{c}q^{-1/2}(q' - q/u), \quad q = q(u),$$

$$\frac{d^2f}{dr^2} = \frac{d}{du}\left(\frac{df}{dr}\right)\frac{du}{dr}$$

$$= \frac{\omega^2}{c^2}(u/q)^{1/2}(2q'' - q'^2/q - q'/u + 2q/u^2).$$

(6.5.8)

As a result, from (6.5.6), using (6.5.7) and (6.5.8), we derive the following equation for the function $q(u)$:

$$qq'' = q'[2qq' - q^2/u - (1-q)(uq'^2/q - 2q' + q/u)], \quad q = q(u).$$

(6.5.9)

Let us now introduce the inverse function $u = u(q)$, satisfying the equality $q(u(q)) = q$. Taking the first and second derivatives of this equality with respect to the variable q, we find

$$q(u(q)) = q, \quad q'(u(q))u'(q) = 1,$$

$$q''(u(q))u'^2(q) + q'(u(q))u''(q) = 0.$$

(6.5.10)

This gives

$$q'(u(q)) = 1/u'(q), \quad q''(u(q)) = -u''(q)/u'^3(q).$$

(6.5.11)

Putting $u = u(q)$ in Eq. (6.5.9) and using (6.5.11), we come to the following equation for the function $u(q)$:

$$qu'' = (1-q)u/q - 2u' + qu'^2/u, \quad u = u(q).$$

(6.5.12)

Introducing the function

$$h(q) = q/u(q),$$

(6.5.13)

we find

$$u = \frac{q}{h}, \quad u' = -\frac{1}{h^2}(qh' - h),$$

$$u'' = -\frac{1}{h^3}(qhh'' - 2qh'^2 + 2hh').$$

(6.5.14)

Substituting now (6.5.14) into Eq. (6.5.12), we get

$$qhh'' - qh'^2 + 2hh' - h^2 = 0, \quad h = h(q). \tag{6.5.15}$$

Let us divide this equation by h^2. Then, it acquires the form

$$q(h'/h)' + 2(h'/h) - 1 = 0, \quad h = h(q). \tag{6.5.16}$$

As is seen, the obtained equation (6.5.16) is a linear differential equation of the first order with respect to the function h'/h. From it, we readily obtain

$$h'/h = 1/2 + \alpha/q^2, \quad \alpha = \text{const.} \tag{6.5.17}$$

From (6.5.17), we find the general form of the function $h(q)$ giving solutions to Eq. (6.5.15):

$$h(q) = \beta \exp(q/2 - \alpha/q), \quad \alpha = \text{const}, \quad \beta = \text{const.} \tag{6.5.18}$$

From (6.5.7) and (6.5.13), we have

$$h = q/u = f^2, \quad u = \omega^2 r^2/c^2. \tag{6.5.19}$$

Substituting (6.5.19) into (6.5.18), we find the following equality for the function $f(r)$:

$$2\ln(f^2/\beta) = f^2\omega^2 r^2/c^2 - 2\alpha c^2/(f^2\omega^2 r^2), \quad f = f(r). \tag{6.5.20}$$

The obtained formula (6.5.20) presents the general solution to the nonlinear differential equation of the second order (6.5.6).

Consider the point $r = 0$. As follows from (6.5.1), this point is at rest relative to the inertial frame of reference with the coordinates $\bar{x}^\mu$. Therefore, in a small vicinity of the immovable point $r = 0$ of the rotating perfect frame, formulas (6.5.1), relating the coordinates $\bar{x}^\mu$ and x^μ of the considered frames of reference, should coincide with the well-known formulas of nonrelativistic physics. To provide this, the value $f(0)$ should be as follows:

$$f(0) = 1. \tag{6.5.21}$$

As follows from (6.5.21), the numbers α and β in formula (6.5.20) should have the values

$$\alpha = 0, \quad \beta = 1. \tag{6.5.22}$$

Therefore, formula (6.5.20) acquires the form

$$4\ln f/f^2 = \omega^2 r^2/c^2. \tag{6.5.23}$$

The obtained formulas (6.5.1)–(6.5.3) and (6.5.23) give the solution of the considered problem.

It is interesting to note that such a simple solution to Eq. (6.2.2) for a freely rotating perfect frame can be obtained only for the value of the parameter $\lambda = 3/4$ determined above.

It follows from formulas (6.5.1) that the value $|f\omega r|$ is the velocity of a point of the rotating frame of reference relative to the considered inertial frame. Since this velocity should be less than the speed of light $|f\omega r| < c$, it readily follows from (6.5.23) that the values of $f(r)$ satisfy the inequalities

$$1 \le f(r) < \exp(1/4). \tag{6.5.24}$$

From (6.5.23) and (6.5.24), we easily obtain that $f(r)$ is a monotonically increasing function running from 1 to $\exp(1/4)$.

Let us turn to expression (6.5.2)–(6.5.3) for the interval ds in the coordinates t, x, y, z of a freely rotating perfect frame of reference.

Let us choose the polar coordinates r, ϕ, related with x, y by the formulas

$$x = r \cos \phi, \quad y = r \sin \phi. \tag{6.5.25}$$

Then, we obtain the following equalities:

$$dx^2 + dy^2 = dr^2 + r^2 d\phi^2, \quad y dx - x dy = -r^2 d\phi,$$
$$x dx + y dy = r dr. \tag{6.5.26}$$

Using (6.5.26) in formulas (6.5.2)–(6.5.3), we come to the following formula for the interval ds in the coordinates t, r, ϕ, z of the perfect frame freely rotating with the angular velocity ω:

$$ds^2 = (c^2 - f^2\omega^2 r^2)dt^2 - 2f^2\omega r^2 dt d\phi - f^2 r^2 d\phi^2$$
$$- (rf' + f)^2 dr^2 - dz^2. \tag{6.5.27}$$

Let us differentiate formula (6.5.23). Then, we find

$$2(1 - 2\ln f)f' = f^3\omega^2 r/c^2. \tag{6.5.28}$$

From formulas (6.5.23) and (6.5.28), we readily obtain

$$rf' + f = f \Big/ \left[1 - \frac{1}{2}f^2\omega^2 r^2/c^2\right]. \tag{6.5.29}$$

Using (6.5.29) and (6.5.23), we can represent formula (6.5.27) for the interval ds in the coordinates t, r, ϕ, z of a freely rotating perfect frame as

$$ds^2 = c^2(1 - f^2\omega^2 r^2/c^2)dt^2 - 2f^2\omega r^2 dt d\phi - f^2 r^2 d\phi^2$$

$$- f^2 dr^2 \Big/ \left(1 - \frac{1}{2}f^2\omega^2 r^2/c^2\right)^2 - dz^2, \tag{6.5.30}$$

$$4\ln f/f^2 = \omega^2 r^2/c^2.$$

Above, we considered perfect frames of reference, in which their internal stresses can be neglected. In the following section, we will investigate elastically deformed frames of reference, in which their internal stresses will be taken into account.

6.6. Description of Elastically Deformed Frames of Reference

Let us pose the problem of describing the space–time metric with respect to the frame of reference associated with an arbitrarily moving elastic body, taking into account the stresses arising in it and, consequently, elastic deformations. We call such a frame of reference elastically deformed.

We will further consider the case in which the deformations in elastically deformed frames are small and obeys Hooke's law.

Consider an arbitrary elastically deformed frame of reference and denote the components of the metric tensor in it by $g^{(e)}_{\mu\nu}$. Along with this frame, we will also consider the perfect frame comoving with it and differing from it only in the absence of elastic deformations. The components of the metric tensor in this comoving perfect frame will be denoted by $g_{\mu\nu}$. The presence of the index (e) will indicate an elastically deformed frame of reference and its absence will indicate the perfect frame comoving with it. The difference between the components $g^{(e)}_{\mu\nu}$ and $g_{\mu\nu}$ should be related in some way to the components of the stress tensor $S_{\mu\nu}$. We will seek this relation later on.

Let $x^{\mu}_{(e)}$ and x^{μ} be rectangular space–time coordinates in an elastically deformed frame of reference and the perfect frame comoving with it, respectively, and introduce a 4-vector of small elastic deformations u^{μ}. For this purpose, we give the following two definitions:

Definition 6.6.1. Let two space–time points $P_{(e)}$ and P have the same coordinates $x^\mu_{(e)}$ and $x^\mu (x^\mu_{(e)} = x^\mu)$ in the considered elastically deformed frame of reference and the perfect frame comoving with it, respectively. Then, we will call the point $P_{(e)}$ a displaced image of the point P.

Definition 6.6.2. Consider an arbitrary coordinate system y^μ and define a vector function of small elastic displacements $u^\mu = u^\mu(y^\nu)(\mu, \nu = 0, 1, 2, 3)$ for points of an elastically deformed frame of reference with respect to points of the perfect frame comoving with it as

$$u^\mu = \overleftarrow{y}^\mu - y^\mu, \qquad (6.6.1)$$

where $\overleftarrow{y}^\mu = \overleftarrow{y}^\mu(y^\nu)$ are the coordinates in the considered coordinate system of the displaced image $P_{(e)}$ of the point P having the coordinates y^μ.

Let us show that the introduced 4-vector u^μ is a relativistic generalization of the vector of small elastic displacements of the nonrelativistic theory of elasticity. For this purpose, let us take the coordinate system x^μ of the considered perfect frame as the coordinates y^μ and choose a point P, connected with this system, that is having constant spatial coordinates in it. In this coordinate system, from (6.6.1), we have

$$\overleftarrow{x}^\mu = x^\mu + u^\mu, \qquad (6.6.2)$$

where u^μ are the components of the vector function of small elastic displacements at the point P of the considered perfect frame and x^μ and $\overleftarrow{x}^\mu$ are the coordinates of the point P and its displaced image $P_{(e)}$ in this frame, respectively.

Owing to Definition 6.6.1, the coordinates $x^\mu_{(e)}$ and x^μ of the space–time points $P_{(e)}$ and P are the same in the elastically deformed and perfect frames, respectively:

$$x^\mu_{(e)} = x^\mu. \qquad (6.6.3)$$

Since we have chosen the point P with unchangeable spatial coordinates relative to the considered perfect frame of reference, from (6.6.3), we obtain that the spatial coordinates of the point $P_{(e)}$ are unchangeable relative to the elastically deformed frame.

When $\mu = 1, 2, 3$, the values $\overset{\leftarrow}{x}^{\mu}$ are the spatial coordinates of the point $P_{(e)}$ in the perfect frame and $x^{\mu}_{(e)}$ are unchangeable spatial coordinates of the considered point $P_{(e)}$ in the elastically deformed frame, which are equal to its spatial coordinates in this frame before the deformation. Therefore, as follows from (6.6.2) and (6.6.3), when $\mu = 1, 2, 3$, the components u^{μ} in the perfect frame of reference are the standard vector of elastic displacement and hence, we have a relativistic generalization of this notion.

Let us give one more definition.

Definition 6.6.3. In an arbitrary frame of reference, let us define the strain tensor $\varepsilon_{\mu\nu}$ as

$$\varepsilon_{\mu\nu} = (\nabla_{\mu} u_{\nu} + \nabla_{\nu} u_{\mu})/2, \qquad (6.6.4)$$

where ∇_{μ} is the covariant derivative.

As will be shown in the following, this definition is a relativistic generalization of the strain tensor of the classical theory of elasticity.

Let us now seek a relativistic relation between the stress and strain tensors. It should satisfy the following two requirements:

(1) The sought relation should be covariant under arbitrary transformations of space–time coordinates.
(2) In the nonrelativistic case, this relation should be the traditional relation between stresses and strains in the classical theory of elasticity.

Let us show that the following relation satisfies the two requirements:

$$\varepsilon_{\mu\nu} = [(1 + \nu_0)/E_0]\{S_{\mu\nu} - [\nu_0/(1 + \nu_0)]S_{\alpha\beta}g^{\alpha\beta}g_{\mu\nu}\}, \qquad (6.6.5)$$

where the strains $\varepsilon_{\mu\nu}$ are defined by formula (6.6.4), $S_{\mu\nu}$ are the stresses in the elastic body, with which the elastically deformed frame under consideration is connected, and E_0 and ν_0 are the modulus of elasticity and the Poisson coefficient of this elastic body, respectively.

Since the components u^μ, introduced above, are a vector, the components $\varepsilon_{\mu\nu}$, defined by formula (6.6.4), are a covariant tensor. Taking into account that $S_{\mu\nu}$ and $g_{\mu\nu}$ are tensors as well, relation (6.6.5) is covariant under arbitrary transformations of space–time coordinates and hence satisfies the first requirement.

Let us turn to the second requirement. In the nonrelativistic case, we have [56]

$$S_{00} \approx 0; \quad \sigma_{kl} = -S_{kl}; \quad e_{kl} = -\varepsilon_{kl} = (\partial u^k/\partial x^l + \partial u^l/\partial x^k)/2;$$

$$g_{kl} = -\delta_{kl}; \quad g_{00} = 1; \quad g_{0k} = 0; \quad k, l = 1, 2, 3, \tag{6.6.6}$$

where δ_{kl} is the Kronecker symbol and σ_{kl} and e_{kl} are the three-dimensional stress and strain tensors, respectively.

From (6.6.5) and (6.6.6), we obtain

$$e_{kl} = [(1+\nu_0)/E_0]\{\sigma_{kl} - [\nu_0/(1+\nu_0)]\sigma\delta_{kl}\}, \quad k, l = 1, 2, 3, \tag{6.6.7}$$

where

$$\sigma = \sigma_{11} + \sigma_{22} + \sigma_{33}. \tag{6.6.8}$$

Formula (6.6.7) presents the well-known relation between strains and stresses of the nonrelativistic theory of elasticity [55].

Therefore, formula (6.6.5) can be regarded as its nonrelativistic generalization. Thus, relation (6.6.5) satisfies the two introduced requirements.

Let us turn to the components $g^{(e)}_{\mu\nu}(x^\gamma)$ of the metric tensor at the space–time point $P_{(e)}$ with coordinates x^μ relative to the considered elastically deformed frame of reference. Let the point $P_{(e)}$ have coordinates $\overset{\leftarrow}{x}{}^\mu$ in the perfect frame comoving with it and $g_{\mu\nu}(\overset{\leftarrow}{x}{}^\gamma)$ denote the components of the metric tensor at the point $P_{(e)}$ in this frame.

The components of the metric tensor $g^{(e)}_{\mu\nu}(x^\gamma)$ and $g_{\mu\nu}(\overset{\leftarrow}{x}{}^\gamma)$ in the coordinate systems x^μ and $\overset{\leftarrow}{x}{}^\mu$ are related by the well-known tensor formula [13]

$$g^{(e)}_{\mu\nu}(x^\gamma) = g_{\alpha\beta}(\overset{\leftarrow}{x}{}^\gamma)\partial\overset{\leftarrow}{x}{}^\alpha/\partial x^\mu \partial\overset{\leftarrow}{x}{}^\beta/\partial x^\nu. \tag{6.6.9}$$

In the considered case of small elastic deformations, from (6.6.2) and (6.6.9), we obtain

$$g_{\mu\nu}^{(e)}(x^\gamma) = g_{\alpha\beta}(\overline{x}^\gamma)(g_\mu^\alpha + \partial u^\alpha/\partial x^\mu)(g_\nu^\beta + \partial u^\beta/\partial x^\nu)$$

$$\approx g_{\mu\nu}(\overline{x}^\gamma) + g_{\mu\beta}\partial u^\beta/\partial x^\nu + g_{\nu\alpha}\partial u^\alpha/\partial x^\mu$$

$$\approx g_{\mu\nu}(x^\gamma) + (\overline{x}^\gamma - x^\gamma)\partial g_{\mu\nu}/\partial x^\gamma + \partial(g_{\mu\beta}u^\beta)/\partial x^\nu$$

$$- u^\beta\partial g_{\mu\beta}/\partial x^\nu + \partial(g_{\nu\alpha}u^\alpha)/\partial x^\mu - u^\alpha\partial g_{\nu\alpha}/\partial x^\mu$$

$$= g_{\mu\nu}(x^\gamma) + u^\gamma\partial g_{\mu\nu}/\partial x^\gamma + \partial u_\mu/\partial x^\nu + \partial u_\nu/\partial x^\mu$$

$$- u^\gamma(\partial g_{\mu\gamma}/\partial x^\nu + \partial g_{\nu\gamma}/\partial x^\mu)$$

$$= g_{\mu\nu}(x^\gamma) + \partial u_\mu/\partial x^\nu + \partial u_\nu/\partial x^\mu - 2\Gamma_{\mu\nu}^\gamma u_\gamma$$

$$= g_{\mu\nu}(x^\gamma) + \nabla_\nu u_\mu + \nabla_\mu u_\nu, \tag{6.6.10}$$

where $\Gamma_{\mu\nu}^\gamma$ are the Christoffel symbols [13].

From (6.6.4) and (6.6.10), we find

$$g_{\mu\nu}^{(e)}(x^\gamma) = g_{\mu\nu}(x^\gamma) + 2\varepsilon_{\mu\nu}(x^\gamma). \tag{6.6.11}$$

This formula relates the metric tensor in the considered elastically deformed frame, taken at the point $P_{(e)}$, with the metric tensor and strain tensor in the perfect frame comoving with this frame, taken at the point P, for which the point $P_{(e)}$ is its displaced image.

It should be noted that for the spatial indices, formula (6.6.11) coincides with the well-known formula of the nonrelativistic theory of elasticity that relates the strain tensor with elastic displacements [56] and, therefore, is its relativistic generalization.

The obtained formulas (6.6.4), (6.6.5) and (6.6.11) present the equations of a relativistic theory of elasticity. To them, one can add the dynamic equations of elastic bodies. When there are no physical fields, they have the form [13]

$$\nabla_\mu T^{\mu\nu} = 0, \quad T^{\mu\nu} = c^2\rho_0\frac{dx^\mu}{ds}\frac{dx^\nu}{ds} + S^{\mu\nu}, \tag{6.6.12}$$

where $T^{\mu\nu}$ is the energy–momentum tensor of an elastic body, ρ_0 is its rest mass density and $S^{\mu\nu}$ is its stress tensor.

Let us use the following formula, which is a consequence of formulas (4.9.8) and (4.9.10) of Section 4.9:

$$\nabla_\mu \left(\rho_0 \frac{dx^\mu}{ds} \frac{dx^\nu}{ds} \right) = \rho_0 \left(\frac{d^2 x^\nu}{ds^2} + \Gamma^\nu_{\alpha\beta} \frac{dx^\alpha}{ds} \frac{dx^\beta}{ds} \right), \qquad (6.6.13)$$

where $\Gamma^\nu_{\alpha\beta}$ are the Christoffel symbols.

Then, from (6.6.12), we find that the dynamic equations for an elastic body can be represented in the form

$$c^2 \rho_0 \left(\frac{d^2 x^\nu}{ds^2} + \Gamma^\nu_{\alpha\beta} \frac{dx^\alpha}{ds} \frac{dx^\beta}{ds} \right) + \nabla_\mu S^{\mu\nu} = 0. \qquad (6.6.14)$$

Let us apply the proposed equations for perfect and elastically deformed frames of reference to study the propagation of strong gravitational waves relative to them.

At first, we will consider the problem of their description relative to perfect frames and then generalize the obtained results for the case of elastically deformed frames, using formula (6.6.11).

6.7. Propagation of Strong Gravitational Waves Relative to a Perfect Frame of Reference

Let us turn to the study of strong gravitational waves in a perfect frame of reference. As follows from Sections 6.2–6.4, the Einstein equations for a gravitational field in a perfect frame can be represented in the form

$$R_{\mu\nu} - \frac{1}{2} R g_{\mu\nu} = \kappa T_{\mu\nu}, \quad g^{\alpha\beta} \frac{\partial g_{\mu\alpha}}{\partial x^\beta} = \lambda g^{\alpha\beta} \frac{\partial g_{\alpha\beta}}{\partial x^\mu}, \qquad (6.7.1)$$

where the parameter λ has the value

$$\lambda = 3/4 \qquad (6.7.2)$$

and x^μ are rectangular coordinates in the perfect frame.

As stated above, there can be two types of perfect frames of reference: local perfect frames and extended perfect frames, which are a set of local perfect frames, in which neighboring frames are at rest relative to each other.

Consider an extended perfect frame that is inertial at infinity. For it, we add the following condition to Eqs. (6.7.1):

$$g_{\mu\nu} \to \bar{g}_{\mu\nu}, \quad (x^1)^2 + (x^2)^2 + (x^3)^2 \to \infty, \tag{6.7.3}$$

where $\bar{g}_{\mu\nu}$ are the components of the Minkowski metric tensor.

Such a system is a large set of local perfect frames, which is a continuation of a local inertial frame located sufficiently far from massive bodies.

Let us turn to the exact solution of the Einstein gravitational equations

$$R_{\mu\nu} = 0 \tag{6.7.4}$$

for gravitational waves propagated in vacuum, which was proposed by I. Robinson and G. Bondi [13]. This solution can be represented in the form

$$ds^2 = 2dx^1 d\eta + \tilde{g}_{ab}(\eta)dx^a dx^b; \quad a, b = 2, 3, \tag{6.7.5}$$

where ds is the four-dimensional interval, η is some time coordinate and x^1, x^2, x^3 are some spatial coordinates. The indices a, b and c, d, used in the following, take the values 2, 3 and $\tilde{g}_{ab}(\eta)$ is a two-dimensional tensor.

For this metric, only the following Christoffel symbols are nonzero:

$$\Gamma^a_{b0} = \frac{1}{2}\kappa^a_b, \quad \Gamma^1_{ab} = -\frac{1}{2}\kappa_{ab}, \tag{6.7.6}$$

where

$$\kappa_{ab} = \dot{\tilde{g}}_{ab}, \quad \kappa^a_b = \tilde{g}^{ac}\kappa_{bc} \tag{6.7.7}$$

and the dot over a letter means the derivative with respect to the time coordinate η.

As for the components of the Ricci tensor, only R_{00} is nonzero. As a result, from (6.7.4), we derive only one equation [13]

$$R_{00} = -\frac{1}{2}\dot{\kappa}^a_a - \frac{1}{4}\kappa^b_a\kappa^a_b = 0. \tag{6.7.8}$$

Equation (6.7.8) can be represented in the form

$$\ddot{\chi} + \frac{1}{8}(\dot{\gamma}_{ac}\gamma^{bc})(\dot{\gamma}_{bd}\gamma^{ad})\chi = 0, \tag{6.7.9}$$

where

$$\tilde{g}_{ab} = -\chi^2 \gamma_{ab}, \quad \det(\gamma_{ab}) = 1. \tag{6.7.10}$$

This solution contains the two arbitrary functions $\gamma_{22}(\eta)$ and $\gamma_{23}(\eta)$. The functions $\gamma_{33}(\eta)$ and $\chi(\eta)$ are determined from the condition $\det(\gamma_{ab}) = 1$ and Eq. (6.7.9).

If we perform the coordinate transformation

$$\eta = (\overset{\leftarrow}{x}{}^0 - \overset{\leftarrow}{x}{}^1)/\sqrt{2}; \quad x^1 = (\overset{\leftarrow}{x}{}^0 + \overset{\leftarrow}{x}{}^1)/\sqrt{2}; \quad x^2 = \overset{\leftarrow}{x}{}^2; \quad x^3 = \overset{\leftarrow}{x}{}^3, \tag{6.7.11}$$

then interval (6.7.5) acquires the form

$$ds^2 = (d\overset{\leftarrow}{x}{}^0)^2 - (d\overset{\leftarrow}{x}{}^1)^2 + \check{g}_{ab} d\overset{\leftarrow}{x}{}^a d\overset{\leftarrow}{x}{}^b, \tag{6.7.12}$$

where

$$\check{g}_{ab} = \check{g}_{ab}(\overset{\leftarrow}{x}{}^0 - \overset{\leftarrow}{x}{}^1) = \tilde{g}_{ab}(\eta). \tag{6.7.13}$$

This exact solution to Eq. (6.7.4) contains two arbitrary functions and describes transverse waves.

It is necessary to note that gravitational waves generated by massive bodies are determined in the general case by the 10 components $T^{\mu\nu}$ of their energy–momentum tensor. Since these components are related by the four differential equations of energy and momentum conservation [13]

$$\nabla_\mu T^{\mu\nu} = 0, \tag{6.7.14}$$

gravitational waves are determined by six independent components of the energy–momentum tensor of the bodies generating the waves. That is why we will seek a class of exact wave solutions to the Einstein gravitational equations in vacuum, which contains six arbitrary wave functions.

For this purpose, let us turn to a special representation of the exact solution (6.7.12)–(6.7.13), in which the following coordinate transformations are made:

$$\overset{\leftarrow}{x}{}^0 = \gamma_0(\xi) + \delta(\xi)x^0; \quad \overset{\leftarrow}{x}{}^1 = \gamma_1(\xi) + \delta(\xi)x^0;$$

$$\overset{\leftarrow}{x}{}^2 = x^2 + \alpha(\xi); \quad \overset{\leftarrow}{x}{}^3 = x^3 + \beta(\xi); \quad \xi = x^0 - x^1. \tag{6.7.15}$$

Then, we easily obtain

$$ds^2 = g_{\mu\nu}dx^\mu dx^\nu,$$
$$g_{00} = \mu_0(\xi) + \theta(\xi)x^0; \quad g_{11} = \mu_1(\xi) + \theta(\xi)x^0; \qquad (6.7.16)$$
$$g_{01} = -(g_{00} + g_{11})/2; \quad g_{a\sigma} = g_{a\sigma}(\xi);$$
$$a = 2,3; \quad \sigma = 0,1,2,3; \quad g_{02} = -g_{12}; \quad g_{03} = -g_{13},$$

where

$$\mu_0 = \dot\gamma_0^2 - \dot\gamma_1^2 + 2\delta(\dot\gamma_0 - \dot\gamma_1) + u\dot\alpha^2 + v\dot\beta^2 + 2w\dot\alpha\dot\beta;$$
$$\mu_1 = \dot\gamma_0^2 - \dot\gamma_1^2 + u\dot\alpha^2 + v\dot\beta^2 + 2w\dot\alpha\dot\beta; \quad \theta = 2\dot\delta(\dot\gamma_0 - \dot\gamma_1);$$
$$g_{02} = -g_{12} = u\dot\alpha + w\dot\beta; \quad g_{03} = -g_{13} = w\dot\alpha + v\dot\beta;$$
$$g_{22} = u; \quad g_{33} = v; \quad g_{23} = w; \quad u = u(\xi); \quad v = v(\xi); \quad w = w(\xi)$$
$$(6.7.17)$$

and

$$u(\xi) = \breve{g}_{22}(\gamma_0 - \gamma_1); \quad v(\xi) = \breve{g}_{33}(\gamma_0 - \gamma_1); \quad w(\xi) = \breve{g}_{23}(\gamma_0 - \gamma_1);$$
$$\xi = x^0 - x^1. \qquad (6.7.18)$$

Since all the values in (6.7.17) are functions of the argument ξ, their partial derivatives with respect to the time coordinate x^0, which are denoted by the dot over them, coincide with their ordinary derivatives with respect to the argument ξ.

The components of the metric tensor (6.7.16) give exact solutions to Eq. (6.7.4), depending on the following seven arbitrary functions of the argument ξ: $\gamma_0, \gamma_1, \delta, \alpha, \beta, \breve{g}_{22}, \breve{g}_{23}$ ($\breve{g}_{33}$ is determined from Eqs. (6.7.9), (6.7.10) and (6.7.13)).

6.8. Gravitational-Wave Solutions in Perfect and Elastically Deformed Frames of Reference

Let us turn to the description of the propagation of gravitational waves relative to perfect frames of reference, in which their own elastic deformations can be neglected. The propagation of gravitational waves relative to elastically deformed frames will be considered at the end of this section.

As follows from Eqs. (6.7.1) and (6.7.2), the gravitational equations (6.7.4) for an empty space in perfect frames of reference acquire the form

$$R_{\mu\nu} = 0, \quad g^{\alpha\beta}\frac{\partial g_{\mu\alpha}}{\partial x^{\beta}} = \lambda g^{\alpha\beta}\frac{\partial g_{\alpha\beta}}{\partial x^{\mu}}, \quad \lambda = 3/4. \tag{6.8.1}$$

Let us seek solutions to Eqs. (6.8.1), corresponding to gravitational waves.

Consider the components $g_{\mu\nu}$ of the metric tensor of form (6.7.16)–(6.7.18), which satisfy Eq. (6.7.4). For their determinant, we have

$$g = \det(g_{\mu\nu}) = -\Delta(\mu_0 - \mu_1)^2/4, \tag{6.8.2}$$

where

$$\Delta = g_{22}g_{33} - g_{23}^2 = \Delta(\xi); \quad g = g(\xi); \quad \xi = x^0 - x^1. \tag{6.8.3}$$

For the contravariant components of the metric tensor $g^{\mu\nu}$, we find

$$g^{00} = d - 4(\mu_1 + \theta x^0)/(\mu_0 - \mu_1)^2;$$
$$g^{11} = d - 4(\mu_0 + \theta x^0)/(\mu_0 - \mu_1)^2;$$
$$g^{01} = (g^{00} + g^{11})/2; \quad d = (2g_{02}g_{03}g_{23} - g_{22}g_{03}^2 - g_{33}g_{02}^2)/g;$$
$$g^{02} = g^{12} = 2(g_{23}g_{03} - g_{02}g_{33})/[\Delta(\mu_0 - \mu_1)];$$
$$g^{03} = g^{13} = 2(g_{23}g_{02} - g_{03}g_{22})/[\Delta(\mu_0 - \mu_1)];$$
$$g^{22} = g_{33}/\Delta; \quad g^{33} = g_{22}/\Delta; \quad g^{23} = -g_{23}/\Delta. \tag{6.8.4}$$

As shown in Section 6.2, the four equations in (6.8.1), describing perfect frames of reference, are equivalent to the following equations containing derivatives of the contravariant components of the metric tensor $g^{\mu\nu}$ and its determinant g:

$$g\partial g^{\mu\nu}/\partial x^{\nu} = -\lambda g^{\mu\nu}\partial g/\partial x^{\nu}; \quad g = \det(g_{\mu\nu}); \quad \lambda = 3/4. \tag{6.8.5}$$

Consider Eq. (6.8.5) when $\mu = 2, 3$. Then, from (6.7.16) and (6.8.2)–(6.8.5), we find

$$g(\dot{g}^{0m} - \dot{g}^{1m}) = -\lambda(g^{0m} - g^{1m})\dot{g}; \quad m = 2, 3, \tag{6.8.6}$$

where the dot over the letters denotes the derivative with respect to x^0, which coincides with the derivative with respect to $\xi = x^0 - x^1$.

Since from (6.8.4), we have

$$g^{0m} = g^{1m}; \quad m = 2, 3, \tag{6.8.7}$$

the two equations (6.8.6) are evident identities.

Therefore, Eq. (6.8.5) when $\mu = 2, 3$ is fulfilled.

Consider Eq. (6.8.5) when $\mu = 0, 1$. Then from (6.7.16) and (6.8.2)–(6.8.5), we find

$$g(\partial g^{00}/\partial x^0 + \partial g^{01}/\partial x^1) = -\lambda(g^{00} - g^{01})\dot{g}, \quad \mu = 0,$$
$$g(\partial g^{01}/\partial x^0 + \partial g^{11}/\partial x^1) = -\lambda(g^{01} - g^{11})\dot{g}, \quad \mu = 1. \tag{6.8.8}$$

As follows from (6.8.4), we can represent the components g^{00}, g^{01}, g^{11} in the form

$$g^{00} = \rho_0 + \nu x^0; \quad g^{11} = \rho_1 + \nu x^0; \quad g^{01} = (\rho_0 + \rho_1)/2 + \nu x^0, \tag{6.8.9}$$

where

$$\rho_i = \rho_i(\xi), \quad i = 0, 1; \quad \nu = \nu(\xi); \quad \xi = x^0 - x^1;$$
$$\rho_0 = d - 4\mu_1/(\mu_0 - \mu_1)^2; \quad \rho_1 = d - 4\mu_0/(\mu_0 - \mu_1)^2; \tag{6.8.10}$$
$$\nu = -4\theta/(\mu_0 - \mu_1)^2.$$

Substituting formulas (6.8.9) into Eqs. (6.8.8) and using (6.8.2), we readily obtain that the two equations (6.8.8) result in the same equation of the form:

$$g(\dot{\rho}_0 - \dot{\rho}_1 + 2\nu) = -\lambda(\rho_0 - \rho_1)\dot{g}; \tag{6.8.11}$$

$$\rho_i = \rho_i(\xi); \quad \nu = \nu(\xi); \quad g = g(\xi); \quad g = -\Delta(\mu_0 - \mu_1)^2/4.$$

From (6.8.10) and (6.8.11), we obtain

$$2\theta = (2\lambda - 1)(\dot{\mu}_0 - \dot{\mu}_1) + \lambda(\mu_0 - \mu_1)\dot{\Delta}/\Delta; \tag{6.8.12}$$

$$\mu_i = \mu_i(\xi); \quad \Delta = \Delta(\xi); \quad \xi = x^0 - x^1.$$

As follows from (6.7.17),

$$\mu_0 - \mu_1 = 2\delta(\dot{\gamma}_0 - \dot{\gamma}_1); \quad \theta = 2\dot{\delta}(\dot{\gamma}_0 - \dot{\gamma}_1). \tag{6.8.13}$$

Therefore, from (6.8.12) and (6.8.13), we find

$$(3 - 2\lambda)\dot{\delta}(\dot{\gamma}_0 - \dot{\gamma}_1) = (2\lambda - 1)\delta(\ddot{\gamma}_0 - \ddot{\gamma}_1) + \lambda\delta(\dot{\gamma}_0 - \dot{\gamma}_1)\dot{\Delta}/\Delta. \tag{6.8.14}$$

Dividing Eq. (6.8.14) by $\lambda\delta(\dot{\gamma}_0 - \dot{\gamma}_1)$, we get

$$(3/\lambda - 2)\dot{\delta}/\delta = \dot{\Delta}/\Delta + (2 - 1/\lambda)(\ddot{\gamma}_0 - \ddot{\gamma}_1)/(\dot{\gamma}_0 - \dot{\gamma}_1). \tag{6.8.15}$$

From (6.8.15), we derive

$$|\delta|^{(3-2\lambda)/\lambda} = C\Delta|\dot{\gamma}_0 - \dot{\gamma}_1|^{(2\lambda-1)/\lambda}, \quad C = \text{const.} \tag{6.8.16}$$

Consider extended perfect frames of reference that are inertial with the Minkowski metric at infinity. For such frames, using (6.7.12) and (6.7.15), we obtain when $x^1 \to \infty$:

$$\overset{\leftharpoonup 0}{x} = x^0 + \text{const}; \quad \overset{\leftharpoonup 1}{x} = x^1 + \text{const};$$

$$\delta = 1; \quad \gamma_0 = \text{const}; \quad \gamma_1 = x^1 - x^0 + \text{const};$$

$$\dot{\gamma}_0 - \dot{\gamma}_1 = 1; \quad \Delta = 1; \quad x^1 \to \infty. \tag{6.8.17}$$

Therefore, from (6.8.16) and (6.8.17), we find

$$\delta = (\dot{\gamma}_0 - \dot{\gamma}_1)^{(2\lambda-1)/(3-2\lambda)}\Delta^{\lambda/(3-2\lambda)}. \tag{6.8.18}$$

Thus, formulas (6.7.16), (6.7.17), (6.8.16) and (6.8.18) give exact wave solutions to the Einstein gravitational equations in vacuum (6.8.1) with respect to perfect frames of reference.

Consider weak gravitational waves. Then, from Eq. (6.7.9), we approximately find [13]

$$\chi = 1 \tag{6.8.19}$$

and from (6.7.10), (6.7.13), (6.8.3) and (6.8.18), we obtain

$$\Delta = \chi^4 = 1; \quad \delta = (\dot{\gamma}_0 - \dot{\gamma}_1)^{(2\lambda-1)/(3-2\lambda)}. \tag{6.8.20}$$

Consider the value $\lambda = 1/2$. Then, Eq. (6.8.5) coincides with the de Donder equations for the harmonic coordinates [53], which were used in many studies of weak gravitational waves [13, 54, 55].

When $\lambda = 1/2$, from (6.8.13), (6.8.20) and (6.7.16)–(6.7.17), we have

$$\Delta = 1; \quad \delta = 1; \quad \theta = 0; \quad g_{\mu\nu} = g_{\mu\nu}(x^0 - x^1); \quad \lambda = 1/2 \quad (6.8.21)$$

and formulas (6.7.15) and (6.7.16) acquire the well-known form of the approximate solution for weak plane gravitational waves, which depend on six arbitrary functions of the argument $x^0 - x^1$ [13].

Therefore, the exact solution to Eqs. (6.8.1), determined by formulas (6.7.16) and (6.8.18) and also dependent on six arbitrary functions of the argument $x^0 - x^1$, presents a generalization of the well-known solution for weak plane gravitational waves.

Since for perfect frames of reference $\lambda = 3/4$, formula (6.8.16) acquires the form

$$\delta = D(\dot{\gamma}_0 - \dot{\gamma}_1)^{1/3}\Delta^{1/2}, \quad \lambda = 3/4, \quad D = \text{const}, \quad (6.8.22)$$

where, as follows from (6.8.18),

$$D = 1 \quad (6.8.23)$$

for extended perfect frames of reference that are inertial with the Minkowski metric at infinity.

The obtained exact wave solution to the Einstein gravitational equations in vacuum with respect to perfect frames of reference contains six arbitrary functions of the argument $x^0 - x^1$ and describes the propagation of gravitational waves in the direction of the axis x^1.

As noted above, in the general case, the gravitational waves radiated from massive bodies should be determined by six independent components $T^{\mu\nu}$ of their energy–momentum tensor due to the presence of the four differential equations of energy–momentum conservation. Thus, the general gravitational-wave solution should contain six arbitrary wave functions. Therefore, the obtained exact gravitational-wave solution for perfect frames of reference, containing six arbitrary functions of the argument $x^0 - x^1$, describes the general case of the propagation of gravitational waves along the direction x^1 in these frames.

Let us note some features of the found solution.

First, consider gravitational waves in the case $\theta = 0$. Then, from (6.7.17) and (6.8.22), we obtain

$$\delta = \text{const}; \quad \theta = 0; \quad \dot{\eta} = A\Delta^{-3/2};$$
$$\eta = (\gamma_0 - \gamma_1)/\sqrt{2}; \quad A = \text{const.}$$

(6.8.24)

As follows from (6.7.10), (6.7.16)–(6.7.18) and (6.8.3), the value Δ can be represented as a function of the argument η:

$$\Delta = \overline{\Delta}(\eta), \quad \overline{\Delta} = \overline{g}_{22}\overline{g}_{33} - \overline{g}_{23}^2 = \chi^4,$$

(6.8.25)

where χ is determined from Eq. (6.7.9).

From (6.8.24), (6.8.25) and (6.7.15), we find

$$\int \chi^6(\eta)d\eta = A\xi + B; \quad \eta = (\gamma_0 - \gamma_1)/\sqrt{2} = (\overline{x}^0 - \overline{x}^1)/\sqrt{2},$$
$$\xi = x^0 - x^1; \quad A = \text{const}, \quad B = \text{const.}$$

(6.8.26)

When $\delta = \text{const}$ and, consequently, relation (6.8.26) is valid, the components of the metric tensor $g_{\mu\nu}$ are functions of the argument $x^0 - x^1$: $g_{\mu\nu} = g_{\mu\nu}(x^0 - x^1)$.

However, when $\delta \neq \text{const}$, we come to another type of gravitational wave. As follows from (6.7.16) and (6.7.17), in this case, $\theta \neq 0$ and the absolute values of the components of the metric tensor g_{00}, g_{01}, g_{11}, which have the form

$$g_{00} = \mu_0(x^0 - x^1) + \theta(x^0 - x^1) \cdot x^0,$$
$$g_{11} = \mu_1(x^0 - x^1) + \theta(x^0 - x^1) \cdot x^0,$$
$$g_{01} = -[\mu_0(x^0 - x^1) + \mu_1(x^0 - x^1)]/2 - \theta(x^0 - x^1) \cdot x^0,$$

(6.8.27)

can significantly increase when such a gravitational wave passes spatial points during a sufficiently long period of time. In this case, the detection of a gravitational wave can be greatly facilitated. Such an anomalous wave can be regarded by an observer as some other, a nongravitational wave. Therefore, the discovered property of a possible significant amplification of gravitational waves may be important for the correct identification of these waves when observing them.

Let us now consider the problem of describing gravitational waves with respect to an elastically deformed frame of reference. It can be solved by applying the results obtained above for perfect frames of reference and formula (6.6.11). This formula just expresses the difference between the components of the metric tensor in an elastically deformed frame and in the perfect frame comoving with it through the strain tensor in the former. Therefore, substituting into formula (6.6.11) the found gravitational-wave solutions for the components of the metric tensor in the perfect frame of reference comoving with the considered elastically deformed frame, together with the components of the strain tensor, we can describe the propagation of gravitational waves with respect to the given deformed frame.

Of particular interest is the application of the solutions obtained above for extended perfect frames of reference. They, together with formula (6.6.11), make it possible to describe the propagation of gravitational waves, taking into account the elastic properties of frames of reference used in astronomical practice.

We supplement the found solution for propagating strong gravitational waves by studying the static gravitational field of a massive spherically symmetric body until the moment it emits gravitational waves.

Let us turn to the Schwarzschild problem of the gravitational field outside a spherically symmetric body, which we will consider in the extended perfect frame associated with this body.

6.9.   The Schwarzschild Problem in a Perfect Frame of Reference

As is well known, the space–time interval ds in the region outside a massive spherically symmetric body generating a gravitational field can be represented in the form [13]

$$ds^2 = c^2(1 - r_g/f)dt^2 - f^2(d\theta^2 + \sin^2\theta d\phi^2) - df^2/(1 - r_g/f),$$
$$(6.9.1)$$

where $f = f(r)$ is some function, r, θ, ϕ are spherical spatial coordinates, t is the time, $r_g = 2f_N m/c^2$ is the gravitational radius, f_N is the Newtonian gravitational constant and m is the mass of the spherically symmetric body, which is the source of the gravitational field.

We will study this gravitational field in the extended perfect frame associated with the considered massive body, which is inertial with the Minkowski metric at infinity. We place its origin at the center of the body.

In the considered frame of reference, let us make the transition from the time coordinate t and the spherical coordinates r, θ, ϕ, using in (6.9.1), to the rectangular coordinates x^μ by the formulas

$$x^0 = ct, \quad x^1 = r\cos\phi\sin\theta, \quad x^2 = r\sin\phi\sin\theta, \quad x^3 = r\cos\theta.$$
$$(6.9.2)$$

From (6.9.2), we readily find

$$(dx^1)^2 + (dx^2)^2 + (dx^3)^2 = dr^2 + r^2(d\theta^2 + \sin^2\theta d\phi^2),$$
$$x^1 dx^1 + x^2 dx^2 + x^3 dx^3 = r dr, \quad r = \sqrt{(x^1)^2 + (x^2)^2 + (x^3)^2}.$$
$$(6.9.3)$$

Formulas (6.9.3) give

$$dr^2 = (x^1 dx^1 + x^2 dx^2 + x^3 dx^3)^2/r^2,$$
$$d\theta^2 + \sin^2\theta d\phi^2 = (1/r^2)[(dx^1)^2 + (dx^2)^2 + (dx^3)^2]$$
$$- (1/r^4)(x^1 dx^1 + x^2 dx^2 + x^3 dx^3)^2. \quad (6.9.4)$$

Substituting (6.9.4) into (6.9.1), we find the components of the metric tensor in the coordinates x^μ:

$$ds^2 = g_{\mu\nu} dx^\mu dx^\nu, \quad g_{00} = 1 - r_g/f, \quad g_{0k} = 0,$$
$$g_{ik} = -f^2 \delta_{ik}/r^2 + [f^2/r^2 - f'^2/(1 - r_g/f)]x^i x^k/r^2, \quad (6.9.5)$$

where $i, k = 1, 2, 3$ and δ_{ik} is the Kronecker symbol.

From (6.9.5), we get the following determinant of the metric tensor:

$$g = \det(g_{\mu\nu}) = -f^4 f'^2/r^4. \quad (6.9.6)$$

The calculations of the contravariant components of the metric tensor result in the following expressions:

$$g^{00} = 1/(1 - r_g/f), \quad g^{0k} = 0,$$
$$g^{ik} = -r^2 \delta_{ik}/f^2 - [(1 - r_g/f)/f'^2 - r^2/f^2]x^i x^k/r^2, \quad (6.9.7)$$
$$i, k = 1, 2, 3.$$

As shown in Section 6.2, the four equations in (6.8.1), describing perfect frames of reference, can be represented in the following equivalent form which contains the derivatives of the contravariant components of the metric tensor and its determinant:

$$g\partial g^{\mu\nu}/\partial x^{\nu} = -\lambda g^{\mu\nu}\partial g/\partial x^{\nu}, \quad \lambda = 3/4. \tag{6.9.8}$$

Substituting expressions (6.9.6) and (6.9.7) for g and $g^{\mu\nu}$ into (6.9.8), we find that the first equation in (6.9.8) ($\mu = 0$) is an evident identity and the other three equations in (6.9.8) ($\mu = 1, 2, 3$) give the same equation of the following form:

$$f''/f' + 2/r - 6f'/f - 2(r_g - 2rf')f'/[f(f - r_g)] = 0,$$
$$\lambda = 3/4. \tag{6.9.9}$$

To solve Eq. (6.9.9), let us introduce the inverse function $r = r(f)$. For it, we have

$$f'(r) = 1/r'(f),$$
$$f''(r) = -r''(f)f'(r)/r'^{2}(f) = -r''(f)/r'^{3}(f). \tag{6.9.10}$$

Substituting (6.9.10) into Eq. (6.9.9), we obtain the following equation for the function $r(f)$:

$$f(r'' - 2r'^{2}/r) + 6r' + 2(r_g r' - 2r)/(f - r_g) = 0, \quad r = r(f). \tag{6.9.11}$$

Let us now introduce the function

$$h(f) = 1/r(f). \tag{6.9.12}$$

From (6.9.12), we have

$$r = 1/h, \quad r' = -h'/h^{2}, \quad r'' = -(hh'' - 2h'^{2})/h^{3}. \tag{6.9.13}$$

Substituting (6.9.13) into Eq. (6.9.11), we find that this equation acquires the form of the following linear equation for the function $h(f)$:

$$f(f - r_g)h'' + 2(3f - 2r_g)h' + 4h = 0, \quad h = h(f). \tag{6.9.14}$$

It is easy to verify by direct substitution into (6.9.14) that this equation has the particular exact solution

$$h(f) = a/(f - r_g), \quad a = \text{const.} \tag{6.9.15}$$

Indeed, in this case,

$$f(f - r_g)h'' + 2(3f - 2r_g)h' + 4h$$
$$= a(f - r_g)^{-2}[2f - 2(3f - 2r_g) + 4(f - r_g)] \equiv 0. \tag{6.9.16}$$

Using the particular solution (6.9.15) of the linear differential equation (6.9.14) of the second order and applying the well-known Liouville formula, we can find another particular solution of this equation.

Let us represent Eq. (6.9.14) in the form

$$h'' + p(f)h' + q(f)h = 0, \quad h = h(f), \tag{6.9.17}$$

where

$$p(f) = \frac{2(3f - 2r_g)}{f(f - r_g)}, \quad q(f) = \frac{4}{f(f - r_g)}, \tag{6.9.18}$$

and denote the particular solution to Eq. (6.9.14), found above, as $h_1(f)$:

$$h_1 = a/(f - r_g). \tag{6.9.19}$$

Then, in accordance with the Liouville formula, another particular solution $h_2(f)$ to Eq. (6.9.14) can be represented in the form

$$h_2 = Ah_1 \int h_1^{-2} \exp\left(-\int p\,df\right)df, \quad A = \text{const.} \tag{6.9.20}$$

Since the function $p(f)$ can be rewritten as

$$p = \frac{2(3f - 2r_g)}{f(f - r_g)} \equiv \frac{4}{f} + \frac{2}{f - r_g}, \tag{6.9.21}$$

from (6.9.19)–(6.9.21), we find

$$h_2 = \frac{A}{a(f - r_g)} \int f^{-4}df = \frac{-A}{3af^3(f - r_g)} + \frac{B}{f - r_g}, \quad A, B = \text{const.} \tag{6.9.22}$$

Since Eq. (6.9.14) is linear of the second order, by using its found two independent solutions (6.9.19) and (6.9.22), we obtain the general solution of this equation:

$$h(f) = (a + b/f^3)/(f - r_g), \qquad (6.9.23)$$

where a and b are arbitrary constants.

From the condition at infinity (6.7.3) for the considered extended perfect frame of reference and expressions (6.9.5) for the metric tensor, we have

$$f(r) = r, \quad r \to \infty. \qquad (6.9.24)$$

Therefore, using (6.9.12), we obtain that as $r \to \infty$,

$$f \to \infty, \quad h(f)f \to 1. \qquad (6.9.25)$$

From (6.9.25), we determine the value of the constant a in (6.9.23):

$$a = 1. \qquad (6.9.26)$$

As a result, formula (6.9.23) acquires the form

$$h(f) = (1 + b/f^3)/(f - r_g), \quad b = \text{const.} \qquad (6.9.27)$$

Using (6.9.12) and (6.9.27), we determine the general solution $f(r)$ of the nonlinear differential equation (6.9.9) that satisfies the condition at infinity (6.7.3):

$$f = r + r_g + br/f^3, \quad \lambda = 3/4. \qquad (6.9.28)$$

The obtained formula (6.9.28) for the function $f(r)$ and formula (6.9.1) for the interval ds give the general solution of the Schwarzschild problem in the considered perfect frame of reference.

As follows from (6.9.28), $f(0) = r_g$. As for the equality $f(r) = r_g$ for some radius $r > 0$, that gives a singularity in metric (6.9.1), it can take place if only $b = -r_g^3$.

Consider the case $b > -r_g^3$. Then, from Eq. (6.9.28), we find that for any $r > 0$, there exists its solution $f(r)$ satisfying the condition $f(r) > r_g$. Indeed, in this case, when $f = r_g$, the left-hand side of

Eq. (6.9.28) is less than its right-hand side and when $f \to +\infty$, its left-hand side is greater than its right-hand side. Therefore, when $r > 0$, there always exists a solution to Eq. (6.9.28) in the interval $(r_g, +\infty)$.

Thus, in the case $b > -r_g^3$, metric (6.9.1) has no singularity when $r > 0$, whereas when $b = -r_g^3$, its singularity appears.

Therefore, we come to the conclusion that physically admissible values of the parameter b are the values $b > -r_g^3$, for which metric (6.9.1) has no singularities at any $r > 0$.

When the number $b = 0$, metric (6.9.1) acquires a particularly simple form:

$$ds^2 = c^2 dt^2/(1 + r_g/r) - (r + r_g)^2(d\theta^2 + \sin^2 \theta d\phi^2)$$
$$- (1 + r_g/r)dr^2, \quad b = 0, \quad \lambda = 3/4, \tag{6.9.29}$$

in which, in accordance with what was said above, a singularity arises only at $r = 0$.

It is interesting to note that such a property takes place only when the value of the parameter $\lambda = 3/4$, corresponding to perfect frames of reference. For example, when $\lambda = 1/2$, the analogous formula for this metric has the form [54]

$$ds^2 = c^2 \frac{r - r_g/2}{r + r_g/2} dt^2 - (r + r_g/2)^2(d\theta^2 + \sin^2 \theta d\phi^2)$$
$$- \frac{r + r_g/2}{r - r_g/2} dr^2, \quad \lambda = 1/2, \tag{6.9.30}$$

in which a singularity arises when $r = r_g/2 > 0$.

It should be stressed that Eqs. (6.7.1) when $\lambda = 1/2$ are the Einstein equations in the harmonic coordinates, which are widely used in many studies [13, 54]. However, as shown above, a clear physical meaning can be given to Eq. (6.7.1) precisely for $\lambda = 3/4$. Then, they present the Einstein equations in perfect frames of reference, in which their internal stresses are small.

Let us turn to the problem of the determination of the energy and momentum of the gravitational field, which is very important to study the properties of gravitational waves.

6.10.   Energy and Momentum of the Gravitational Field

The problem of describing the energy and momentum of the gravitational field has not yet been considered resolved and continues to be debatable since the formulas proposed for them contain certain defects and have been subjected to justified criticism. We will study this problem in detail.

The initial equation for its solution, when the sources of gravity are a charged matter and an electromagnetic field, has, as is well known, the form [13]

$$\nabla_\nu(T_m^{\mu\nu} + T_e^{\mu\nu}) = 0, \tag{6.10.1}$$

where ∇_ν is the covariant derivative and $T_m^{\mu\nu}, T_e^{\mu\nu}$ are the energy–momentum tensors of the matter and electromagnetic field, respectively.

Consider the case of a dust-like matter. Then, we have [13]

$$T_m^{\mu\nu} = c^2\rho_0\frac{dx^\mu}{ds}\frac{dx^\nu}{ds}, \quad T_e^{\mu\nu} = \frac{1}{4\pi}\left(-F^{\mu\alpha}F^\nu{}_\alpha + \frac{1}{4}g^{\mu\nu}F_{\alpha\beta}F^{\alpha\beta}\right),$$

$$F_{\mu\nu} = \frac{\partial A_\nu}{\partial x^\mu} - \frac{\partial A_\mu}{\partial x^\nu}, \quad ds^2 = g_{\mu\nu}dx^\mu dx^\nu, \quad \mu,\nu = 0,1,2,3, \tag{6.10.2}$$

where x^μ are space–time coordinates, ρ_0 is the rest mass density of the matter in a local inertial frame of reference comoving with it, $F_{\mu\nu}$ is the tensor of strengths of an electromagnetic field and A_μ are its potentials.

By means of the approach, proposed by L.D. Landau and E.M. Lifshitz, Eq. (6.10.1) can be represented in the following form [13]:

$$\frac{\partial}{\partial x^\nu}[\phi(-g)(T_m^{\mu\nu} + T_e^{\mu\nu} + T_g^{\mu\nu})] = 0, \tag{6.10.3}$$

where ϕ is an arbitrary function, defined for positive values of its argument and twice differentiable, $g = \det(g_{\mu\nu}) < 0$ and $T_g^{\mu\nu}$ is an energy–momentum pseudotensor of the gravitational field dependent on the choice of the function $\phi(-g)$.

In [13], the choice $\phi(-g) = -g$ was made and an expression for $T_g^{\mu\nu}$, which is symmetrical in indices μ, ν, was obtained. This expression has not become generally accepted due to a shortcoming, which will be discussed in the following.

Equations like (6.10.3) are usually considered to be energy–momentum conservation equations. Of course, there is no doubt that this view is correct if the expression in the square brackets in (6.10.3) is continuous and piecewise differentiable. Strictly speaking, it is under this condition that one can apply the Gauss theorem to Eq. (6.10.3) and obtain the integral energy–momentum conservation law.

However, let us pose the following question: Will the integral law of conservation of energy–momentum follow from equation (6.10.3) if the rest mass density of matter is a discontinuous function of spatial coordinates? After all, as a rule, it is such a situation that one has to deal with. For example, when the gravitational field is generated by a body occupying a finite spatial volume V. Then, the density ρ_0 is discontinuous:

$$\rho_0 \neq 0, (x^1, x^2, x^3) \in V; \quad \rho_0 = 0, (x^1, x^2, x^3) \notin V. \qquad (6.10.4)$$

To answer the above question, consider a surface S in the three-dimensional space dependent, generally speaking, on time, at which the density ρ_0 is discontinuous and, therefore, at it the expression

$$\theta^{\mu\nu} = \phi(-g)(T_m^{\mu\nu} + T_e^{\mu\nu} + T_g^{\mu\nu}) \qquad (6.10.5)$$

is also a discontinuous function.

Let us introduce the following denotation:

$$\Delta\theta^{\mu\nu} = \theta_+^{\mu\nu} - \theta_-^{\mu\nu}, (x^1, x^2, x^3) \in S, \qquad (6.10.6)$$

where the indices '+' and '−' denote the limits of the considered values as we tend to the point $(x^1, x^2, x^3) \in S$ from the external and internal sides, respectively, with respect to the surface S.

Then, using (6.10.3) and the Gauss theorem, assuming that the values $\theta^{\mu\nu}$ tend to zero fast enough at spatial infinity, we obtain

$$\frac{d}{dx^0} \int \theta^{\mu 0} dV = \int \frac{\partial \theta^{\mu 0}}{\partial x^0} dV + \int_S \Delta\theta^{\mu 0} \frac{dx^k}{dx^0} n_k dS$$

$$= -\int \frac{\partial \theta^{\mu k}}{\partial x^k} dV + \int_S \Delta\theta^{\mu 0} \frac{dx^k}{dx^0} n_k dS$$

$$= \int_S \left(\Delta\theta^{\mu 0} \frac{dx^k}{dx^0} - \Delta\theta^{\mu k} \right) n_k dS, \qquad (6.10.7)$$

where $k = 1, 2, 3, dV = dx^1 dx^2 dx^3$, the volume integrals are taken over the entire three-dimensional space and n_k are the orthogonal projections onto the spatial axes of the unit vector of the normal to the surface S, directed inside the volume V.

Note that when differentiating with respect to time the volume integral in the left-hand side of (6.10.7), it has been presented as the sum of the two integrals taken over the time-dependent external and internal volumes with respect to S, and then the well-known formula for differentiating these integrals has been applied.

From (6.10.2), we get

$$T_m^{\mu 0} \frac{dx^k}{dx^0} = T_m^{\mu k}. \tag{6.10.8}$$

Taking into account that $T_e^{\mu\nu}$ are continuous functions of the coordinates x^μ, since $T_e^{\mu\nu}$ depend only on the first derivatives of the electromagnetic potentials A_μ, from formulas (6.10.5), (6.10.7) and (6.10.8), we obtain

$$\frac{d}{dx^0} \int \phi(-g)(T_m^{\mu 0} + T_e^{\mu 0} + T_g^{\mu 0})dV$$

$$= \int_S \phi(-g) \left(\Delta T_g^{\mu 0} \frac{dx^k}{dx^0} - \Delta T_g^{\mu k} \right) n_k dS. \tag{6.10.9}$$

From this formula, we can draw the following conclusion: In order for the integral law of conservation of energy–momentum to follow from the differential equations (6.10.3) in the case of an arbitrary discontinuous density of matter, it is necessary that $\Delta T_g^{\mu\nu} = 0$, that is, that the pseudotensor $T_g^{\mu\nu}$ of the gravitational field is always continuous.

It should be noted that in a number of works, this circumstance was not given due attention, and they considered formulas for the pseudotensor $T_g^{\mu\nu}$ of the energy–momentum of the gravitational field, depending on the second derivatives of the metric tensor $g_{\mu\nu}$. However, in view of what has been said above, such formulas cannot be considered correct.

Indeed, if the density of matter ρ_0 is a discontinuous function of coordinates, then, as follows from the equations of the gravitational field, the second derivatives of the metric tensor $g_{\mu\nu}$ are, generally speaking, also discontinuous. Consequently, the pseudotensors $T_g^{\mu\nu}$ of the energy–momentum of the gravitational field, containing the second derivatives of the metric tensor, can also become discontinuous.

Thus, as follows from formula (6.10.9), the pseudotensors $T_g^{\mu\nu}$ depending on the second derivatives of the metric tensor do not generally ensure the validity of the integral energy–momentum conservation law of the form

$$\int \phi(-g)(T_m^{\mu 0} + T_e^{\mu 0} + T_g^{\mu 0})dV = \text{const.} \qquad (6.10.10)$$

In order for formula (6.10.10) to be correct in the general case, we should choose a pseudotensor $T_g^{\mu\nu}$, which not only satisfies the differential equation (6.10.3) but also depends only on the metric tensor $g_{\mu\nu}$ and its first derivatives. Then, the pseudotensor $T_g^{\mu\nu}$ will always be continuous and, therefore, formula (6.10.10) will follow from (6.10.3).

Let us now formulate the following requirements for the choice of the function $\phi(-g)$ in Eq. (6.10.3) and the pseudotensor $T_g^{\mu\nu}$, which, as will be seen from what follows, will make it possible to uniquely determine these values.

Requirement 1. *The integral law (6.10.10) of energy–momentum conservation should be fulfilled in both cases of continuous and discontinuous density of matter ρ_0.*

Requirement 2. *In this law, the 4-vectors of momentum of the matter particles should be proportional to their 4-vectors of velocity. At the same time, their relativistic masses should not change with different choices of purely spatial coordinates, that is, they should not change with coordinate transformations of the form*

$$\bar{x}^0 = x^0, \quad \bar{x}^k = \bar{x}^k(x^1, x^2, x^3), \quad k = 1, 2, 3. \qquad (6.10.11)$$

The necessity of the first requirement is quite obvious, and the condition under which it is fulfilled was obtained above.

As for the second requirement, it corresponds to the physical essence of the concepts of energy and momentum for particles of matter and is relativistically covariant. Let us show that this requirement is satisfied with the following choice of function $\phi(-g)$ in formula (6.10.3):

$$\phi(-g) = \sqrt{-g}. \qquad (6.10.12)$$

We will use the invariance of the following expression dI, in which the integral is taken over a small four-dimensional volume ε [13]:

$$dI = \sqrt{-g}\int_\varepsilon dV dt, \quad dt = dx^0/c, \quad dV = dx^1 dx^2 dx^3. \quad (6.10.13)$$

Let us choose a local inertial frame of reference, relative to which the volume dV is at rest. In it, we have

$$dV = dV_0, \quad dt = d\tau, \quad d\tau = ds/c, \quad \sqrt{-g} = 1, \quad (6.10.14)$$

where V_0 and τ denote a volume and time in this frame of reference.
From the invariance of dI, we obtain

$$\sqrt{-g}dV dt = dV_0 d\tau = dV_0 ds/c. \quad (6.10.15)$$

Using expression (6.10.12) for the function $\phi(-g)$, from formula (6.10.15), we find the following expression for the first term on the left-hand side of (6.10.10):

$$\int \sqrt{-g}T_m^{\mu0}dV = \int \sqrt{-g}c^2\rho_0\frac{dx^\mu}{ds}\frac{dx^0}{ds}dV$$

$$= \int c^2\rho_0\frac{dx^\mu}{ds}dV_0 = \int c^2\frac{dx^\mu}{ds}dm_0. \quad (6.10.16)$$

where $dm_0 = \rho_0 dV_0$ is the rest mass of the small volume dV_0.

Taking into account (6.10.12) and (6.10.16), we can represent formula (6.10.10) in the form

$$\int c^2\frac{dx^\mu}{ds}dm_0 + \int \sqrt{-g}(T_e^{\mu0} + T_g^{\mu0})dV = \text{const}. \quad (6.10.17)$$

The first integral in (6.10.17) presents the energy and momentum of matter and the second integral gives the energy and momentum of electromagnetic and gravitational fields.

As follows from the first term in (6.10.17), the kinetic energy E and the momentum $\mathbf{p}$ of a matter particle are determined by the formulas

$$E = mc^2, \quad \mathbf{p} = m\mathbf{v},$$

$$m = \frac{m_0}{\left(\frac{g_{\mu\nu}}{c^2}\frac{dx^\mu}{dt}\frac{dx^\nu}{dt}\right)^{1/2}}, \quad (6.10.18)$$

in which m_0 is the particle rest mass and $\mathbf{v}$ is its velocity with the components $dx^k/dt(k = 1, 2, 3, \ t = x^0/c)$. In addition, as is seen from (6.10.18), the relativistic mass m does not change under the purely spatial coordinate transformations (6.10.11).

Thus, the second requirement is fulfilled by our choice $\phi(-g) = \sqrt{-g}$ in the differential equation (6.10.3).

It should be noted that if we had used the following differential equation, instead of (6.10.3):

$$\frac{\partial}{\partial x^\nu}[\phi(-g)(T^\nu_{m\mu} + T^\nu_{e\mu} + T^\nu_{g\mu})] = 0, \qquad (6.10.19)$$

the second requirement could not be fulfilled since the components of the momentum of matter particles would not be proportional to the components dx^k/dt of their velocities.

As has been shown above, in order to satisfy both requirements under consideration, we should find an expression for the pseudotensor $T^{\mu\nu}_g$ of the energy–momentum of the gravitational field, which would satisfy Eq. (6.10.3) when $\phi(-g) = \sqrt{-g}$ and which would depend on the derivatives of the metric tensor not higher than the first order.

We can uniquely determine such an expression using the method proposed by L.D. Landau and E.M. Lifshitz [13]. Note that in [13], the choice $\phi(-g) = -g$ was made that, as we saw, does not satisfy our second requirement.

We apply this method to find the pseudotensor $T^{\mu\nu}_g$ satisfying equation (6.10.3) for $\phi(-g) = \sqrt{-g}$.

Let us define the desired pseudotensor $T^{\mu\nu}_g$ using the following formula:

$$T^{\mu\nu}_g = \frac{1}{\sqrt{-g}}\frac{\partial h^{\mu\nu\alpha}}{\partial x^\alpha} - (T^{\mu\nu}_m + T^{\mu\nu}_e), \qquad (6.10.20)$$

$$h^{\mu\nu\alpha} = \frac{c^4}{16\pi f_N}\frac{1}{\sqrt{-g}}\frac{\partial}{\partial x^\beta}[(-g)(g^{\mu\nu}g^{\alpha\beta} - g^{\mu\alpha}g^{\nu\beta})], \qquad (6.10.21)$$

where f_N is the Newtonian gravitational constant.

Then, due to the evident antisymmetry of the values $h^{\mu\nu\alpha}$ in the indices ν, α:

$$h^{\mu\nu\alpha} = -h^{\mu\alpha\nu}, \qquad (6.10.22)$$

we have

$$\frac{\partial^2 h^{\mu\nu\alpha}}{\partial x^\nu \partial x^\alpha} = \frac{\partial^2 h^{\mu\alpha\nu}}{\partial x^\alpha \partial x^\nu} = -\frac{\partial^2 h^{\mu\nu\alpha}}{\partial x^\nu \partial x^\alpha}, \qquad (6.10.23)$$

and hence, we find

$$\frac{\partial^2 h^{\mu\nu\alpha}}{\partial x^\nu \partial x^\alpha} = 0. \qquad (6.10.24)$$

From (6.10.20) and (6.10.24), we derive the equality

$$\frac{\partial}{\partial x^\nu}[\sqrt{-g}(T_m^{\mu\nu} + T_e^{\mu\nu} + T_g^{\mu\nu})] = 0, \qquad (6.10.25)$$

which means the validity of formula (6.10.3) when $\phi(-g) = \sqrt{-g}$ and is just one of the necessary requirements for the desired energy–momentum pseudotensor $T_g^{\mu\nu}$ of the gravitational field.

We now show that, in addition, expression (6.10.20) for the pseudotensor $T_g^{\mu\nu}$ contains derivatives of the components of the metric tensor not higher than the first order, that is, that it satisfies all the necessary requirements and, therefore, is the desired expression for the pseudotensor.

To do this, first, we use the following well-known equalities [13] for the case under consideration, when the sources of the gravitational field are the energy–momentum tensor $T_m^{\mu\nu}$ of matter and the energy–momentum tensor $T_e^{\mu\nu}$ of the electromagnetic field:

$$\frac{16\pi f_N}{c^4}(T_m^{\mu\nu} + T_e^{\mu\nu}) = 2R^{\mu\nu} - Rg^{\mu\nu}$$

$$= (g^{\mu\alpha}g^{\nu\beta} - \frac{1}{2}g^{\mu\nu}g^{\alpha\beta})g^{\rho\sigma}\left[\frac{\partial^2 g_{\rho\beta}}{\partial x^\alpha \partial x^\sigma} + \frac{\partial^2 g_{\alpha\sigma}}{\partial x^\rho \partial x^\beta} - \frac{\partial^2 g_{\rho\sigma}}{\partial x^\alpha \partial x^\beta}\right.$$

$$\left. - \frac{\partial^2 g_{\alpha\beta}}{\partial x^\rho \partial x^\sigma} + 2g_{\omega\theta}(\Gamma^\omega_{\alpha\sigma}\Gamma^\theta_{\rho\beta} - \Gamma^\omega_{\alpha\beta}\Gamma^\theta_{\rho\sigma})\right], \qquad (6.10.26)$$

in which $R^{\mu\nu}$ is the Ricci tensor, $R = g_{\mu\nu}R^{\mu\nu}$ and $\Gamma^\omega_{\alpha\beta}$ are the Christoffel symbols.

Then, performing identical transformations in the last equality in formula (6.10.26), we can represent the expressions in its first equality in the following form:

$$\frac{16\pi f_N}{c^4}(T_m^{\mu\nu} + T_e^{\mu\nu})$$

$$= \frac{\partial}{\partial x^\rho}\left\{\frac{1}{(-g)}\frac{\partial}{\partial x^\alpha}[(-g)(g^{\mu\nu}g^{\rho\alpha} - g^{\mu\rho}g^{\nu\alpha})]\right\}$$

$$+ \left(\frac{\partial g_{\rho\sigma}}{\partial x^\beta} - \frac{\partial g_{\rho\beta}}{\partial x^\sigma}\right)\frac{\partial}{\partial x^\alpha}\left[g^{\rho\sigma}\left(g^{\mu\alpha}g^{\nu\beta} - \frac{1}{2}g^{\mu\nu}g^{\alpha\beta}\right)\right]$$

$$+ \left(\frac{\partial g_{\alpha\beta}}{\partial x^\sigma} - \frac{\partial g_{\alpha\sigma}}{\partial x^\beta}\right)\frac{\partial}{\partial x^\rho}\left[g^{\rho\sigma}\left(g^{\mu\alpha}g^{\nu\beta} - \frac{1}{2}g^{\mu\nu}g^{\alpha\beta}\right)\right]$$

$$+ 2g_{\omega\theta}g^{\rho\sigma}\left(g^{\mu\alpha}g^{\nu\beta} - \frac{1}{2}g^{\mu\nu}g^{\alpha\beta}\right)(\Gamma^\omega_{\alpha\sigma}\Gamma^\theta_{\rho\beta} - \Gamma^\omega_{\alpha\beta}\Gamma^\theta_{\rho\sigma}). \quad (6.10.27)$$

From (6.10.20), (6.10.21) and (6.10.27), we get

$$\frac{16\pi f_N}{c^4}T_g^{\mu\nu} = \frac{1}{2g^2}\frac{\partial(-g)}{\partial x^\rho}\frac{\partial}{\partial x^\alpha}[(-g)(g^{\mu\nu}g^{\rho\alpha} - g^{\mu\rho}g^{\nu\alpha})]$$

$$+ \frac{\partial}{\partial x^\alpha}\left[g^{\rho\sigma}\left(g^{\mu\alpha}g^{\nu\beta} - \frac{1}{2}g^{\mu\nu}g^{\alpha\beta}\right)\right]\left(\frac{\partial g_{\rho\beta}}{\partial x^\sigma} - \frac{\partial g_{\rho\sigma}}{\partial x^\beta}\right)$$

$$+ \frac{\partial}{\partial x^\rho}\left[g^{\rho\sigma}\left(g^{\mu\alpha}g^{\nu\beta} - \frac{1}{2}g^{\mu\nu}g^{\alpha\beta}\right)\right]\left(\frac{\partial g_{\alpha\sigma}}{\partial x^\beta} - \frac{\partial g_{\alpha\beta}}{\partial x^\sigma}\right)$$

$$+ 2g_{\omega\theta}g^{\rho\sigma}\left(g^{\mu\alpha}g^{\nu\beta} - \frac{1}{2}g^{\mu\nu}g^{\alpha\beta}\right)(\Gamma^\omega_{\alpha\beta}\Gamma^\theta_{\rho\sigma} - \Gamma^\omega_{\alpha\sigma}\Gamma^\theta_{\rho\beta}).$$

$$(6.10.28)$$

Consequently, the proposed expression for the pseudotensor $T_g^{\mu\nu}$ does not contain derivatives of the components of the metric tensor above the first order, which was to be proved. Thus, it is indeed the desired expression for $T_g^{\mu\nu}$.

We now represent formula (6.10.28) in a different form, expressing the derivatives of the metric tensor components in terms of the Christoffel symbols.

To do this, we apply the following well-known formulas for the derivatives of the covariant and contravariant components of the metric tensor and its determinant [13]:

$$\frac{\partial g_{\rho\beta}}{\partial x^\sigma} = g_{\alpha\beta}\Gamma^\alpha_{\rho\sigma} + g_{\alpha\rho}\Gamma^\alpha_{\beta\sigma}, \quad \frac{\partial g^{\rho\beta}}{\partial x^\sigma} = -g^{\alpha\beta}\Gamma^\rho_{\alpha\sigma} - g^{\alpha\rho}\Gamma^\beta_{\alpha\sigma},$$

$$\frac{1}{g}\frac{\partial g}{\partial x^\sigma} = 2\Gamma^\alpha_{\alpha\sigma}. \tag{6.10.29}$$

Replacing in formula (6.10.28) the derivatives of the covariant and contravariant components of the metric tensor and its determinant through the Christoffel symbols using formulas (6.10.29), we finally obtain the following expression for the gravitational field energy–momentum pseudotensor:

$$\begin{aligned}
\frac{16\pi f_N}{c^4}T_g^{\mu\nu} &= (g^{\mu\rho}g^{\nu\alpha} - g^{\mu\nu}g^{\rho\alpha})(\Gamma^\sigma_{\rho\alpha}\Gamma^\beta_{\sigma\beta} - \Gamma^\sigma_{\rho\beta}\Gamma^\beta_{\alpha\sigma}) \\
&\quad + g^{\mu\rho}g^{\alpha\sigma}(\Gamma^\nu_{\rho\beta}\Gamma^\beta_{\alpha\sigma} - \Gamma^\nu_{\sigma\beta}\Gamma^\beta_{\rho\alpha}) \\
&\quad + g^{\nu\rho}g^{\alpha\sigma}(\Gamma^\mu_{\rho\beta}\Gamma^\beta_{\alpha\sigma} + \Gamma^\mu_{\alpha\sigma}\Gamma^\beta_{\rho\beta} \\
&\quad - \Gamma^\mu_{\sigma\beta}\Gamma^\beta_{\rho\alpha} - \Gamma^\mu_{\rho\alpha}\Gamma^\beta_{\sigma\beta}) \\
&\quad + g^{\rho\alpha}g^{\sigma\beta}(\Gamma^\mu_{\rho\sigma}\Gamma^\nu_{\alpha\beta} - \Gamma^\mu_{\rho\alpha}\Gamma^\nu_{\sigma\beta}). \tag{6.10.30}
\end{aligned}$$

This is the desired expression for $T_g^{\mu\nu}$, which satisfies the two requirements imposed on it.

Let us now turn to formula (6.10.18) for the relativistic mass and, as its application, consider the motion of a particle of matter with rest mass m_0 and electric charge q in gravitational and electromagnetic fields. As is known, this motion in an arbitrary coordinate system x^μ is described by the following equation [13]:

$$m_0 c\left[\frac{d^2 x^\mu}{ds^2} + \Gamma^\mu_{\alpha\beta}\frac{dx^\alpha}{ds}\frac{dx^\beta}{ds}\right] = \frac{q}{c}F^\mu_{\ \alpha}\frac{dx^\alpha}{ds}, \tag{6.10.31}$$

where $F_{\mu\nu}$ is the tensor of strengths of an electromagnetic field.

Taking into account formula (6.10.18) for the relativistic mass m of a matter particle in the presence of a gravitational field, we can represent the equation of its motion (6.10.31) in the form

$$\frac{d}{dt}\left(m\frac{dx^\mu}{dt}\right) = \frac{m}{c^2}G^\mu_{\alpha\beta}\frac{dx^\alpha}{dt}\frac{dx^\beta}{dt} + \frac{q}{c}F^\mu_{\ \alpha}\frac{dx^\alpha}{dt}, \tag{6.10.32}$$

in which

$$G^\mu_{\alpha\beta} = -c^2\Gamma^\mu_{\alpha\beta}, \quad t = x^0/c. \qquad (6.10.33)$$

As follows from (6.10.33), the right-hand side of Eq. (6.10.32) with $\mu = 1, 2, 3$ can be interpreted as the components of the force acting on a particle of matter, and when $\mu = 0$ — as the power of this force divided by the speed of light.

In addition, as can be seen from Eq. (6.10.32), the values $G^\mu_{\alpha\beta}$, similar to the values F^μ_α for the electromagnetic field, acquire the meaning of the strengths of the gravitational field.

Chapter 7

Weyl's Principle of Scale Invariance and a New Cosmology

The generally accepted cosmological ΛCDM model, based on Einstein's theory of gravity, is currently experiencing a deep crisis [57–60]. There are many cosmological questions that it has not yet answered. Among the most difficult questions for standard cosmology, a special place is occupied by the problem of the cosmological singularity and the mystery of dark matter and dark energy. Many attempts have been made to solve these problems. However, they were unsuccessful. A lot of expectations are connected with the James Webb Space Telescope launched in December 2021. As for the first data obtained from it, some of them show that orderly disk galaxies already existed more than 13.4 billion years ago, contrary to the standard cosmological model [61].

The existing crisis in cosmology is of a systemic nature, and in order to overcome it, it is necessary to revise some of the fundamental foundations affecting Einstein's theory of gravity. This point of view accompanies the last chapter. It proposes a change in the geometric basis of Einstein's theory: the transition from the Riemannian geometry used in it to its generalization — Weyl's conformal geometry.

A great advantage of Weyl's geometry is the equivalence of metrics in it, differing in the choice of the scale factor at different space–time points, which presents Weyl's principle of scale invariance. This is achieved by introducing four potentials in addition to the ten components of the metric tensor. Weyl himself interpreted his potentials as electromagnetic, seeking to create a unified geometric theory of gravity and electromagnetism. However, the equations he came up with did not fit well with the available experimental data, and they had to be abandoned. At the same time, the Weyl geometry itself is a very attractive and deep theory, which constantly maintains the interest in it of many modern physicists involved in gravitation and cosmology.

This chapter presents a new view of Weyl's geometry and ways of constructing on its basis a conformally invariant theory of gravity generalizing Einstein's theory. One of the key points of the proposed generalization is a different, in comparison with Weyl, interpretation of the four potentials introduced by him: they are treated as small potentials caused by the physical vacuum.

The gravitational theory that this idea led to contains equations of the Einstein type but with the Weyl connection that is invariant under the choice of the scale factor in the metric tensor. An important link in the development of this approach is the finding of differential equations for the Weyl potentials based on the requirement of consistency of the system of gravitational equations.

In the proposed generalization of Einstein's theory, due to the smallness of the Weyl potentials, their influence on the gravitational interaction can be neglected when considering not-too-long time intervals. Then, with a high degree of accuracy, Einstein's theory of gravity is correct. However, as it turns out, a cumulative effect manifests itself in cosmological processes that take place over billions of years. It consists in the gradual accumulation of a small action of Weyl's potentials. As a result, their influence can become significant during cosmological time. Therefore, the new gravitational theory, which practically coincides with Einstein's theory for not-too-long time intervals, leads to a new cosmology when considering astronomical processes lasting billions of years.

The studies performed show that the proposed cosmology based on Weyl's principle of scale invariance is consistent with the available observational data and is a real alternative to the standard cosmological models based on Einstein's theory. At the same time, unlike standard cosmology, the new cosmological theory does not have a physically unacceptable singularity and gives natural explanations for the mysterious dark matter and a number of amazing properties of spiral galaxies.

In this chapter, the results of our publications [62,63] are used.

7.1. Unexplained Gravitational and Cosmological Phenomena

This chapter is devoted to anomalous gravitational and cosmological phenomena that remain still unexplained within the framework of the Einstein theory. To them, one can attribute a number of puzzling properties of stars and galaxies and mysteries of standard cosmology.

Let us note the following of them [64,65]:

(1) The mysterious spiral structure of many galaxies.
(2) The unusual distribution of young and old stars in spiral galaxies: Most young stars are observed in the spiral arms, whereas old stars are mainly concentrated near the galaxy center.
(3) The analogous property of clusters of galaxies: As a rule, old galaxies are much closer to the cluster center than young galaxies in clusters.
(4) The surprising property of star axial rotations: As a rule, angular velocities of rotation of old stars are much slower than those of young stars.
(5) The mystery of the cosmological singularity.
(6) The mysterious dark matter and dark energy. The absence of reasonable candidates for them within the framework of standard cosmology.

Recently, the following difficult problem of cosmology has arisen:

(7) How to explain the existence of orderly disk galaxies more than 13.4 billion years ago [61], which contradicts the standard cosmological model?

The absence of satisfactory models for the above anomalous astronomical phenomena shows that the Einstein gravitational theory may need some modification.

In order to solve difficult cosmological problems, a number of approaches to generalize the Einstein gravitational theory were proposed [66]. Among them, gravitational theories based on Weyl's geometry occupy a prominent place [67].

Let us turn to the geometry proposed by H. Weyl in 1918 [68]. Weyl's theory appeared to overcome a difficulty of the Einstein definition of the invariant interval between two space–time points close to each other. In the Einstein theory, this interval is uniquely determined by 10 components of the metric tensor. The difficulty is associated with the question: Why not multiply the Einstein interval by an arbitrary positive scalar function? Then, the new interval is also invariant and is as acceptable as the previous one.

Trying to answer this question, Weyl proposed the principle of scale invariance and came to the wonderful idea to create a conformal gravitational theory that is invariant under gauge multiplications of the metric tensor. To realize the idea, he added new four potentials to ten gravitational potentials which are components of the metric tensor. Weyl's potentials play the following role. In the gauge-invariant Weyl geometry, the multiplication of the metric tensor by an arbitrary positive function is accompanied by a corresponding gauge transformation of Weyl's potentials. This gauge transformation is chosen so that the connection and curvature tensor of Weyl's geometry become independent of the choice of a multiplier for the metric tensor.

Weyl noted that the gauge transformations of his four potentials are quite analogous to those of four electromagnetic potentials. This led him to the idea to identify his potentials with electromagnetic ones. Developing this idea, Weyl came to a unified geometrical theory of gravitational and electromagnetic fields.

However, his attempt to unify such different fields was not successful. In particular, the gravitational equations of the Weyl theory differ significantly from the Einstein equations and do not accord with experimental data. That is why the remarkable attempt to unify gravitation and electromagnetism was rejected by most of the physicists. Nevertheless, though the physical theory proposed by Weyl is not satisfactory, the Weyl gauge-invariant geometry looks very attractive and promising. For this reason, in this chapter, we undertake a new attempt to generalize the Einstein gravitational theory on the basis of Weyl's geometry. One of the main differences of our generalization from the Weyl unified theory consists in another interpretation of Weyl's potentials.

In order to have a new gravitational theory close to the Einstein theory, we regard the Weyl four potentials as small functions. We interpret these small four potentials as a 4-vector describing the influence of the physical vacuum on a matter moving in it.

Using Weyl's principle of scale invariance, we propose generalized gravitational and electromagnetic equations which give small corrections to the Einstein and Maxwell field theories. These equations are applied to a number of unsolved problems of cosmology.

One of the serious difficulties of modern cosmology is the problem of the cosmological singularity. As is known, the modern cosmological models based on the Einstein gravitational theory show that the Universe was born 13.8 billion years (13.8 Gyr) ago and at this initial moment, its density and temperature were infinite. However, it is very difficult to accept this conclusion since the infinite values of physical parameters indicate that there are some defects in standard cosmology. Moreover, the Universe's age of 13.8 Gyr brings a number of very great difficulties.

For example, it is very difficult to explain why orderly disk galaxies already existed more than 13.4 years ago [61].

Besides, consider the unsolved problem of dark matter. As follows from astronomical data, the invisible (dark) part of the mass of clusters of galaxies is an order of magnitude greater than the total mass of their observed stars. The simplest way to explain this phenomenon is that the dark matter consists of old, faintly glowing and extinct stars. However, this idea contradicts the accepted age of the Universe: this age is too small for old galaxies to have so large number of faintly glowing and extinct stars. Then, the question arises: Is the age of 13.8 Gyr really true for the Universe?

Let us now turn to the cosmology that follows from the generalized Einstein gravitational theory based on Weyl's geometry. It is shown in this chapter that in contrast to standard cosmology, the proposed new theory gives a nonsingular cosmological solution. At the same time, the new cosmology is a real alternative to the generally accepted ΛCDM model and allows us to answer a number of unresolved cosmological questions.

7.2. Generalization of the Equations of General Relativity Based on Weyl's Principle of Scale Invariance

As stated above, in 1918, H. Weyl proposed a generalization of the Einstein gravitational theory. In it, the connection $\Gamma_{\mu\nu}^{\gamma}$ was

defined as [68]

$$\Gamma^{\gamma}_{\mu\nu} = \frac{1}{2}g^{\gamma\alpha}(\partial_{\mu}g_{\alpha\nu} + \partial_{\nu}g_{\alpha\mu} - \partial_{\alpha}g_{\mu\nu})$$
$$+ \frac{1}{2}(\lambda^{\gamma}g_{\mu\nu} - \lambda_{\mu}\delta^{\gamma}_{\nu} - \lambda_{\nu}\delta^{\gamma}_{\mu}), \Gamma^{\gamma}_{\mu\nu} = \Gamma^{\gamma}_{\nu\mu}, \qquad (7.2.1)$$

where $\partial_{\mu} \equiv \partial/\partial x^{\mu}$, x^{μ} are space–time coordinates, $\mu = 0, 1, 2, 3$, $g_{\mu\nu}$ are components of the metric tensor, δ^{μ}_{ν} is the Kronecker symbol and λ^{μ} are components of a four-dimensional vector.

Weyl's formula (7.2.1) for the connection $\Gamma^{\gamma}_{\mu\nu}$ gives the following expressions for the covariant derivatives ∇_{μ} of the metric tensor:

$$\nabla_{\mu}g_{\nu\gamma} = \lambda_{\mu}g_{\nu\gamma}, \nabla_{\mu}g^{\nu\gamma} = -\lambda_{\mu}g^{\nu\gamma}. \qquad (7.2.2)$$

Weyl's connection (7.2.1) is invariant under the gauge transformations

$$g_{\mu\nu} \to \exp(\phi)g_{\mu\nu}, \lambda_{\mu} \to \lambda_{\mu} + \partial_{\mu}\phi, \qquad (7.2.3)$$

where ϕ is an arbitrary differentiable function of the coordinates x^{μ}. This property of connection (7.2.1) plays an important role in the Weyl theory.

In Weyl's geometry, the space–time interval ds is defined as the invariant differential expression $(g_{\mu\nu}dx^{\mu}dx^{\nu})^{1/2}$, which gives the kinematic equation for beams of light [68]

$$g_{\mu\nu}dx^{\mu}dx^{\nu} = 0. \qquad (7.2.4)$$

Hermann Weyl (1885–1955)

Then, the gauge transformations (7.2.3), consistent with Eq. (7.2.4), give physically admissible components $g_{\mu\nu}$ and λ_{μ}, which do not change the connection $\Gamma^{\gamma}_{\mu\nu}$.

Weyl regarded the components λ_{μ} in formula (7.2.1) as potentials of an electromagnetic field, and his goal was to construct a unified theory of gravitation and electromagnetism. However, despite all its attractiveness, this theory was not supported by either

Einstein or other famous physicists since it did not give satisfactory results.

Therefore, our goal is to study such a conformally invariant generalization of Einstein's gravitational theory with Weyl's geometry, in which the components λ_μ are not potentials of the electromagnetic field. In order to have a theory consistent with the well-known experimental data, we will interpret the components λ_μ as potentials that owe their existence to the physical vacuum and give only small corrections to the Einstein gravitational equations.

Although we will use Weyl's geometry to generalize Einstein's equations of general relativity, we will follow a way different from that chosen by Weyl.

We will adhere to the following principles:

(1) Weyl's principle of scale invariance. *In vacuum, the equations for gravitational fields should be invariant under the scale transformations (7.2.3).*

(2) Additional principle. *The generalized gravitational equations can be reduced to the Einstein equations with an additional energy–momentum tensor corresponding to Weyl's vector field with potentials λ_μ.*

It should be noted that the theory proposed by Weyl is based on a second-order conformally invariant Lagrangian with respect to curvature [68]. The Weyl Lagrangian and a number of other second-order conformally invariant Lagrangians with respect to curvature lead to fourth-order gravitational equations with respect to derivatives of the metric [69], in contrast to Einstein's equations, which are second order with respect to them. Consequently, such generalizations of Einstein's theory do not satisfy the second of the requirements formulated above, and therefore the desired gravitational equations must be of the second order with respect to derivatives of the metric tensor.

Let us now turn to the following relativistically covariant generalization of the Einstein gravitational equations, which is based on the Weyl connection (7.2.1) and is different from the Weyl equations:

$$R_{\mu\nu} + R_{\nu\mu} - g_{\mu\nu}R = (16\pi f_N/c^4)T_{\mu\nu},$$

$$R_{\mu\nu} = \partial_\alpha \Gamma^\alpha_{\mu\nu} - \partial_\nu \Gamma^\alpha_{\mu\alpha} + \Gamma^\alpha_{\mu\nu}\Gamma^\beta_{\alpha\beta} - \Gamma^\beta_{\mu\alpha}\Gamma^\alpha_{\nu\beta},$$

$$R = g^{\mu\nu}R_{\mu\nu}, \tag{7.2.5}$$

where $R_{\mu\nu}$ is the Ricci tensor, the connection $\Gamma^\gamma_{\mu\nu}$ is determined by formula (7.2.1), $T_{\mu\nu}$ is the energy–momentum tensor of matter and nongravitational fields, λ^μ is the 4-vector of Weyl's potentials and f_N is the Newtonian gravitational constant. In the first equation in (7.2.5), we have taken into account that the energy–momentum tensor $T_{\mu\nu}$ should be symmetric: $T_{\mu\nu} = T_{\nu\mu}$ [13].

As will be shown later on, the gravitational equations (7.2.5) accord with the two requirements formulated above.

It should be noted that since the connection $\Gamma^\gamma_{\mu\nu}$ is invariant under the gauge transformations (7.2.3), the tensors $R_{\mu\nu}$ and $g_{\mu\nu}R = g_{\mu\nu}g^{\alpha\beta}R_{\alpha\beta}$ contained in the first equation in (7.2.5) are also conformally invariant. Therefore, the left-hand side of this equation is conformally invariant.

Consider now gravitational and electromagnetic fields in vacuum. Then, we come to Eqs. (7.2.5) with the following energy–momentum tensor $T_{\mu\nu}$:

$$T_{\mu\nu} = \frac{1}{4\pi}\left(-F_\mu{}^\alpha F_{\nu\alpha} + \frac{1}{4}g_{\mu\nu}F_{\alpha\beta}F^{\alpha\beta}\right), \qquad (7.2.6)$$

where $F_{\mu\nu}$ is the antisymmetric tensor of electromagnetic field strengths.

Let us require that Eqs. (7.2.5) and (7.2.6) should be invariant under the gauge transformations (7.2.3). As noted above, the Ricci tensor $R_{\mu\nu}$ and the left-hand side of the first equation in (7.2.5) are invariant under the gauge transformations (7.2.3). It can be readily verified that expression (7.2.6) is invariant under the following gauge transformations:

$$g_{\mu\nu} \to \exp(\phi)g_{\mu\nu}, \ F_{\mu\nu} \to \exp(\phi/2)F_{\mu\nu}, \qquad (7.2.7)$$

where ϕ is an arbitrary differentiable function of space–time coordinates.

Thus, in vacuum, Eqs. (7.2.5) and (7.2.6) for gravitational and electromagnetic fields are invariant under the following gauge transformations:

$$g_{\mu\nu} \to \exp(\phi)g_{\mu\nu}, \ \lambda_\mu \to \lambda_\mu + \partial_\mu\phi, \ F_{\mu\nu} \to \exp(\phi/2)F_{\mu\nu}. \quad (7.2.8)$$

Let us find the equations for the electromagnetic field strengths $F_{\mu\nu}$ that generalize Maxwell's equations and are covariant in vacuum under the gauge transformations (7.2.8). For this purpose, consider the following equations:

$$\left(\nabla^\mu - \frac{1}{2}\lambda^\mu\right)F_{\mu\nu} = 4\pi\sigma_0 dx_\nu/ds, \qquad (7.2.9)$$

$$\left(\nabla_\gamma - \frac{1}{2}\lambda_\gamma\right)F_{\mu\nu} + \left(\nabla_\nu - \frac{1}{2}\lambda_\nu\right)F_{\gamma\mu} + \left(\nabla_\mu - \frac{1}{2}\lambda_\mu\right)$$
$$F_{\nu\gamma} = 0, \qquad (7.2.10)$$

where $F_{\mu\nu} = -F_{\nu\mu}$, σ_0 is the charge density of matter in a comoving local inertial frame and dx_ν/ds are the covariant components of the matter 4-vector of velocity.

Let us show that they are the desired equations. Actually, when $\lambda_\mu = 0$, Eqs. (7.2.9) and (7.2.10) represent the well-known covariant generalization of the Maxwell equations [13]. Besides, from (7.2.8), we obtain the following gauge transformation:

$$\left(\nabla_\gamma - \frac{1}{2}\lambda_\gamma\right)F_{\mu\nu} \to \exp(\phi/2)\left(\nabla_\gamma - \frac{1}{2}\lambda_\gamma\right)F_{\mu\nu}. \qquad (7.2.11)$$

Taking into account (7.2.11), we find that when $\sigma_0 = 0$, Eqs. (7.2.9) and (7.2.10) are covariant under the gauge transformations (7.2.8).

Therefore, Eqs. (7.2.9) and (7.2.10) are the desired equations for the electromagnetic field strengths $F_{\mu\nu}$.

As will be shown later on, in the case of choosing the standard gauge for the components $g_{\mu\nu}$, the Weyl potentials λ_μ are usually very small quantities: $\sqrt{\lambda^\mu\lambda_\mu} \sim 1/A$ where A is the radius of the spatial curvature of the Universe. However, we will see that these small potentials can have a significant effect on cosmological processes that take place over billions of years.

Consider the gravitational equations (7.2.5). From them, we find

$$R_{\nu\mu} - R_{\mu\nu} = \partial_\nu\Gamma^\alpha_{\mu\alpha} - \partial_\mu\Gamma^\alpha_{\nu\alpha} = 2(\partial_\mu\lambda_\nu - \partial_\nu\lambda_\mu). \qquad (7.2.12)$$

Let us introduce the antisymmetric tensor

$$\Lambda_{\mu\nu} = \nabla_\mu\lambda_\nu - \nabla_\nu\lambda_\mu \equiv \partial_\mu\lambda_\nu - \partial_\nu\lambda_\mu. \qquad (7.2.13)$$

The tensor $\Lambda_{\mu\nu}$, which is invariant under the gauge transformations (7.2.3), can be interpreted as the tensor of strengths of the Weyl field.

From (7.2.12) and (7.2.13), we obtain

$$R_{\nu\mu} - R_{\mu\nu} = 2\Lambda_{\mu\nu}. \tag{7.2.14}$$

Let us put

$$\Gamma^\beta_{\mu\nu} = \bar{\Gamma}^\beta_{\mu\nu} + \gamma^\beta_{\mu\nu}, \bar{\Gamma}^\beta_{\mu\nu} = \frac{1}{2}g^{\beta\alpha}(\partial_\mu g_{\alpha\nu} + \partial_\nu g_{\alpha\mu} - \partial_\alpha g_{\mu\nu}),$$

$$\gamma^\beta_{\mu\nu} = \frac{1}{2}(\lambda^\beta g_{\mu\nu} - \lambda_\mu \delta^\beta_\nu - \lambda_\nu \delta^\beta_\mu),$$

$$R_{\mu\nu} = \bar{R}_{\mu\nu} + W_{\mu\nu}, \bar{R} = g^{\mu\nu}\bar{R}_{\mu\nu},$$

$$\bar{R}_{\mu\nu} = \partial_\alpha \bar{\Gamma}^\alpha_{\mu\nu} - \partial_\nu \bar{\Gamma}^\alpha_{\mu\alpha} + \bar{\Gamma}^\alpha_{\mu\nu}\bar{\Gamma}^\beta_{\alpha\beta} - \bar{\Gamma}^\beta_{\mu\alpha}\bar{\Gamma}^\alpha_{\nu\beta}, \tag{7.2.15}$$

$$W_{\mu\nu} = \partial_\alpha \gamma^\alpha_{\mu\nu} - \partial_\nu \gamma^\alpha_{\mu\alpha} + \bar{\Gamma}^\alpha_{\mu\nu}\gamma^\beta_{\alpha\beta} + \gamma^\alpha_{\mu\nu}\bar{\Gamma}^\beta_{\alpha\beta}$$

$$- \bar{\Gamma}^\beta_{\mu\alpha}\gamma^\alpha_{\nu\beta} - \gamma^\beta_{\mu\alpha}\bar{\Gamma}^\alpha_{\nu\beta} + \gamma^\alpha_{\mu\nu}\gamma^\beta_{\alpha\beta} - \gamma^\beta_{\mu\alpha}\gamma^\alpha_{\nu\beta}.$$

Taking into account (7.2.15), the first equation in (7.2.5) can be rewritten in the form

$$\bar{R}_{\mu\nu} - \frac{1}{2}g_{\mu\nu}\bar{R} = \frac{8\pi f_N}{c^4}(T_{\mu\nu} + \Theta_{\mu\nu}),$$

$$\Theta_{\mu\nu} = -\frac{c^4}{8\pi f_N}\left(W_{\mu\nu} - \frac{1}{2}Wg_{\mu\nu} + \Lambda_{\mu\nu}\right), W = g^{\mu\nu}W_{\mu\nu},$$

$$\tag{7.2.16}$$

which presents the Einstein gravitational equations with the additional energy–momentum tensor $\Theta_{\mu\nu}$ corresponding to Weyl's field with the potentials λ_μ.

Thus, the proposed gravitational equations (7.2.5) accord with both principles formulated above.

As is well-known, Eq. (7.2.16) gives the following differential relations [13]:

$$\bar{\nabla}_\mu(T^{\mu\nu} + \Theta^{\mu\nu}) \equiv \partial_\mu(T^{\mu\nu} + \Theta^{\mu\nu}) + \bar{\Gamma}^\mu_{\alpha\mu}(T^{\alpha\nu} + \Theta^{\alpha\nu})$$

$$+ \bar{\Gamma}^\nu_{\alpha\mu}(T^{\alpha\mu} + \Theta^{\alpha\mu}) = 0, \tag{7.2.17}$$

where $\bar{\nabla}_\mu$ denotes the covariant derivative, defined by the Christoffel symbols $\bar{\Gamma}^\beta_{\mu\nu}$.

Let us choose the standard gauge for the components $g_{\mu\nu}$. This means that in local inertial frames of reference, the components $g_{\mu\nu} = \bar{g}_{\mu\nu}$, where $\bar{g}_{\mu\nu}$ is the Minkowski metric tensor. Then, from (7.2.17), we obtain the following differential equations of energy–momentum conservation in a local inertial frame:

$$\partial_\mu(T^{\mu\nu} + \Theta^{\mu\nu}) = 0, \quad g_{\mu\nu} = \bar{g}_{\mu\nu}, \qquad (7.2.18)$$

which contain the additional energy–momentum tensor $\Theta^{\mu\nu}$ corresponding to the Weyl vectorial field.

Further, we will study the gravitational equations (7.2.5) that include the Weyl potentials λ_μ and use the Bianchi identities [13] to obtain a generalized differential relation for the matter energy–momentum tensor $T^{\mu\nu}$.

This relation will be applied to obtain covariant kinematic equations of motion of charged dust-like matter in the gravitational and electromagnetic fields generated by it. We will come to four differential equations of the second order with respect to three functions $x^k(x^0), k = 1, 2, 3$, describing the time dependence of the spatial coordinates of material points of the dust-like matter. The condition of consistency of these equations will allow us to find four differential equations of the second order for the four vacuum potentials λ_μ. These four equations and ten gravitational equations (7.2.5) will give us fourteen second-order differential equations for the fourteen functions λ_μ and $g_{\mu\nu}$.

Within the framework of the proposed gravitational equations (7.2.5), we will consider a number of cosmological problems. We will find a nonsingular cosmological solution for a homogeneous and isotropic physical vacuum with Weyl's potentials λ_μ and determine the effect of these potentials on particles moving in the physical vacuum. This solution will be applied to explain a number of astrophysical and cosmological phenomena.

7.3.   Generalized Equation for the Energy–Momentum Tensor of Matter

Consider the curvature tensor $R^\nu_{\alpha\beta\gamma}$, defined by the following formula [13]:

$$R^\nu_{\alpha\beta\gamma} = \partial_\beta\Gamma^\nu_{\alpha\gamma} - \partial_\gamma\Gamma^\nu_{\alpha\beta} + \Gamma^\mu_{\alpha\gamma}\Gamma^\nu_{\mu\beta} - \Gamma^\mu_{\alpha\beta}\Gamma^\nu_{\mu\gamma}, \qquad (7.3.1)$$

where the connection $\Gamma^\nu_{\alpha\beta}$ is given by formula (7.2.1) and, since $\Gamma^\nu_{\alpha\beta} = \Gamma^\nu_{\beta\alpha}$, the well-known Bianchi identities take place [13]

$$\nabla_\mu R^\nu_{\alpha\beta\gamma} + \nabla_\gamma R^\nu_{\alpha\mu\beta} + \nabla_\beta R^\nu_{\alpha\gamma\mu} = 0. \tag{7.3.2}$$

From identities (7.3.2), we obtain the following equality:

$$g^{\alpha\beta}(\nabla_\mu R^\mu_{\alpha\beta\gamma} + \nabla_\gamma R^\mu_{\alpha\mu\beta} + \nabla_\beta R^\mu_{\alpha\gamma\mu}) = 0. \tag{7.3.3}$$

Let us represent equality (7.3.3) in another form containing the Ricci tensor $R_{\alpha\beta}$, instead of the curvature tensor $R^\nu_{\alpha\beta\gamma}$, using the well-known relations [13]

$$R^\nu_{\alpha\nu\beta} = R_{\alpha\beta}, R^\nu_{\alpha\beta\gamma} = -R^\nu_{\alpha\gamma\beta}. \tag{7.3.4}$$

With this aim, first consider the following tensor:

$$R_{\nu\alpha\beta\gamma} \equiv g_{\nu\mu} R^\mu_{\alpha\beta\gamma} = g_{\nu\mu}(\partial_\beta \Gamma^\mu_{\alpha\gamma} - \partial_\gamma \Gamma^\mu_{\alpha\beta} + \Gamma^\sigma_{\alpha\gamma}\Gamma^\mu_{\sigma\beta} - \Gamma^\sigma_{\alpha\beta}\Gamma^\mu_{\sigma\gamma}). \tag{7.3.5}$$

As noted above, Weyl's connection satisfies the equality $\Gamma^\mu_{\alpha\beta} = \Gamma^\mu_{\beta\alpha}$. As is known [13], it gives that in a small vicinity of an arbitrary point x^μ, one can choose a local coordinate system in which at this point $\Gamma^\mu_{\alpha\beta} = 0$. Then, in the chosen coordinate system at the considered point, expression (7.3.5) acquires the following form:

$$\Gamma^\mu_{\alpha\beta} = 0, R_{\nu\alpha\beta\gamma} = g_{\nu\mu}(\partial_\beta \Gamma^\mu_{\alpha\gamma} - \partial_\gamma \Gamma^\mu_{\alpha\beta})$$

$$= \partial_\beta(g_{\nu\mu}\Gamma^\mu_{\alpha\gamma}) - \partial_\gamma(g_{\nu\mu}\Gamma^\mu_{\alpha\beta}). \tag{7.3.6}$$

Formulas (7.2.1) and (7.3.6) give

$$R_{\nu\alpha\beta\gamma} = \frac{1}{2}\partial_\beta(\partial_\alpha g_{\nu\gamma} - \partial_\nu g_{\alpha\gamma} + \lambda_\nu g_{\alpha\gamma} - \lambda_\alpha g_{\nu\gamma} - \lambda_\gamma g_{\nu\alpha})$$

$$- \frac{1}{2}\partial_\gamma(\partial_\alpha g_{\nu\beta} - \partial_\nu g_{\alpha\beta} + \lambda_\nu g_{\alpha\beta} - \lambda_\alpha g_{\nu\beta} - \lambda_\beta g_{\nu\alpha}),$$

$$\Gamma^\mu_{\alpha\beta} = 0. \tag{7.3.7}$$

In the considered local coordinate system, from (7.3.7), we obtain

$$R_{\nu\alpha\beta\gamma} = -R_{\nu\alpha\gamma\beta}, \tag{7.3.8}$$

$$R_{\nu\alpha\beta\gamma} + R_{\alpha\nu\beta\gamma} = \nabla_\gamma(\lambda_\beta g_{\nu\alpha}) - \nabla_\beta(\lambda_\gamma g_{\nu\alpha})$$

$$= g_{\nu\alpha}(\nabla_\gamma\lambda_\beta - \nabla_\beta\lambda_\gamma) = g_{\nu\alpha}\Lambda_{\gamma\beta}, \tag{7.3.9}$$

where formulas (7.2.2) and (7.2.13) are taken into account.

Since the left-hand and right-hand sides of equalities (7.3.8) and (7.3.9) are tensors, these equalities are valid in an arbitrary coordinate system.

Using relations (7.3.8), (7.3.9) and (7.3.4), we find

$$g^{\alpha\beta}R_{\nu\alpha\gamma\beta} = -g^{\alpha\beta}R_{\nu\alpha\beta\gamma} = -g^{\alpha\beta}(-R_{\alpha\nu\beta\gamma} + g_{\nu\alpha}\Lambda_{\gamma\beta})$$

$$= g^{\alpha\beta}g_{\alpha\mu}R^{\mu}_{\nu\beta\gamma} + \delta^{\beta}_{\nu}\Lambda_{\beta\gamma}$$

$$= R^{\mu}_{\nu\mu\gamma} + \Lambda_{\nu\gamma} = R_{\nu\gamma} + \Lambda_{\nu\gamma}. \tag{7.3.10}$$

From (7.3.5) and (7.3.10), we find

$$g^{\alpha\beta}R^{\mu}_{\alpha\gamma\beta} = g^{\mu\nu}g^{\alpha\beta}R_{\nu\alpha\gamma\beta} = R^{\mu}_{\gamma} + \Lambda^{\mu}_{\gamma}. \tag{7.3.11}$$

Let us turn to equality (7.3.3) and represent it in the form

$$\nabla_{\mu}(g^{\alpha\beta}R^{\mu}_{\alpha\beta\gamma}) + \nabla_{\gamma}(g^{\alpha\beta}R^{\mu}_{\alpha\mu\beta}) + \nabla_{\beta}(g^{\alpha\beta}R^{\mu}_{\alpha\gamma\mu})$$

$$- R^{\mu}_{\alpha\beta\gamma}\nabla_{\mu}g^{\alpha\beta} - R^{\mu}_{\alpha\mu\beta}\nabla_{\gamma}g^{\alpha\beta} - R^{\mu}_{\alpha\gamma\mu}\nabla_{\beta}g^{\alpha\beta} = 0. \tag{7.3.12}$$

Using (7.2.2) and (7.3.4), from (7.3.12), we find

$$(\nabla_{\mu} + \lambda_{\mu})(g^{\alpha\beta}R^{\mu}_{\alpha\gamma\beta}) - (\nabla_{\gamma} + \lambda_{\gamma})R + (\nabla_{\beta} + \lambda_{\beta})R^{\beta}_{\gamma} = 0,$$

$$R = g^{\alpha\beta}R_{\alpha\beta}. \tag{7.3.13}$$

Taking into account relation (7.3.11), from equality (7.3), we obtain

$$(\nabla_{\mu} + \lambda_{\mu})\left(R^{\mu}_{\gamma} - \frac{1}{2}\delta^{\mu}_{\gamma}R + \frac{1}{2}\Lambda^{\mu}_{\gamma}\right) = 0. \tag{7.3.14}$$

Multiply Eq. (7.3.14) by $g^{\gamma\nu}$ and use the following formula, taking into account the second formula in (7.2.2):

$$g^{\gamma\nu}\nabla_{\mu}Q^{\mu}_{\gamma} = \nabla_{\mu}(g^{\gamma\nu}Q^{\mu}_{\gamma}) - Q^{\mu}_{\gamma}\nabla_{\mu}g^{\gamma\nu} = (\nabla_{\mu} + \lambda_{\mu})Q^{\mu\nu}, \tag{7.3.15}$$

where we denote

$$Q^{\mu}_{\gamma} = R^{\mu}_{\gamma} - \frac{1}{2}\delta^{\mu}_{\gamma}R + \frac{1}{2}\Lambda^{\mu}_{\gamma}. \tag{7.3.16}$$

Then, we obtain

$$(\nabla_{\mu} + 2\lambda_{\mu})Q^{\mu\nu} = 0, \quad Q^{\mu\nu} = R^{\mu\nu} - \frac{1}{2}g^{\mu\nu}R + \frac{1}{2}\Lambda^{\mu\nu}. \tag{7.3.17}$$

Using (7.2.2), we find

$$
\begin{aligned}
\nabla^\mu \lambda^\nu &= g^{\mu\sigma} \nabla_\sigma (g^{\nu\gamma} \lambda_\gamma) = g^{\mu\sigma} g^{\nu\gamma} (\nabla_\sigma \lambda_\gamma - \lambda_\sigma \lambda_\gamma) \\
&= g^{\mu\sigma} g^{\nu\gamma} \nabla_\sigma \lambda_\gamma - \lambda^\mu \lambda^\nu .
\end{aligned}
\tag{7.3.18}
$$

From (7.3.18) and (7.2.13), we obtain

$$
\nabla^\mu \lambda^\nu - \nabla^\nu \lambda^\mu = g^{\mu\sigma} g^{\nu\gamma} (\nabla_\sigma \lambda_\gamma - \nabla_\gamma \lambda_\sigma) = g^{\mu\sigma} g^{\nu\gamma}
$$

$$
\Lambda_{\sigma\gamma} = \Lambda^{\mu\nu}.
\tag{7.3.19}
$$

Therefore, Eq. (7.3.17) can be represented as

$$
(\nabla_\mu + 2\lambda_\mu) \left[R^{\mu\nu} - \frac{1}{2} g^{\mu\nu} R + \frac{1}{2}(\nabla^\mu \lambda^\nu - \nabla^\nu \lambda^\mu) \right] = 0.
\tag{7.3.20}
$$

From (7.2.5) and (7.2.14), we find

$$
R^{\mu\nu} - \frac{1}{2} g^{\mu\nu} R = (8\pi f_N / c^4) T^{\mu\nu} - \Lambda^{\mu\nu}.
\tag{7.3.21}
$$

Therefore, Eqs. (7.3.19)–(7.3.21) give the differential relation for the energy–momentum tensor $T^{\mu\nu}$

$$
(\nabla_\mu + 2\lambda_\mu)[(16\pi f_N / c^4) T^{\mu\nu} + \nabla^\nu \lambda^\mu - \nabla^\mu \lambda^\nu] = 0.
\tag{7.3.22}
$$

This equation represents conditions of consistency for the gravitational equations (7.2.5) with Weyl's connection. When the energy–momentum tensor $T^{\mu\nu} = 0$, Eq. (7.3.22) gives differential equations of the second order for the Weyl vector λ^μ. Further, we will use Eq. (7.3.22) to find differential equations for λ^μ when $T^{\mu\nu} \neq 0$.

7.4. Differential Relation for the Energy–Momentum Tensor of Charged Particles and Their Electromagnetic Field

Consider an arbitrary system of charged particles interacting by means of gravitational and electromagnetic forces. Then, for its energy–momentum tensor $T^{\mu\nu}$, we have [13]

$$
T^{\mu\nu} = T^{\mu\nu}_{(m)} + T^{\mu\nu}_{(e)},
\tag{7.4.1}
$$

where $T^{\mu\nu}_{(m)}$ and $T^{\mu\nu}_{(e)}$ correspond to the charged particles and the electromagnetic field generated by them, respectively, and are described by the following expressions [13]:

$$T^{\mu\nu}_{(m)} = c^2 \rho_0 \frac{dx^\mu}{ds} \frac{dx^\nu}{ds}, \quad ds^2 = g_{\mu\nu} dx^\mu dx^\nu, \qquad (7.4.2)$$

where ρ_0 is the mass density in a comoving local inertial frame of reference, and

$$T^{\mu\nu}_{(e)} = \frac{1}{4\pi} \left(-F^{\mu\alpha} F^\nu{}_\alpha + \frac{1}{4} g^{\mu\nu} F_{\alpha\beta} F^{\alpha\beta} \right). \qquad (7.4.3)$$

Here, $F_{\mu\nu}$ is the tensor of electromagnetic field strengths.

Consider the energy–momentum tensor $T^{\mu\nu}_{(m)}$ of the charged matter. In a local inertial frame with $x^\mu = \bar{x}^\mu$ and $g_{\mu\nu} = \bar{g}_{\mu\nu}$, where $\bar{g}_{\mu\nu}$ is the Minkowski metric tensor, we have the differential equation of rest mass conservation [13]

$$\partial_\mu \left(\rho_0 \frac{dx^\mu}{ds} \right) = 0. \qquad (7.4.4)$$

In this frame, from (7.2.1) and (7.4.4), we find

$$\nabla_\mu \left(\rho_0 \frac{dx^\mu}{ds} \right) = \partial_\mu \left(\rho_0 \frac{dx^\mu}{ds} \right) + \Gamma^\nu_{\mu\nu} \rho_0 \frac{dx^\mu}{ds} = -2\lambda_\mu \rho_0 \frac{dx^\mu}{ds},$$
$$x^\mu = \bar{x}^\mu, \quad g_{\mu\nu} = \bar{g}_{\mu\nu}. \qquad (7.4.5)$$

Using (7.4.5), we come to the following covariant equation of rest mass conservation:

$$(\nabla_\mu + 2\lambda_\mu) \left(\rho_0 \frac{dx^\mu}{ds} \right) = 0. \qquad (7.4.6)$$

From Eqs. (7.4.2), we obtain

$$(\nabla_\mu + 2\lambda_\mu) T^{\mu\nu}_{(m)} = c^2 (\nabla_\mu + 2\lambda_\mu) \left(\rho_0 \frac{dx^\mu}{ds} \right) \frac{dx^\nu}{ds}$$
$$+ c^2 \rho_0 \frac{dx^\mu}{ds} \nabla_\mu \left(\frac{dx^\nu}{ds} \right). \qquad (7.4.7)$$

Using (7.4.6), we can represent Eq. (7.4.7) in the form

$$(\nabla_\mu + 2\lambda_\mu)T^{\mu\nu}_{(m)} = c^2\rho_0 \frac{dx^\mu}{ds}\nabla_\mu\left(\frac{dx^\nu}{ds}\right). \qquad (7.4.8)$$

Since we have

$$\frac{dx^\mu}{ds}\nabla_\mu\left(\frac{dx^\nu}{ds}\right) = \frac{dx^\mu}{ds}\left[\partial_\mu\left(\frac{dx^\nu}{ds}\right) + \Gamma^\nu_{\mu\gamma}\frac{dx^\gamma}{ds}\right]$$

$$= \frac{d^2x^\nu}{ds^2} + \Gamma^\nu_{\mu\gamma}\frac{dx^\mu}{ds}\frac{dx^\gamma}{ds}, \qquad (7.4.9)$$

from (7.4.8), we obtain

$$(\nabla_\mu + 2\lambda_\mu)T^{\mu\nu}_{(m)} = c^2\rho_0\left(\frac{d^2x^\nu}{ds^2} + \Gamma^\nu_{\mu\gamma}\frac{dx^\mu}{ds}\frac{dx^\gamma}{ds}\right). \qquad (7.4.10)$$

Let us now turn to Eq. (7.4.3) for the energy–momentum tensor of an electromagnetic field and study it in this theory with Weyl's connection.

From (7.4.3), we find

$$\nabla_\mu T^{\mu\nu}_{(e)} = \frac{1}{4\pi}\left(- F^\nu{}_\alpha\nabla_\mu F^{\mu\alpha} - F^{\mu\alpha}\nabla_\mu F^\nu{}_\alpha - \frac{1}{4}\lambda^\nu F_{\alpha\beta}F^{\alpha\beta}\right.$$

$$\left. + \frac{1}{4}g^{\mu\nu}F^{\alpha\beta}\nabla_\mu F_{\alpha\beta} + \frac{1}{4}g^{\mu\nu}F_{\alpha\beta}\nabla_\mu F^{\alpha\beta}\right). \quad (7.4.11)$$

Using (7.2.2) and (7.2.9), we obtain

$$\nabla_\mu F^{\mu\alpha} = \nabla_\mu(g^{\mu\gamma}g^{\alpha\delta}F_{\gamma\delta}) = g^{\alpha\delta}(\nabla^\gamma F_{\gamma\delta} - 2\lambda^\gamma F_{\gamma\delta})$$

$$= 4\pi\sigma_0 dx^\alpha/ds - \frac{3}{2}g^{\alpha\delta}\lambda^\gamma F_{\gamma\delta}$$

$$= 4\pi\sigma_0 dx^\alpha/ds - \frac{3}{2}\lambda_\mu F^{\mu\alpha}, \qquad (7.4.12)$$

$$\nabla_\mu F^\nu{}_\alpha = \nabla_\mu(g^{\nu\gamma}F_{\gamma\alpha}) = g^{\nu\gamma}\nabla_\mu F_{\gamma\alpha} - \lambda_\mu F^\nu{}_\alpha, \quad (7.4.13)$$

$$g^{\mu\nu}F_{\alpha\beta}\nabla_\mu F^{\alpha\beta} = g^{\mu\nu}F_{\alpha\beta}\nabla_\mu(g^{\alpha\gamma}g^{\beta\delta}F_{\gamma\delta})$$

$$= g^{\mu\nu}F^{\gamma\delta}\nabla_\mu F_{\gamma\delta} - 2\lambda^\nu F^{\gamma\delta}F_{\gamma\delta}. \qquad (7.4.14)$$

Substituting formulas (7.4.12)–(7.4.14) into Eq. (7.4.11), we find

$$
\nabla_\mu T^{\mu\nu}_{(e)} = \frac{1}{4\pi}\left(-4\pi\sigma_0 F^{\nu\alpha}dx_\alpha/ds + \frac{5}{2}\lambda_\mu F^{\mu\alpha}F^\nu{}_\alpha - \frac{3}{4}\lambda^\nu F_{\alpha\beta}F^{\alpha\beta} \right.
$$
$$
\left. -g^{\nu\gamma}F^{\mu\alpha}\nabla_\mu F_{\gamma\alpha} + \frac{1}{2}g^{\mu\nu}F^{\alpha\beta}\nabla_\mu F_{\alpha\beta} \right). \tag{7.4.15}
$$

Using Eq. (7.2.10), we obtain

$$
\frac{1}{2}g^{\mu\nu}F^{\alpha\beta}\nabla_\mu F_{\alpha\beta} = -\frac{1}{2}g^{\mu\nu}F^{\alpha\beta}\left[\nabla_\beta F_{\mu\alpha} + \nabla_\alpha F_{\beta\mu} \right.
$$
$$
\left. -\frac{1}{2}(\lambda_\mu F_{\alpha\beta} + \lambda_\beta F_{\mu\alpha} + \lambda_\alpha F_{\beta\mu}) \right]
$$
$$
= -g^{\mu\nu}F^{\alpha\beta}\nabla_\beta F_{\mu\alpha} + \frac{1}{4}\lambda^\nu F^{\alpha\beta}F_{\alpha\beta}
$$
$$
+\frac{1}{2}g^{\mu\nu}\lambda_\beta F^{\alpha\beta}F_{\mu\alpha}
$$
$$
= g^{\nu\gamma}F^{\mu\alpha}\nabla_\mu F_{\gamma\alpha} + \frac{1}{4}\lambda^\nu F_{\alpha\beta}F^{\alpha\beta}
$$
$$
-\frac{1}{2}g^{\nu\gamma}\lambda_\mu F^{\mu\alpha}F_{\gamma\alpha}. \tag{7.4.16}
$$

Substituting (7.4.16) into Eq. (7.4.15), we find

$$
\nabla_\mu T^{\mu\nu}_{(e)} = -\sigma_0 F^{\nu\alpha}dx_\alpha/ds + \frac{1}{4\pi}\left(2\lambda_\mu F^{\mu\alpha}F^\nu{}_\alpha - \frac{1}{2}\lambda^\nu F_{\alpha\beta}F^{\alpha\beta} \right). \tag{7.4.17}
$$

Using formula (7.4.3), Eq. (7.4.17) can be rewritten as

$$
(\nabla_\mu + 2\lambda_\mu)T^{\mu\nu}_{(e)} = -\sigma_0 F^{\nu\alpha}dx_\alpha/ds. \tag{7.4.18}
$$

Formulas (7.4.1), (7.4.10) and (7.4.18) give the following differential relation for the energy–momentum tensor $T^{\mu\nu}$:

$$
(\nabla_\mu + 2\lambda_\mu)T^{\mu\nu} = c^2\rho_0\left(\frac{d^2x^\nu}{ds^2} + \Gamma^\nu_{\alpha\beta}\frac{dx^\alpha}{ds}\frac{dx^\beta}{ds} \right) - \sigma_0 F^{\nu\alpha}\frac{dx_\alpha}{ds}. \tag{7.4.19}
$$

7.5. Differential Equations for Components of Weyl's Vector

From Eqs. (7.3.22) and (7.4.19), we derive

$$(\nabla_\mu + 2\lambda_\mu)(\nabla^\nu \lambda^\mu - \nabla^\mu \lambda^\nu) + \frac{16\pi f_N}{c^2}\rho_0 \left(\frac{d^2 x^\nu}{ds^2} + \Gamma^\nu_{\alpha\beta} \frac{dx^\alpha}{ds} \frac{dx^\beta}{ds} \right)$$

$$- \frac{16\pi f_N}{c^4}\sigma_0 F^{\nu\alpha} \frac{dx_\alpha}{ds} = 0. \tag{7.5.1}$$

The Weyl connection $\Gamma^\nu_{\alpha\beta}$ is defined by Eq. (7.2.1). It can be represented in the form

$$\Gamma^\nu_{\alpha\beta} = \bar\Gamma^\nu_{\alpha\beta} + \gamma^\nu_{\alpha\beta}, \tag{7.5.2}$$

where $\bar\Gamma^\nu_{\alpha\beta}$ are the Christoffel symbols defined as

$$\bar\Gamma^\nu_{\alpha\beta} = \frac{1}{2}g^{\nu\gamma}(\partial_\alpha g_{\gamma\beta} + \partial_\beta g_{\alpha\gamma} - \partial_\gamma g_{\alpha\beta}) \tag{7.5.3}$$

and $\gamma^\nu_{\alpha\beta}$ are as follows:

$$\gamma^\nu_{\alpha\beta} = \frac{1}{2}(\lambda^\nu g_{\alpha\beta} - \lambda_\alpha \delta^\nu_\beta - \lambda_\beta \delta^\nu_\alpha). \tag{7.5.4}$$

Let us multiply Eq. (7.5.1) by dx_ν/ds and sum it over ν. Then, using (7.5.2) and (7.5.4) and taking into account the antisymmetry of $F^{\nu\alpha}$, we obtain

$$\frac{dx_\nu}{ds}(\nabla_\mu + 2\lambda_\mu)(\nabla^\nu \lambda^\mu - \nabla^\mu \lambda^\nu)$$

$$+ \frac{16\pi f_N}{c^2}\rho_0 \frac{dx_\nu}{ds}\left(\frac{d^2 x^\nu}{ds^2} + \bar\Gamma^\nu_{\alpha\beta} \frac{dx^\alpha}{ds} \frac{dx^\beta}{ds} \right)$$

$$+ \frac{8\pi f_N}{c^2}\rho_0 \frac{dx_\nu}{ds}\frac{dx^\alpha}{ds}\frac{dx^\beta}{ds}(\lambda^\nu g_{\alpha\beta} - \lambda_\alpha \delta^\nu_\beta - \lambda_\beta \delta^\nu_\alpha) = 0. \tag{7.5.5}$$

Let us now use the following identity:

$$\frac{dx_\nu}{ds}\left(\frac{d^2 x^\nu}{ds^2} + \bar\Gamma^\nu_{\alpha\beta} \frac{dx^\alpha}{ds} \frac{dx^\beta}{ds} \right) = 0. \tag{7.5.6}$$

Indeed, in a local inertial frame with $x^\mu = \bar{x}^\mu$ and $g_{\mu\nu} = \bar{g}_{\mu\nu}$, where $\bar{g}_{\mu\nu}$ is the Minkowski metric tensor, Eq. (7.5.6) is evident since

$$\bar{\Gamma}^\nu_{\alpha\beta} = 0, \quad \frac{dx_\nu}{ds}\frac{d^2 x^\nu}{ds^2} = \frac{1}{2}\frac{d}{ds}\left(\frac{dx_\nu}{ds}\frac{dx^\nu}{ds}\right) \equiv 0, \quad g_{\mu\nu} = \bar{g}_{\mu\nu}. \quad (7.5.7)$$

Since the left-hand side of Eq. (7.5.6) is a scalar [13], Eq. (7.5.6) is also true in arbitrary coordinate systems.

It can be readily verified that the following identity is true:

$$\frac{dx_\nu}{ds}\frac{dx^\alpha}{ds}\frac{dx^\beta}{ds}(\lambda^\nu g_{\alpha\beta} - \lambda_\alpha \delta^\nu_\beta - \lambda_\beta \delta^\nu_\alpha) = -\lambda^\nu \frac{dx_\nu}{ds}. \quad (7.5.8)$$

From (7.5.2), (7.5.4), (7.5.6) and (7.5.8), we derive

$$\frac{dx_\nu}{ds}\left(\frac{d^2 x^\nu}{ds^2} + \Gamma^\nu_{\alpha\beta}\frac{dx^\alpha}{ds}\frac{dx^\beta}{ds}\right) = -\frac{1}{2}\lambda^\nu \frac{dx_\nu}{ds}. \quad (7.5.9)$$

Substituting Eqs. (7.5.6) and (7.5.8) into (7.5.5), we obtain

$$\frac{dx_\nu}{ds}\left[(\nabla_\mu + 2\lambda_\mu)(\nabla^\nu \lambda^\mu - \nabla^\mu \lambda^\nu) - \frac{8\pi f_N}{c^2}\rho_0 \lambda^\nu\right] = 0. \quad (7.5.10)$$

Since dx_ν can be arbitrary, Eq. (7.5.10) gives the following field equations for the Weyl vector λ^ν:

$$(\nabla_\mu + 2\lambda_\mu)(\nabla^\nu \lambda^\mu - \nabla^\mu \lambda^\nu) = \frac{8\pi f_N}{c^2}\rho_0 \lambda^\nu. \quad (7.5.11)$$

As follows from (7.4.2) and (7.4.3),

$$T^\mu_{\mu(m)} = c^2 \rho_0, \quad T^\mu_{\mu(e)} = 0. \quad (7.5.12)$$

Therefore, from (7.4.1), we obtain

$$T = c^2 \rho_0, \quad T \equiv T^\mu_\mu. \quad (7.5.13)$$

Using formula (7.5.13), we can rewrite Eq. (7.5.11) in the form

$$(\nabla_\mu + 2\lambda_\mu)(\nabla^\nu \lambda^\mu - \nabla^\mu \lambda^\nu) = \frac{8\pi f_N}{c^4}T\lambda^\nu. \quad (7.5.14)$$

Substituting Eq. (7.5.11) into (7.5.1), we come to the following dynamic equations for matter:

$$c^2 \rho_0 \left(\frac{d^2 x^\nu}{ds^2} + \Gamma^\nu_{\alpha\beta} \frac{dx^\alpha}{ds} \frac{dx^\beta}{ds} + \frac{1}{2} \lambda^\nu \right) - \sigma_0 F^\nu{}_\alpha \frac{dx^\alpha}{ds} = 0. \quad (7.5.15)$$

From (7.3.19) and (7.5.14), we find

$$(\nabla_\mu + 2\lambda_\mu)\Lambda^{\nu\mu} = \frac{8\pi f_N}{c^4} T\lambda^\nu. \quad (7.5.16)$$

Using (7.2.2), we obtain

$$\nabla_\mu \Lambda^{\nu\mu} = \nabla_\mu (g^{\nu\alpha} g^{\mu\beta} \Lambda_{\alpha\beta}) = g^{\nu\alpha} \nabla^\beta \Lambda_{\alpha\beta} - 2\lambda_\mu \Lambda^{\nu\mu}. \quad (7.5.17)$$

Therefore, Eq. (7.5.16) acquires the form

$$g^{\nu\alpha} \nabla^\beta \Lambda_{\alpha\beta} = \frac{8\pi f_N}{c^4} T\lambda^\nu, \quad (7.5.18)$$

This equation gives

$$\nabla^\mu \Lambda_{\nu\mu} = \frac{8\pi f_N}{c^4} T\lambda_\nu, \quad (7.5.19)$$

where $\Lambda_{\nu\mu}$ are defined by Eq. (7.2.13).

Thus, Eq. (7.5.19) is equivalent to Eq. (7.5.14).

Equation (7.5.19) represents four differential equations for the components λ_ν of Weyl's vector which should be added to the gravitational equations (7.2.5).

As follows from (7.2.13) and (7.5.13), in vacuum with $\rho_0 = 0$ and $T = c^2 \rho_0 = 0$, the field equations (7.5.19) and hence (7.5.14) for the components λ_ν are invariant under the gauge transformations (7.2.3). Therefore, Eq. (7.5.19) and hence Eq. (7.5.14) satisfy the above-stated Weyl's principle of scale invariance in vacuum.

Let us turn to the dynamic equations (7.5.15). Using equality (7.5.9) and the antisymmetry of the tensor $F_{\nu\mu}$, we find

$$\frac{dx_\nu}{ds} \left(\frac{d^2 x^\nu}{ds^2} + \Gamma^\nu_{\alpha\beta} \frac{dx^\alpha}{ds} \frac{dx^\beta}{ds} + \frac{1}{2} \lambda^\nu \right) = 0,$$

$$\frac{dx_\nu}{ds} F^\nu{}_\alpha \frac{dx^\alpha}{ds} = F_{\nu\alpha} \frac{dx^\nu}{ds} \frac{dx^\alpha}{ds} = 0. \quad (7.5.20)$$

Therefore, multiplying the left-hand side of Eq. (7.5.15) by dx_ν/ds and summing it over ν, we get zero. This means that the four differential equations (7.5.15) are not independent: the first equation

($\nu = 0$) is a consequence of the other three equations ($\nu = 1, 2, 3$). Hence, it suffices to solve the differential equations (7.5.15) only for $\nu = 1, 2, 3$.

7.6. Homogeneous and Isotropic Metric in Vacuum

Examine the space–time geometry of a big spatial region of physical vacuum located sufficiently far from massive bodies. We will assume that this region is homogeneous and isotropic and describe its geometry by the well-known Robertson–Walker metric. This metric can be represented as [70]

$$ds^2 = (dx^0)^2 - A^2[dr^2/(1 - Kr^2) + r^2(d\theta^2 + \sin^2\theta d\varphi^2)],$$
$$A = A(x^0), \tag{7.6.1}$$

where A is the scale factor, x^0/c is the time, r, θ, φ are some spherical coordinates and the parameter K takes the values $-1, 1, 0$ corresponding to the cases of negative space curvature, positive space curvature and to the flat case, respectively.

Let us require that the components λ_ν of the Weyl vector should be independent of rotations and translations of rectangular space axes in the considered homogeneous and isotropic space. Then, for the components λ_μ, we get

$$\lambda_1 = \lambda_2 = \lambda_3 = 0, \lambda_0 = \lambda_0(x^0). \tag{7.6.2}$$

From (7.2.13) and (7.6.2), we find

$$\Lambda_{\mu\nu} = 0 \tag{7.6.3}$$

and substituting (7.6.3) into Eq. (7.5.19), we obtain

$$T \equiv T^\mu_\mu = 0. \tag{7.6.4}$$

Let us choose the dimensionless time coordinate η defined by the formula

$$d\eta = dx^0/A(x^0). \tag{7.6.5}$$

Then, from (7.6.1), we find

$$ds^2 = A^2[d\eta^2 - dr^2/(1 - Kr^2) - r^2(d\theta^2 + \sin^2\theta d\varphi^2)], \quad A = A(\eta). \tag{7.6.6}$$

Let λ_μ^* describe the components of the Weyl vector in the coordinate system $(\eta, r, \theta, \varphi)$. Then, since λ_μ and λ_μ^* are components of a covariant vector in the coordinate systems $(x^0, r, \theta, \varphi)$ and $(\eta, r, \theta, \varphi)$, respectively, from (7.6.2) and (7.6.5), we find

$$\lambda_1^* = \lambda_2^* = \lambda_3^* = 0, \ \lambda_0^* = \lambda_0 dx^0/d\eta = A\lambda_0, \ \lambda_0^* = \lambda_0^*(\eta). \quad (7.6.7)$$

In the coordinate system $(\eta, r, \theta, \varphi)$, from (7.2.1), (7.2.5), (7.6.6) and (7.6.7), we obtain the following nonzero components $\Gamma_{\mu\nu}^\gamma$ and R_μ^ν[62]:

$$\Gamma_{00}^0 = \Gamma_{01}^1 = \Gamma_{02}^2 = \Gamma_{03}^3 = \alpha, \ \alpha = \dot{A}/A - \lambda_0^*/2, \ \dot{A} \equiv dA/d\eta,$$

$$\Gamma_{11}^0 = \alpha/(1 - Kr^2), \ \Gamma_{22}^0 = \alpha r^2, \ \Gamma_{33}^0 = \alpha r^2 \sin^2\theta,$$

$$\Gamma_{11}^1 = Kr/(1 - Kr^2), \ \Gamma_{22}^1 = -r(1 - Kr^2),$$

$$\Gamma_{33}^1 = -r(1 - Kr^2)\sin^2\theta,$$

$$\Gamma_{12}^2 = \Gamma_{13}^3 = 1/r, \ \Gamma_{33}^2 = -\sin\theta\cos\theta, \ \Gamma_{23}^3 = \cot\theta, \quad (7.6.8)$$

$$R_0^0 = -3A^{-2}\dot{\alpha}, \ R_1^1 = R_2^2 = R_3^3 = -A^{-2}\left(2K + \dot{\alpha} + 2\alpha^2\right),$$

$$\dot{\alpha} \equiv d\alpha/d\eta. \quad (7.6.9)$$

From the gravitational equations (7.2.5) and formulas (7.6.9), we find

$$3A^{-2}\dot{\alpha} = -\left(4\pi f_N/c^4\right)\left(\bar{T}_0^0 - 3\bar{T}_1^1\right),$$

$$\bar{T}_1^1 = \bar{T}_2^2 = \bar{T}_3^3,$$

$$A^{-2}\left(2K + \dot{\alpha} + 2\alpha^2\right) = \left(4\pi f_N/c^4\right)\left(\bar{T}_0^0 + \bar{T}_1^1\right), \quad (7.6.10)$$

where $T_\mu^\nu = \bar{T}_\mu^\nu$ is the energy–momentum tensor of the physical vacuum.

From (7.6.4) and (7.6.10), we obtain

$$\dot{\alpha} = -\left(8\pi f_N A^2/(3c^4)\right)\bar{T}_0^0, \ \dot{\alpha} + 2\left(\alpha^2 + K\right)$$

$$= (8\pi f_N A^2/(3c^4))\bar{T}_0^0. \quad (7.6.11)$$

Taking the sum of the two equations (7.6.11), we find

$$\dot{\alpha} + \alpha^2 + K = 0, \ \dot{\alpha} \equiv \partial\alpha/\partial\eta, \quad (7.6.12)$$

where, as noted above, $K = \pm 1, \ 0$.

7.7. Nonsingular Cosmological Solution to the Generalized Einstein Gravitational Equations with Weyl's Connection

From Eq. (7.6.12), we obtain the following solutions:

(1a) $K = -1, \alpha = \tanh(\eta + \eta_0)$, (1b) $K = -1, \alpha = \coth(\eta + \eta_0)$,

(2) $K = 1, \alpha = -\tan(\eta + \eta_0)$, (3) $K = 0, \alpha = 1/(\eta + \eta_0)$,

$$(7.7.1)$$

where $\eta_0 = \text{const}$.

As follows from (7.7.1), case (1a) is the only one in which there is no singularity.

Let us choose this case. Then, from (7.6.11) and (7.7.1), we find

$$\alpha = \tanh(\eta + \eta_0), K = -1, \ \bar{T}_0^0 = -\frac{3c^4}{8\pi f_N A^2} \cosh^{-2}(\eta + \eta_0). \ (7.7.2)$$

From (7.6.8), we have $\alpha = \dot{A}/A - \lambda_0^*/2$. Hence, using (7.7.2), we obtain

$$\dot{A}/A - \lambda_0^*/2 = \tanh(\eta + \eta_0). \tag{7.7.3}$$

Let us represent the scale factor A in the form

$$A = A_0 \exp(\chi/2), \chi = \chi(\eta), \ A_0 = \text{const} > 0. \tag{7.7.4}$$

Then, from (7.7.3), we get

$$\lambda_0^* = -2\tanh(\eta + \eta_0) + \dot{\chi}. \tag{7.7.5}$$

Here, $\chi = \chi(\eta)$ is a differentiable function which can be arbitrary.

Using (7.6.6) and (7.7.4) and taking into account that $K = -1$, as indicated in (7.7.2), we find

$$ds^2 = (A_0)^2 \exp(\chi)[d\eta^2 - dr^2/(1+r^2) - r^2(d\theta^2 + \sin^2\theta d\phi^2)]. \ (7.7.6)$$

Let us now perform the gauge transformations (7.2.3) with $\phi = -\chi$. Then, formulas (7.7.5), (7.7.6) and the expression for $\bar{T}_0^0$ in

(7.7.2) acquire the forms

$$ds^2 = (A_0)^2[d\eta^2 - dr^2/(1 + r^2) - r^2(d\theta^2 + \sin^2\theta d\varphi^2)], \quad (7.7.7)$$

$$\lambda_0^* = -2\tanh(\eta + \eta_0), \ \bar{T}_0^0 = -\frac{3c^4}{8\pi f_N(A_0)^2}\cosh^{-2}(\eta + \eta_0). \quad (7.7.8)$$

Let us choose the time coordinate $x^0 = A_0(\eta + \eta_0)$ instead of η and the radial coordinate $\bar{r} = A_0 r$ instead of r. Then, formulas (7.7.7), (7.7.8) and (7.6.7) give

$$ds^2 = (dx^0)^2 - d\bar{r}^2/[1 + (\bar{r}/A_0)^2] - \bar{r}^2(d\theta^2 + \sin^2\theta d\varphi^2), \quad (7.7.9)$$

$$\lambda_0 = -(2/A_0)\tanh(x^0/A_0), \ \lambda_1 = \lambda_2 = \lambda_3 = 0,$$

$$A_0 = \text{const} > 0. \quad (7.7.10)$$

Nikolai Lobachevsky
(1792–1856)

The spatial part of metric (7.7.9) corresponds to the Lobachevsky geometry with the radius A_0, which coincides with the Euclidean geometry for small values of $\bar{r}$. Therefore, in the chosen gauge, the considered homogeneous and isotropic metric corresponds in small spatial regions to the Minkowski metric. Thus, the chosen gauge gives the Minkowski geometry in local frames of references that are inertial relative to the considered homogeneous and isotropic physical vacuum. Further, we will use this gauge.

In the following section, we will apply formulas (7.7.9) and (7.7.10) to propagating electromagnetic waves and moving free particles in vacuum. It will be shown that when the time x^0 is positive, the component λ_0 of the Weyl vector causes cosmological redshift of photon's frequency.

7.8. Influence of Weyl's Field on Propagating Electromagnetic Waves and Moving Free Particles in Vacuum

Consider a plane electromagnetic wave propagating in vacuum along the axis x^1. Near a straight line, the Lobachevsky geometry, described by the spatial part of metric (7.7.9), coincides with the Euclidean geometry. Therefore, for the considered electromagnetic wave in a rectangular coordinate system, we have $g_{\mu\nu} = \bar{g}_{\mu\nu}$, where $\bar{g}_{\mu\nu}$ is the Minkowski metric tensor.

Taking this into account, we get, using (7.2.1), that in the region where the considered wave propagates

$$ds^2 = \bar{g}_{\mu\nu}dx^\mu dx^\nu, \quad \Gamma^\gamma_{\mu\nu} = \frac{1}{2}(\lambda^\gamma \bar{g}_{\mu\nu} - \lambda_\mu \delta^\gamma_\nu - \lambda_\nu \delta^\gamma_\mu). \quad (7.8.1)$$

In the examined case in which $\sigma_0 = 0$ and $g_{\mu\nu} = \bar{g}_{\mu\nu}$, Eqs. (7.2.9) and (7.2.10) acquire the form

$$\bar{g}^{\mu\gamma}\left[\left(\partial_\gamma - \frac{1}{2}\lambda_\gamma\right)F_{\mu\nu} - \Gamma^\alpha_{\mu\gamma}F_{\alpha\nu} - \Gamma^\alpha_{\nu\gamma}F_{\mu\alpha}\right] = 0, \quad (7.8.2)$$

$$\left(\partial_\gamma - \frac{1}{2}\lambda_\gamma\right)F_{\mu\nu} + \left(\partial_\nu - \frac{1}{2}\lambda_\nu\right)F_{\gamma\mu} + \left(\partial_\mu - \frac{1}{2}\lambda_\mu\right)F_{\nu\gamma}$$
$$-\Gamma^\alpha_{\mu\gamma}(F_{\alpha\nu} + F_{\nu\alpha}) - \Gamma^\alpha_{\nu\gamma}(F_{\mu\alpha} + F_{\alpha\mu})$$
$$-\Gamma^\alpha_{\mu\nu}(F_{\gamma\alpha} + F_{\alpha\gamma}) = 0. \quad (7.8.3)$$

Since $F_{\mu\nu} = -F_{\nu\mu}$, from Eq. (7.8.3), we derive

$$\left(\partial_\gamma - \frac{1}{2}\lambda_\gamma\right)F_{\mu\nu} + \left(\partial_\nu - \frac{1}{2}\lambda_\nu\right)F_{\gamma\mu} + \left(\partial_\mu - \frac{1}{2}\lambda_\mu\right)F_{\nu\gamma} = 0. \quad (7.8.4)$$

Let us seek solutions to Eqs. (7.8.2) and (7.8.4) in the form

$$F_{\mu\nu} = \cosh^{-1}(x^0/A_0)G_{\mu\nu}, \quad (7.8.5)$$

where $G_{\mu\nu}$ are some antisymmetric differentiable functions of x^γ. Then, using (7.7.10), we have

$$\left(\partial_\gamma - \frac{1}{2}\lambda_\gamma\right)F_{\mu\nu} = \cosh^{-1}(x^0/A_0)\partial_\gamma G_{\mu\nu}. \quad (7.8.6)$$

Substituting (7.8.5) and (7.8.6) into Eqs. (7.8.2) and (7.8.4), we obtain

$$\bar{g}^{\mu\gamma}(\partial_\gamma G_{\mu\nu} - \Gamma^\alpha_{\mu\gamma}G_{\alpha\nu} - \Gamma^\alpha_{\nu\gamma}G_{\mu\alpha}) = 0, \qquad (7.8.7)$$

$$\partial_\gamma G_{\mu\nu} + \partial_\nu G_{\gamma\mu} + \partial_\mu G_{\nu\gamma} = 0, \qquad (7.8.8)$$

where $\Gamma^\gamma_{\mu\nu}$ are given in (7.8.1).

From (7.7.10) and (7.8.1), we find that the only nonzero components of $\Gamma^\gamma_{\mu\nu}$ are as follows:

$$\Gamma^0_{00} = \Gamma^0_{11} = \Gamma^0_{22} = \Gamma^0_{33} = \Gamma^1_{01} = \Gamma^2_{02} = \Gamma^3_{03} = -\frac{1}{2}\lambda_0,$$

$$\lambda_0 = -(2/A_0)\tanh(x^0/A_0). \qquad (7.8.9)$$

Using (7.8.9) and the antisymmetry of $G_{\mu\nu}$: $G_{\mu\nu} = -G_{\nu\mu}$, we find

$$\bar{g}^{\mu\gamma}\Gamma^\alpha_{\mu\gamma}G_{\alpha\nu} = \lambda_0 G_{0\nu}, \bar{g}^{\mu\gamma}\Gamma^\alpha_{\nu\gamma}G_{\mu\alpha} = -\lambda_0 G_{0\nu}. \qquad (7.8.10)$$

Therefore, Eq. (7.8.7) acquires the form

$$\partial^\mu G_{\mu\nu} = 0, g^{\mu\nu} = \bar{g}^{\mu\nu}. \qquad (7.8.11)$$

Let us put

$$G_{\mu\nu} = \partial_\mu U_\nu - \partial_\nu U_\mu, \qquad (7.8.12)$$

where U_μ are some differentiable functions. Then, Eq. (7.8.8) is identically satisfied and Eq. (7.8.11) acquires the form

$$\partial^\mu \partial_\mu U_\nu - \partial_\nu \partial^\mu U_\mu = 0. \qquad (7.8.13)$$

Let us now choose the Lorentz gauge for the potentials U_μ in (7.8.12):

$$\partial^\mu U_\mu = 0. \qquad (7.8.14)$$

Then, from (7.8.13), we get

$$\partial^\mu \partial_\mu U_\nu = 0. \qquad (7.8.15)$$

Consider plane waves running along the abscissa and satisfying (7.8.15). They have the form

$$U_\nu = \Phi_\nu(x^0 - x^1), \qquad (7.8.16)$$

where Φ_ν are arbitrary twice differentiable functions.

From (7.8.14), we readily find

$$\Phi_0 = -\Phi_1 + \text{const.} \tag{7.8.17}$$

Substituting (7.8.16) and (7.8.17) into formula (7.8.12), we obtain the following expressions for the antisymmetric functions $G_{\mu\nu}$:

$$G_{01} = G_{23} = 0, G_{02} = -G_{12} = P(x^0 - x^1),$$
$$G_{03} = -G_{13} = Q(x^0 - x^1), \tag{7.8.18}$$

where P and Q are arbitrary differentiable functions of the argument $x^0 - x^1$.

Substituting (7.8.18) into formula (7.8.5), we find that the antisymmetric functions $F_{\mu\nu}$ are as follows:

$$F_{01} = F_{23} = 0, F_{02} = -F_{12} = \cosh^{-1}(x^0/A_0)P(x^0 - x^1),$$
$$F_{03} = -F_{13} = \cosh^{-1}(x^0/A_0)Q(x^0 - x^1). \tag{7.8.19}$$

Substituting now expressions (7.8.19) into formula (7.4.3) and taking into account that $g_{\mu\nu} = \bar{g}_{\mu\nu}$ in the considered case, we come to the following formula for the energy density T_{00} of the electromagnetic plane wave under examination:

$$T_{00} = \frac{1}{4\pi}\cosh^{-2}(x^0/A_0)[P^2(x^0 - x^1) + Q^2(x^0 - x^1)]. \tag{7.8.20}$$

Consider a part of the examined plane wave with $a \le x^0 - x^1 \le b$, where a and b are arbitrary constants, $a < b$. Then from (7.8.20), we find that the energy E of the plane wave part is as follows:

$$E = E_0 \cosh^{-2}(x^0/A_0), E_0 = \text{const.} \tag{7.8.21}$$

Therefore, for the frequency ν of a photon moving along the abscissa, we find

$$\nu = h_0 \cosh^{-2}(x^0/A_0), h_0 = \text{const.} \tag{7.8.22}$$

Formula (7.8.22) describes photon's redshift when $x^0 > 0$ caused by the component λ_0 of the Weyl vector.

Consider now a free movement of a material point in vacuum along the abscissa x^1 and use the dynamic equations (7.5.15). As noted at the end of Section 7.5, the first equation ($\nu = 0$) of (7.5.15) is a consequence of the other three equations ($\nu = 1, 2, 3$) of it. Therefore, it suffices to solve Eq. (7.5.15) for $\nu = 1, 2, 3$.

Let us use formulas (7.8.1) and (7.7.10). Then, for the considered free material point, for which $F_{\mu\nu} = 0$ and $x^2 = x^3 = 0$, the differential equations (7.5.15) give

$$\frac{d^2x^1}{ds^2} - \lambda_0 \frac{dx^0}{ds}\frac{dx^1}{ds} = 0, \quad x^2 = x^3 = 0, \quad ds^2 = (dx^0)^2 - (dx^1)^2,$$

$$\lambda_0 = -(2/A_0)\tanh\left(x^0/A_0\right), \quad A_0 = \text{const.} \tag{7.8.23}$$

Let us introduce the time

$$\tau = x^0/c. \tag{7.8.24}$$

Then, from (7.8.23), we derive

$$d\bar{V}/d\tau = c\lambda_0\bar{V}, \quad \bar{V} \equiv V(1 - V^2/c^2)^{-1/2},$$
$$V \equiv dx^1/d\tau, \quad \lambda_0 = -(2/A_0)\tanh(c\tau/A_0). \tag{7.8.25}$$

From (7.8.25), we obtain

$$\bar{V} \equiv V(1 - V^2/c^2)^{-1/2} = b_0\cosh^{-2}(c\tau/A_0), \quad b_0 = \text{const.} \tag{7.8.26}$$

From this equation, we easily find V and get

$$(1 - V^2/c^2)^{-1} = 1 + (b_0/c)^2\cosh^{-4}(c\tau/A_0), \quad V = dx^1/d\tau. \tag{7.8.27}$$

Formula (7.8.27) gives

$$E^2 = (m_0c^2)^2 + d_0\cosh^{-4}(c\tau/A_0),$$
$$E = m_0c^2(1 - V^2/c^2)^{-1/2}, \tag{7.8.28}$$

where $d_0 = \text{const} \geq 0$ and E and m_0 are the energy and rest mass of the considered material point, respectively.

When $\tau \gg A_0/c$ and $m_0 > 0$, formulas (7.8.28) give

$$E_{\text{kin}}(\tau) = \frac{8d_0}{m_0c^2}\exp(-4c\tau/A_0), \quad \tau \gg A_0/c,$$

$$A_0 > 0, m_0 > 0, E_{\text{kin}} \equiv E - m_0c^2 = E_{\text{kin}}(\tau), \tag{7.8.29}$$

where E_{kin} is the kinetic energy of the material point.

From (7.8.28) and (7.8.29), we get

$$E_{\text{kin}}(0)/E_{\text{kin}}(\tau) = D_0 \exp(4c\tau/A_0), \tau \gg A_0/c,$$

$$A_0 > 0, \ m_0 > 0, \ D_0 = (\sqrt{1 + \delta_0} - 1)/(8\delta_0),$$

$$\delta_0 = d_0/(m_0 c^2)^2. \tag{7.8.30}$$

Let us apply formulas (7.8.28) to a photon. Then, we get

$$E = \sqrt{d_0}\cosh^{-2}(c\tau/A_0), m_0 = 0. \tag{7.8.31}$$

It is interesting to note that formula (7.8.31) coincides with formula (7.8.21), which was obtained by studying the influence of the Weyl potential λ_0 on the propagation of a plane electromagnetic wave.

When $\tau \gg A_0/c$, from (7.8.31) and (7.8.23), we obtain

$$E = 4\sqrt{d_0} \exp(-2c\tau/A_0), \nu = \nu_0 \exp(-2l/A_0),$$

$$\lambda_0 = -2/A_0, A_0 > 0, \tau \gg A_0/c, \tag{7.8.32}$$

where $\nu = \nu(l)$ is the photon frequency, $\nu_0 = \nu(0)$ and l is the distance gone by the photon.

When $l \ll A_0$, from (7.8.32), we come to the following formula for the redshift z of the photon frequency:

$$z = (\nu_0 - \nu)/\nu = 2l/A_0, \ l \ll A_0, \ \tau \gg A_0/c, \tag{7.8.33}$$

where A_0 is the spatial curvature radius of the Universe.

As is well known [70], the cosmological redshift z in the spectrum of a galaxy is proportional to the distance from it when $z \ll 1$. At the same time, formula (7.8.33) also implies that the values z and l are proportional to each other and, besides, it describes a cosmological law since A_0 is the Universe radius.

Therefore, formula (7.8.33) can give a new explanation for the cosmological redshift in the spectra of galaxies without the hypothesis of the expansion of the Universe. This redshift can be interpreted as a result of the influence of Weyl's field on photons.

It should be stressed that formula (7.8.33) can be represented in the form of Hubble's law [70]

$$cz = Hl, H = 2c/A_0, l \ll A_0, \ \tau \gg A_0/c, \tag{7.8.34}$$

where H can be identified with the Hubble constant.

Relying on modern astronomical data, the Hubble constant is estimated as

$$H \approx 70 \, \mathrm{km} \cdot \mathrm{sec}^{-1} \cdot \mathrm{Mpc}^{-1} \approx 0.072 \, \mathrm{Gyr}^{-1}. \qquad (7.8.35)$$

From this estimate and (7.8.34), we obtain the following estimate for the Universe radius A_0:

$$A_0 \approx 2.64 \cdot 10^{23} \, \mathrm{km} = 8.55 \cdot 10^3 \, \mathrm{Mpc}. \qquad (7.8.36)$$

It is interesting to note that in our paper [71], formula (7.8.32) for the function $\nu(l)$ was derived in a different way: as a result of studying the wave function of a photon covariant in the Lobachevsky space.

Let us now apply Eqs. (7.8.23) to the nonrelativistic case. Then, for a free movement along the axis x^1 of a material point in vacuum, from (7.8.24) and (7.8.25), we obtain

$$d^2 x^1/d\tau^2 = c\lambda_0 dx^1/d\tau, \; x^2 = x^3 = 0, \; \left|dx^1/d\tau\right| \ll c,$$

$$\tau = x^0/c, \lambda_0 = -(2/A_0)\tanh(c\tau/A_0). \qquad (7.8.37)$$

From formulas (7.8.37), we find that the following small force $\mathbf{F}_{\mathrm{vac}}$ acts on the considered material point, which is caused by Weyl's field in vacuum:

$$\mathbf{F}_{\mathrm{vac}} = \lambda_0 c m_0 \mathbf{V}, \; \lambda_0 = -(2/A_0)\tanh(c\tau/A_0), \qquad (7.8.38)$$

where m_0 and $\mathbf{V}$ are the rest mass and the three-dimensional vector of velocity of the material point, respectively, and $A_0 > 0$ is the spatial curvature radius of the Universe.

When the time $\tau < 0$, this small force accelerates the material point and when $\tau > 0$, this force decelerates it.

It should be stressed that the vector $\mathbf{V}$ presents the velocity of the material point relative to the frame of reference under consideration in which the physical vacuum is homogeneous and isotropic.

From (7.8.38), we come to the following asymptotic values of the force $\mathbf{F}_{\mathrm{vac}}$:

$$\mathbf{F}_{\mathrm{vac}} = (2c/A_0)m_0\mathbf{V}, \quad \tau \to -\infty, \qquad (7.8.39)$$

$$\mathbf{F}_{\mathrm{vac}} = -(2c/A_0)m_0\mathbf{V}, \quad \tau \to +\infty, \quad A_0 = \mathrm{const} > 0. \quad (7.8.40)$$

As can be seen from (7.8.38), the asymptotic formulas (7.8.39) and (7.8.40) correspond to the stationary values $\lambda_0 = 2/A_0$ and $\lambda_0 = -2/A_0$ of the Weyl potential λ_0.

It is interesting to note the following. As our study [71] showed, formulas (7.8.39) and (7.8.40) can also be derived from the following axiom concerning the free translational motion of a rigid body in the Lobachevsky space: The points of such a body must move along straight lines, and its straight line segments must remain straight when moving.

7.9. Description of Cosmological Data in the New Gravitational Theory

Above, the generalized gravitational equations (7.2.5) were proposed with the Weyl connection depending on the metric tensor and Weyl's potentials. The field equations (7.2.5) are covariant, conformally invariant in vacuum, and second order with respect to the derivatives of the metric tensor. From them, after using the Bianchi identities, the differential relation (7.3.22) for the energy–momentum tensor $T^{\mu\nu}$ was obtained. In the case of charged dust-like matter and electromagnetic fields, from relation (7.3.22) and identity (7.5.6), we have found four differential equations (7.5.14) of the second order for the Weyl potentials λ_μ, consistent with the gravitational equations (7.2.5). The differential relation (7.3.22) and equations (7.5.14) allowed us to find the differential equations (7.5.15) describing the kinematics of dust-like matter in gravitational and electromagnetic fields.

It should be said that the second-order gravitational differential equations (7.2.5) can be reduced to the Einstein gravitational equations (7.2.16) with an additional energy–momentum tensor corresponding to the Weyl vector field. As can be seen from Eq. (7.2.18), the proposed gravitational equations (7.2.5) lead to the differential law of conservation of the total energy and momentum of matter, including the energy and momentum components of the Weyl vector field.

The proposed gravitational equations (7.2.5) have been applied above to cosmological problems. For them, a nonsingular cosmological solution was found for a homogeneous and isotropic physical vacuum. This solution is described by metric (7.7.6) containing an arbitrary time-dependent function χ. The form of this function is related to the choice of gauge for the metric tensor. The chosen gauge gives the Minkowski geometry in local frames of reference that are inertial relative to the considered homogeneous and isotropic physical vacuum. This gauge leads to metric (7.7.7) corresponding to the Lobachevsky space with a constant radius of curvature $A = A_0$.

As follows from formulas (7.7.8), the energy density $\bar{T}_0^0$ of the physical vacuum and the Weyl field potential λ_0 depend on time.

Therefore, the obtained cosmological solution describes a nonstationary Universe.

The proposed generalization (7.2.9) and (7.2.10) of Maxwell's electrodynamic equations, based on Weyl's principle of scale invariance, was applied to study the propagation of a plane electromagnetic wave. As a result, we came to formula (7.8.21) describing the influence of Weyl's field on this wave.

Then, we applied the kinematic equations (7.5.15) and formulas (7.7.10) for the Weyl field potentials to a moving photon and came to formula (7.8.31) coinciding with formula (7.8.21). As a result, we arrived at formula (7.8.32) for the redshift of the photon frequency. This formula gives a new interpretation to the cosmological redshift in the spectra of galaxies as the influence of Weyl's field on photons moving in vacuum.

It should be emphasized that formula (7.8.32), which results in the Hubble law, is valid for cosmological time $\tau \gg A_0/c$. Therefore, we come to the conclusion that the cosmological time $\tau = \tau_0$, corresponding to the modern epoch, just satisfies the relation $\tau \gg A_0/c$.

It should also be noted that in view of formulas (7.8.32), the energy density $\bar{T}_0^0$ of the physical vacuum and the Weyl field potential λ_0 are finite at any time and, therefore, the obtained cosmological solution is applicable for arbitrary values of the time coordinate τ: $-\infty < \tau < +\infty$. In this solution, the absolute value of the energy density of the physical vacuum reaches its maximum at the cosmological time $\tau = 0$ and tends to zero as time $\tau \to \pm\infty$.

From (7.7.10), we find that the vacuum potential $\lambda_0 = \lambda_0(\tau)$ is a decreasing function and $\lambda_0(-\infty) = 2/A_0$, $\lambda_0(0) = 0$, $\lambda_0(+\infty) = -2/A_0$, where $A_0 > 0$.

Thus, the proposed cosmological model has no singularities and is applicable both for positive and negative values of cosmological time.

It should be noted that in our model, the Weyl field affects the energies of particles, but their rest masses remain unchanged, which follows from the differential equation of rest mass conservation (7.4.4). This is a significant difference between our model and the Hoyle–Narlikar cosmological model, which is also applicable for negative values of cosmological time. In the Hoyle–Narlikar model, in contrast to our model, the rest masses of electrons and nucleons change with time [72].

Let us now turn to the study of the thermal history of the Universe. As follows from formula (7.8.28), due to the influence of Weyl's field, the kinetic energy $E_{\text{kin}} \equiv E - m_0 c^2$ of a free particle moving in vacuum increases when the cosmological time $\tau < 0$ and reaches a maximum at $\tau = 0$. When $\tau > 0$, the kinetic energy E_{kin} decreases and tends to zero as $\tau \to +\infty$.

Consider the time $\tau = \tau_0$ corresponding to the modern era. As shown by the above analysis, $\tau_0 \gg A_0/c$. From formula (7.8.29), we obtain that the kinetic energy $E_{\text{kin}}(\tau)$ of a moving free particle with a nonzero rest mass decreases exponentially when $\tau \gg A_0/c$. Using formula (7.8.30), we find that in view of the inequality $\tau_0 \gg A_0/c$, for such a particle, the relation $E_{\text{kin}}(0) \gg E_{\text{kin}}(\tau_0)$ is fulfilled.

Consequently, at the moment $\tau = 0$, the kinetic energies of particles inside cosmic bodies must have had very high values and hence, at this moment, the temperatures of cosmic bodies must have been very high compared to those in the modern era.

Therefore, an epoch near the moment of time $\tau = 0$ can be identified with the stage of the early Universe in standard cosmology [70].

Thus, we conclude that the early Universe, for which the cosmological time τ is near zero, was very hot, just as in standard cosmology. As will be shown in the following section, the well-known results of standard cosmology for the early Universe concerning the cosmic microwave background radiation and the primordial nucleosynthesis [70] should also take place in our cosmology.

7.10.   New Interpretation of the Cosmic Microwave Background Radiation and the Nature of Dark Matter

Let us turn to the differential equations (7.2.18) of energy–momentum conservation in a local inertial frame of reference and give an interpretation to the components $\Theta^{\mu\nu}$ of the energy–momentum tensor corresponding to the Weyl vectorial field. Our main goal is to determine the components of this energy–momentum tensor that could correspond to the well-known homogeneous and isotropic cosmic microwave background radiation [70].

Consider the components $\Theta^{\mu\nu}$ in a local inertial frame of reference with the Minkowski metric tensor $g_{\mu\nu} = \bar{g}_{\mu\nu}$, relative to which the physical vacuum is homogeneous and isotropic and the vacuum potentials λ_μ are determined by formula (7.7.10). In this frame, from (7.2.15), (7.2.16) and (7.7.10), after simple calculations, we obtain the following components $\Theta^{\mu\nu}$:

$$\Theta^{00} = -\frac{3c^4}{32\pi f_N}\lambda_0^2, \; \Theta^{11} = \Theta^{22} = \Theta^{33} = -\frac{c^4}{8\pi f_N}\left(\dot{\lambda}_0 - \frac{1}{4}\lambda_0^2\right),$$

$$\Theta^{\mu\nu} = 0, \mu \neq \nu, \; \lambda_0 = \lambda_0(x^0), \lambda_1 = \lambda_2 = \lambda_3 = 0, \qquad (7.10.1)$$

$$\Lambda_{\mu\nu} = 0, \; \dot{\lambda}_0 \equiv d\lambda_0/dx^0, \; g_{\mu\nu} = \bar{g}_{\mu\nu}.$$

Using the formula for λ_0 in (7.7.10), from (7.10.1), we find

$$\Theta^{00} = \frac{3c^4}{8\pi f_N A_0^2}[\cosh^{-2}(c\tau/A_0) - 1],$$

$$\Theta^{11} = \Theta^{22} = \Theta^{33} = \frac{c^4}{8\pi f_N A_0^2}[\cosh^{-2}(c\tau/A_0) + 1], \quad (7.10.2)$$

$$\Theta^{\mu\nu} = 0, \mu \neq \nu.$$

From formula (7.7.8) for the vacuum energy density $\bar{T}_0^0$ and the formula for the cosmological time τ in (7.8.24), we get in the considered local inertial frame

$$\bar{T}^{00} = -\frac{3c^4}{8\pi f_N A_0^2}\cosh^{-2}(c\tau/A_0). \qquad (7.10.3)$$

Therefore, from (7.10.2) and (7.10.3), we obtain the following equality presenting the law of energy conservation in a vacuum region:

$$\bar{T}^{00} + \Theta^{00} = -\frac{3c^4}{8\pi f_N A_0^2} = \text{const.} \qquad (7.10.4)$$

Expressions (7.10.2) for the energy–momentum tensor $\Theta^{\mu\nu}$ of the Weyl vectorial field can be represented as the sum of their constant part $\Theta^{\mu\nu}_{(0)}$ and variable nonnegative part $\Theta^{\mu\nu}_{(1)}$:

$$\Theta^{\mu\nu} = \Theta^{\mu\nu}_{(0)} + \Theta^{\mu\nu}_{(1)}, \Theta^{00}_{(0)} = -\frac{3c^4}{8\pi f_N A_0^2},$$

$$\Theta^{kk}_{(0)} = \frac{c^4}{8\pi f_N A_0^2}, \ k = 1, 2, 3, \Theta^{\mu\nu}_{(0)} = 0, \ \mu \neq \nu, \quad (7.10.5)$$

$$\Theta^{00}_{(1)} = \frac{3c^4}{8\pi f_N A_0^2}\cosh^{-2}(c\tau/A_0), \Theta^{kk}_{(1)} = \frac{c^4}{8\pi f_N A_0^2}\cosh^{-2}(c\tau/A_0),$$

$$\Theta_{(1)} \equiv \bar{g}_{\mu\nu}\Theta^{\mu\nu}_{(1)} = 0, \Theta^{\mu\nu}_{(1)} = 0, \mu \neq \nu. \quad (7.10.6)$$

Consider the variable part $\Theta^{\mu\nu}_{(1)}$ of the energy–momentum tensor of the Weyl vectorial field. Comparing the time-dependent energy density $\Theta^{00}_{(1)}$ in (7.10.6) with formula (7.8.31) for the energy of a particle with zero rest mass, we find that $\Theta^{00}_{(1)}$ is proportional to the energy of a photon. Besides, formulas (7.10.6) give that the energy density $\Theta^{00}_{(1)}$ is positive and the spur $\Theta_{(1)}$ of the tensor $\Theta^{\mu\nu}_{(1)}$ is equal to zero. Therefore, we come to the conclusion that the part $\Theta^{\mu\nu}_{(1)}$ of the considered energy–momentum tensor describes a radiation in vacuum.

This radiation is homogeneous and isotropic. That is why we could interpret it as the well-known **cosmic microwave background radiation**.

Let us apply this interpretation to estimate the cosmological time $\tau = \tau_0$ corresponding to the present epoch. With this aim, let us use the well-known value of the temperature θ of the cosmic microwave background (CMB) radiation in the present epoch, which is approximately equal to 2.75°K [70].

Then, using the Stefan–Boltzmann law, we find the energy density $\Theta^{00}_{(1)}(\tau_0)$ of CMB radiation in the present epoch:

$$\theta = 2.75°K, \ \Theta^{00}_{(1)}(\tau_0) = (4\sigma/c)\theta^4 = 4.33 \cdot 10^{-13} \text{ erg/cm}^3, \quad (7.10.7)$$

where σ is the Stefan–Boltzmann constant.

On the other hand, from formula (7.10.6) at the time $\tau = 0$, using the value (7.8.36) of A_0, we obtain

$$\Theta^{00}_{(1)}(0) = \frac{3c^4}{8\pi f_N A_0^2} = 2.07 \cdot 10^{-9} \text{ erg/cm}^3. \quad (7.10.8)$$

From formulas (7.10.6)–(7.10.8), we find

$$\cosh(c\tau_0/A_0) = \sqrt{\Theta^{00}_{(1)}(0)/\Theta^{00}_{(1)}(\tau_0)} = 69.1. \qquad (7.10.9)$$

This formula and the value (7.8.36) of A_0 give the value of the cosmological time τ_0 corresponding to the present epoch:

$$\tau_0 = 4.93 A_0/c = 4.34 \cdot 10^{18}\,\text{sec} = 138\,\text{Gyr}. \qquad (7.10.10)$$

The obtained time turns out to be an order of magnitude longer than that accepted in standard cosmology. As will be shown later on, the new estimate (7.10.10) for the time τ_0 allows us to explain the nature of dark matter.

To check formula (7.10.10), let us consider the helium ^4He nucleosynthesis in our galaxy for the found time τ_0. As is known, the stars of our galaxy radiate 10^{44} erg/sec [73], which represents their luminosity. Therefore, during the time τ_0 given by (7.10.10), the stars of our galaxy can radiate the following energy $E_{\text{rad}}(\tau_0)$:

$$E_{\text{rad}}(\tau_0) \approx 10^{44}\,\text{erg/sec} \cdot \tau_0 = 4.34 \cdot 10^{62}\,\text{erg}. \qquad (7.10.11)$$

As is known, during the formation of a helium nucleus inside a star, the energy $2.5 \cdot 10^{-5}$ erg is emitted [73]. Therefore, from (7.10.11), we find the following number $N(^4\text{He})$ of helium nuclei formed in our galaxy during the time τ_0:

$$N(^4\text{He}) \approx 1.74 \cdot 10^{67}. \qquad (7.10.12)$$

The mass M_{gal} of our galaxy is $4 \cdot 10^{44}$ g [73]. Consequently, from (7.10.12), we obtain the mass fraction of helium ^4He in our galaxy, which could have been formed in it during the time τ_0:

$$Y(^4\text{He}) = N(^4\text{He}) \cdot m(^4\text{He})/M_{\text{gal}} \approx 0.29, \qquad (7.10.13)$$

where $m(^4\text{He}) = 6.65 \cdot 10^{-24}$ g is the rest mass of the helium nucleus ^4He.

On the other hand, as follows from observational data, the mass fraction of the helium nucleus ^4He in our galaxy is $Y(^4\text{He}) \approx 0.28$ [73]. Consequently, formula (7.10.13), in which the obtained value (7.10.10) for the cosmological time τ_0 has been used, allows us to give a new explanation of the mass fraction of helium in our galaxy known from observations.

Let us now apply the proposed cosmology to explain the nature of dark matter. As follows from astronomical observations, the invisible (dark) mass in galaxy clusters is an order of magnitude greater than the total mass of the stars observed in them. This property of dark matter still remains unexplained within the framework of standard cosmology.

Consider the question of dark matter within the framework of our concept. According to the obtained estimate (7.10.10), the cosmological time τ_0 is an order of magnitude longer than that accepted in standard cosmology. This leads us to the following explanation of the nature of dark matter, which is impossible in standard cosmology: The mysterious dark matter can consist of old, faintly glowing and extinct stars.

Indeed, as follows from estimate (7.10.10) for the cosmological time τ_0, the total mass of old, faintly glowing and extinct stars in galaxy clusters can be an order of magnitude greater than the total mass of the stars observed in them, which accords with astronomical observations.

It is worth noting that recent studies using New Horizons' Long Range Reconnaissance Imager (LORRI) images have returned the most precise measurement of the cosmic optical background to date. It turned out that the cosmic optical background is about two times brighter than the theoretical models based on the estimation of the brightness of known objects in the Universe suggest [74]. These unexpected observational data can just be explained by weak radiation of old, faintly glowing stars in our interpretation of dark matter, which has a mass in galaxy clusters an order of magnitude greater than the total mass of the stars observed in them. Thus, this measurement of the cosmic optical background can be regarded as a good confirmation of our interpretation of dark matter.

Let us now compare the proposed gravitational equations with Weyl's potentials with the frequently considered Einstein equations containing the time-dependent cosmological term Λ. With this aim, we note that instead of the standard cosmological term Λ, Eqs. (7.6.11) contain the expressions $(\lambda_0^*/A)dA/d\eta$, $(\lambda_0^*)^2$ and $d\lambda_0^*/d\eta$, taking into account the formula for α in (7.6.8).

Thus, there is a significant difference between the proposed gravitational equations and Einstein's equations with a time-dependent cosmological term.

It should be noted that in standard cosmology, the cosmological term Λ describes dark energy. In our cosmology, instead of this term, we use the Weyl potentials λ_μ.

7.11. Astronomical Applications of the New Cosmology

Let us now apply the proposed cosmological theory to the unsolved problems of the evolution and structure of spiral galaxies.

First, as follows from formula (7.8.38), when $\tau > 0$, small decelerating forces act on the stars of a galaxy. That is why the stars rotating about the center of a galaxy have to move in spiral orbits slowly, over billions of years, approaching the galaxy's center.

Therefore, formula (7.8.38) allows one to explain the visible spiral structure of many galaxies [64].

Second, old stars approaching the galaxy center for a sufficiently long time could be near the center. Hence, the proposed cosmological model allows one to explain the well-known fact that the galaxy's central condensation is mostly composed of old stars (Population II stars), whereas the galaxy's spiral arms contain a large number of young stars (Population I stars) [64].

Besides, according to our theory, the stars of the spiral arms of a galaxy gradually, over billions of years, approach its center, the earlier the spiral arms of a galaxy were formed in the past, the closer they are at present to the galaxy center.

It should be noted that this conclusion corresponds with observational data. Namely, in passing from subclass **c** of the spiral galaxies to subclass **b** and then to **a**, we observe an increasing percentage of old stars and, at the same time, a decreasing spreading of the spiral arms [64].

Consider formula (7.8.38) again. It follows from it that small decelerating forces should act on the points of a star rotating around its axis. Therefore, the angular velocity of its rotation should slowly decrease with time. This effect just makes it possible to explain the observed slow rotation of old stars, which was noted in Section 7.1.

Let us apply formula (7.8.38) to the solar system. As follows from (7.10.10), $\tanh(c\tau_0/A_0) \approx 1$. Therefore, for the solar system, we can use the asymptotic expression (7.8.40) of formula (7.8.38). From (7.8.40) and the formula for H in (7.8.34), we arrive at the following equations for the motions of the Sun and one of its planets:

$$\mathbf{a_P} = \mathbf{a_{gr}} - H\mathbf{v_P}, \mathbf{a_S} = -H\mathbf{v_S}, H = 2c/A_0, \qquad (7.11.1)$$

where $\mathbf{a_{gr}}$ is the classical gravitational acceleration of the planet caused by the Sun, H is the Hubble constant and $\mathbf{v_S}$, $\mathbf{a_S}$ and $\mathbf{v_P}$, $\mathbf{a_P}$ are the vectors of velocities and accelerations of the Sun and planet, respectively, relative to the inertial frame of reference in which the physical vacuum is homogeneous and isotropic.

From (7.11.1), we obtain

$$\mathbf{a} = \mathbf{a_{gr}} - H\mathbf{v}, \ \mathbf{v} = \mathbf{v_P} - \mathbf{v_S}, \ \mathbf{a} = \mathbf{a_P} - \mathbf{a_S}, \qquad (7.11.2)$$

where $\mathbf{v}$ and $\mathbf{a}$ are the vectors of velocity and acceleration of the planet relative to the Sun.

Therefore, planets of the solar system get a small acceleration $\delta\mathbf{a} = -H\mathbf{v}$, in addition to $\mathbf{a_{gr}}$, caused by the influence of the Weyl field.

Let $\mathbf{r}$ be the radius vector for which the initial point is the center of the Sun. Then, since for a planet of the Sun, the vector product $\mathbf{a_{gr}} \times \mathbf{r} = 0$, from (7.11.2), we find that for the planet,

$$\mathbf{a} \times \mathbf{r} = -H\mathbf{v} \times \mathbf{r}, \qquad (7.11.3)$$

where $H = 2c/A_0$.

It is easy to verify that formula (7.11.3) can be rewritten in the form

$$d\mathbf{q}/dt = -H\mathbf{q}, \ \mathbf{q} = \mathbf{v} \times \mathbf{r}, \qquad (7.11.4)$$

where t is the time. Therefore,

$$\mathbf{q} = \mathbf{v} \times \mathbf{r} = \mathbf{q_0} \exp(-Ht), \qquad (7.11.5)$$

where $\mathbf{q_0}$ is a constant vector.

Consider the rotation of a planet around the Sun over a sufficiently long period of time and take into account that its orbit is close to circular. Let r and v be, respectively, the average values for one revolution of the values $|\mathbf{r}|$ and $|\mathbf{v}|$ for the planet under consideration. Then, since $v = (f_N M_S/r)^{1/2}$, where M_S is the Sun mass and f_N is the Newtonian gravitational constant, from (7.11.5), we obtain

$$r = r_0 \exp(-2Ht), v = v_0 \exp(Ht), \qquad (7.11.6)$$

where $r_0 = \text{const}$, $v_0 = \text{const}$.

From (7.11.6), we get

$$\omega = d\varphi/dt = v/r = \omega_0 \exp(3Ht), \quad \omega_0 = \text{const}, \qquad (7.11.7)$$

where ω is the average angular velocity of the planet for its one revolution and φ is the central angle that it passes during its revolution around the Sun in time t.

Formula (7.11.7) gives

$$\varphi = \frac{\omega_0}{3H} \exp(3Ht) + \varphi_0, \quad \varphi_0 = \text{const}, \qquad (7.11.8)$$

Consider now formula (7.11.8) when $|t| \ll 1/H$. Then, expanding it into a series in powers of t, we approximately obtain

$$\varphi = \varphi_1 + \omega_0 t + \delta\varphi, \ \delta\varphi = 1.5H\omega_0 t^2, \quad \varphi_1 = \text{const}. \qquad (7.11.9)$$

The value $\delta\phi$ is a correction to the classical formula for the angle of rotation of the planet around the Sun, which is given by the proposed gravitational theory based on Weyl's principle of scale invariance in vacuum.

Let us return to the formula in (7.11.6) for the average distance for one revolution of a planet between it and the Sun. From this formula, we find that the planets of the solar system should slowly approach the Sun. Similarly, the satellites of Sun's planets should also slowly approach them. This allows us to explain the formation of rings around Saturn, Jupiter, Uranus and Neptune. Such formation becomes possible after the moment when a satellite reaches the Roche radius of its planet. Then, as is known, tidal forces begin to destroy the satellite of the planet [75].

Since the Mars satellite Phobos rotates at a distance from the center of the planet close to its Roche radius [75], a rarefied ring around Mars could have arisen. It is important to note that the impact of the particles of this hypothetical ring could explain the unexpected loss of communication with a number of spacecraft and probes approaching the Martian surface. For example, such communication losses occurred with the Soviet spacecraft Phobos-2 in 1989, with the American Mars Observer in 1993 and Mars Polar Lander in 1999 and with the British Beagle 2 in 2003.

Let us now turn to the applications of the proposed generalization of Einstein's gravitational theory to the problems of astronomical dating of events in ancient history. In particular, one of these problems concerns the mysterious jump in the second derivative of the elongation of the Moon in the period from approximately 700 AD to 1300 AD [76,77]. To this end, consider a correction to the location of geographic places where total solar eclipses were observed in ancient history. This correction must be taken into account in view of formula (7.11.9).

Using this formula, we find the following shift value δs for geographic locations where total solar eclipses were observed in ancient history, which is due to the correction $\delta\varphi$ for the angle φ of revolution of the Earth around the Sun:

$$\delta s \approx R_{\mathrm{M}}\delta\varphi = 3\pi T_{\mathrm{E}} R_{\mathrm{M}} H (n_{\mathrm{yr}})^2 \approx 0.00026(n_{\mathrm{yr}})^2 \,\mathrm{km},$$

$$(7.11.10)$$

where $R_{\mathrm{M}} \approx 384400\,\mathrm{km}$ is the distance between the Earth and the Moon, $T_{\mathrm{E}} \approx 3.16 \times 10^7$ sec is the period of the revolution of the Earth around the Sun, $H \approx 70\,\mathrm{km/sec/Mpc} = 2.27 \times 10^{-18}\,\mathrm{sec}^{-1}$ and n_{yr} is the number of years that have passed since the event in question in ancient history. This shift should be made in the direction of the apparent annual motion of the Sun.

When $n_{\mathrm{yr}} = 500$, from (7.11.10), we obtain $\delta s \approx 65\,\mathrm{km}$.

As is known, the width of the band on the Earth's surface, where a total solar eclipse can be observed, is less than 200 km. Consequently, the found correction (7.11.10) is essential for astronomical dating of events that occurred more than 500 years ago.

References

1. Ryder, L. H. *Quantum Field Theory*. Cambridge University Press, Cambridge, 1985.
2. Faddeev, L. D., Slavnov, A. A. *Gauge Fields: Introduction to Quantum Theory*. Benjamin, New York, 1961.
3. Frampton, P. *Gauge Field Theories*. Wiley-VCH, Weinheim, 2008.
4. Rabinowitch, A. S. Yang-Mills fields and nonlinear electrodynamics. *Russ. J. Math. Phys.*, 2005, **12**, 379–385.
5. Rabinowitch, A. S. Yang-Mills fields of nonstationary spherical objects with big charges. *Russ. J. Math. Phys.*, 2008, **15**, 389–394.
6. Rabinowitch, A. S. On new solutions of classical Yang-Mills equations with cylindrical sources. *Appl. Math.*, 2010, **1**, 1–7.
7. Rabinowitch, A. S. New wave solutions of the Yang-Mills equations with axially symmetric sources. *Int. J. Adv. Math. Sci.*, 2013, **1**, 109–121.
8. Ueda, S. *The New View of the Earth*. Freeman, San Francisco, 1978.
9. Singer, S. *The Nature of Ball Lightning*. Plenum, New York, 1971.
10. Tamm, I. E. *Fundamentals of the Theory of Electricity*. Mir, Moscow, 1979.
11. Feynman, R. P., Leighton, R. B., Sands, M. *The Feynman Lectures on Physics*, vol. 2. Addison-Wesley, Reading-London, 1964.
12. Bragin, Y. A., Kocheev, A. A., Kichtenko, V. N., *et al. Radiowave Propagation and Ionospheric Physics*. Nauka, Novosibirsk, 1981, pp. 165–183.
13. Landau, L. D., Lifshitz, E. M. *The Classical Theory of Fields*. Pergamon, Oxford, 1971.
14. Smirnov, B. M. Analysis of the nature of ball lightning. *Sov. Phys. Usp.*, 1975, **18**, 636–640.
15. Rakov, V. A., Uman, M. A. *Lightning: Physics and Effects*. Cambridge University Press, Cambridge, 2003.

16. Cowen, R. Signature of antimatter detected in lightning. *Science News*, 2009, **176**, No. 12, p. 9.

17. Chace, W. G., Moore H. K. (Eds.) *Exploding Wires*. Plenum Press, New York, 1962.

18. Rabinowitch, A. S. Modified Yang-Mills theory and electroweak interactions. *Int. J. Theor. Phys.*, 2000, **39**, 2457–2466.

19. Actor, A. Classical solutions of SU(2) Yang-Mills theories. *Rev. Mod. Phys.*, 1979, **51**, 461–525.

20. Biró, T. S., Matinyan, S. G., Müller, B. *Chaos and Gauge Field Theory*. World Scientific, Singapore, 1994.

21. Coleman, S. R. Non-abelian plane waves. *Phys. Lett. B*, 1977, **70**, 59–60.

22. Campbell, W. B., Morgan, T. A. Nonabelian plane fronted waves. *Phys. Lett. B*, 1979, **84**, 87–88.

23. Gueven, R. Solution for gravity coupled to nonabelian plane waves. *Phys. Rev. D*, 1979, **19**, 471–472.

24. Lo, S.-Y., Desmond, P., Kovacs, E. General self-dual non-abelian plane waves. *Phys. Lett. B*, 1980, **90**, 419–421.

25. Basler, M., Hädicke, A. On non-abelian SU(2) plane waves. *Phys. Lett. B*, 1984, **144**, 83–86.

26. Oh, C. H., Teh, R. Nonabelian progressive waves. *J. Math. Phys.*, 1985, **26**, 841–844.

27. Raczka, P. A. Colliding plane waves in nonabelian gauge theory. *Phys. Lett. B*, 1986, **177**, 60–62.

28. Rabinowitch, A. S. Exact axially symmetric wave solutions of the Yang-Mills equations. *Theor. Math. Phys.*, 2006, **148**, 1081–1085.

29. Rabinowitch, A. S. On non-abelian expanding waves. *J. Phys. A: Math. Theor.*, 2007, **40**, 14575–14579.

30. Rabinowitch, A. S. On a new class of non-abelian expanding waves. *Phys. Lett. B*, 2008, **664**, 295–300.

31. Rabinowitch, A. S. On transverse progressive waves in Yang-Mills fields. *Eur. Phys. J. Plus*, 2021, **136**, 574.

32. Rabinowitch, A. S. On the propagation of local perturbations in Yang-Mills fields with SU(2) symmetry. *Russ. J. Math. Phys.*, 2022, **29**, 576–580.

33. Rabinowitch, A. S. Relativistic theory of nuclear forces. *Int. J. Theor. Phys.*, 1994, **33**, 2049–2056.

34. Rabinowitch, A. S. Binding energies of nuclei. *Int. J. Theor. Phys.*, 1997, **36**, 533–544.

35. Rabinowitch, A. S. Nuclear forces and neutron stars. *Int. J. Theor. Phys.*, 1998, **37**, 1477–1489.

36. Rabinowitch, A. S. On nonlinear dynamical equations for relativistic nucleons moving near atomic nuclei. *Eur. Phys. J. Plus*, 2020, **135**, 695.

37. Acosta, V, Cowan, C. L., Graham, B. J. *Essentials of Modern Physics.* Harper & Row, New York, 1973.
38. Ericson, T., Weise, W. *Pions and Nuclei.* Clarendon Press, Oxford, 1988.
39. Feynman, R. P. *The Theory of Fundamental Processes.* Benjamin, New York, 1961.
40. Schiff, L. I. Nonlinear meson theory of nuclear forces. *Phys. Rev.*, 1951, **84**, 1–9.
41. Skyrme, T. H. R. A unified field theory of mesons and baryons. *Nucl. Phys.*, 1962, **31**, 556–569.
42. Yndurráin, F. J. *Quantum Chromodynamics.* Springer-Verlag, Heidelberg, 1983.
43. Bogolubov, N. N., Shirkov, D. V. *Introduction to the Theory of Quantized Fields.* John Wiley & Sons, New York, 1959.
44. Oppenheimer, J. R., Volkoff, G. M. On massive neutron cores. *Phys. Rev.*, 1939, **55**, 374–381.
45. Rabinowitch, A. S. Relativistic quantum physics equation for number of electrons. *Int. J. Theor. Phys.*, 1993, **32**, 791–799.
46. Rabinowitch, A. S. On the anomalous magnetic moment of nucleons. *Hadronic J.*, 1996, **19**, 375–384.
47. Rabinowitch, A. S. On a generalization of the Dirac equation for a description of the quark structure of nucleons. *Russ. Phys. J.*, 2008, **51**, 822–830.
48. Rabinowitch, A. S. Generalized Dirac equation describing the quark structure of nucleons. In Studenikin A. I. (Ed.) *Particle Physics on the EVE of LHC.* World Scientific, Singapore, 2009, pp. 431–434.
49. Bjorken, J. D., Drell, S. D. *Relativistic Quantum Mechanics.* McGraw Hill, New York, 1964.
50. Arfken, G., Weber, H. *Mathematical Methods for Physicists.* Academic Press, New York, 2000.
51. Rabinowitch, A. S. Energy-momentum pseudotensor of the gravitational field. *Phys. Essays*, 1993, **6**, 572–575.
52. Rabinowitch, A. S. Noninertial frames of reference in general relativity. *Phys. Essays*, 1996, **9**, 387–392.
53. de Donder, T. *Theorie des Champs Gravifiques.* Gauthier-Villars, Paris, 1926.
54. Fock, V. A. Three lectures on relativity theory. *Rev. Mod. Phys.*, 1957, **29**, 325–333.
55. Weber, J. *General Relativity and Gravitational Waves.* Dover, New York, 2004.
56. Sneddon, J. N., Berry, D. S. *The Classical Theory of Elasticity.* Springer-Verlag, Berlin, 1958.

57. Li, M., Li, X., Wang, S., Wang, Y. *Dark Energy.* World Scientific, Singapore, 2015.

58. Lerner, E. J. Observations contradict galaxy size and surface brightness predictions that are based on expanding universe hypothesis. *Mon. Not. Roy. Astron. Soc.*, 2018, **477**, 3185–3196.

59. Risality, G., Lusso, E. Cosmological constraints from the Hubble diagram of quasars at high redshifts. *Nat. Astron.*, 2019, **3**, 272–277.

60. Freedman, W. L. Measurements of the Hubble constant: Tensions in perspective. *Astrophys. J.*, 2021, **919**, 16.

61. Naidu, R. P., Oesch, P. A., van Dokkum, P., *et al.* Two remarkably luminous galaxy candidates at $z \approx 10 - 12$ Revealed by JWST. *Astrophys. J. Lett.*, 2022, **940**, L14.

62. Rabinowitch, A. S. Generalized Einstein gravitational theory with vacuum vectorial field. *Class. Quantum Grav.*, 2003, **20**, 1389–1402.

63. Rabinowitch, A. S. On a generalization of the equations of general relativity based on Weyl's principle of scale invariance. *Gravit. Cosmol.*, 2021, **27**, 202–211.

64. Oster, L. *Modern Astronomy.* Holden-Day, San Francisco, 1973.

65. Clegg, B. *Dark Matter and Dark Energy. The Hidden 95% of the Universe.* Icon Books, London, 2019.

66. Clifton, T., Ferreira, P. G., Padilla, A., Scordis, C. Modified gravity and Cosmology. *Phys. Rep.*, 2012, **513**, 1–189.

67. Maeder, A. An alternative to the ΛCDM model: The case of scale invariance. *Astroph. J.*, 2017, **834**, 194.

68. Weyl, H. *Space—Time—Matter.* Dover, New York, 1952.

69. Alonso, J. C., Barbero, F., Julve, J., Tiemblo, A. Particle contents of higher-derivative gravity. *Class. Quantum Grav.*, 1994, **11**, 865–882.

70. Kolb, E. W., Turner M. S. *The Early Universe.* Addison-Wesley, New York, 1990.

71. Rabinowitch, A. S. Lobachevsky geometry and unsolved problems of solar cosmogony. *Int. J. Theor. Phys.*, 1991, **30**, 521–529.

72. Hoyle, F., Narlikar, J. *The Physics-Astronomy Frontier.* Freeman, San Francisco, 1980.

73. Klimishin, I. A. *Relativistic Astronomy.* Nauka, Moscow, 1983.

74. Bernal, J. L., Sato-Polito, G., Kamionkowski, M. Cosmic optical background excess, dark matter, and line-intensity mapping. *Phys. Rev. Lett.*, 2022, **129**, 231301.

75. Sinclair, A. T. *Mon. Not. Roy. Astron. Soc.*, 1972, **155**, 249–274.

76. Fomenko, A. T. The jump of the second derivative of the moon's elongation. *Celest. Mech.*, 1981, **29**, 33–40.

77. Newton, R. R. *Ancient Astronomical Observations and the Accelerations of the Earth and Moon.* John Hopkins Press, Baltimore, 1970.

Index

Printed in the United States
by Baker & Taylor Publisher Services